Studies in Computational Intelligence

Volume 1246

The series "Studies in Computational Intelligence" (SCI) publishes new developments and advances in the various areas of computational intelligence—quickly and with a high quality. The intent is to cover the theory, applications, and design methods of computational intelligence, as embedded in the fields of engineering, computer science, physics and life sciences, as well as the methodologies behind them. The series contains monographs, lecture notes and edited volumes in computational intelligence spanning the areas of neural networks, connectionist systems, genetic algorithms, evolutionary computation, artificial intelligence, cellular automata, self-organizing systems, soft computing, fuzzy systems, and hybrid intelligent systems. Of particular value to both the contributors and the readership are the short publication timeframe and the world-wide distribution, which enable both wide and rapid dissemination of research output.

Indexed by SCOPUS, DBLP, WTI AG (Switzerland), zbMATH, SCImago.

All books published in the series are submitted for consideration in Web of Science.

Hang Chen • Zhengjun Liu
Editors

Artificial Intelligence in Digital Image Processing: Theories, Methods, and Applications

Springer

Editors
Hang Chen
School of Space Information
Space Engineering University
Beijing, China

Zhengjun Liu
School of Physics
Harbin Institute of Technology
Harbin, Heilongjiang, China

ISSN 1860-949X ISSN 1860-9503 (electronic)
Studies in Computational Intelligence
ISBN 978-3-032-12890-4 ISBN 978-3-032-12891-1 (eBook)
https://doi.org/10.1007/978-3-032-12891-1

This Springer imprint is published by the registered company Springer Nature Switzerland AG
The registered company address is: Gewerbestrasse 11, 6330 Cham, Switzerland

Preface

In the era of digital transformation, the intersection of artificial intelligence (AI) and digital image processing has emerged as a transformative force, reshaping industries, advancing scientific discovery, and redefining human-computer interaction. Digital images, as one of the most prevalent forms of information carriers, permeate every aspect of modern life—from medical diagnostics and satellite remote sensing to smart surveillance and creative media. However, the exponential growth in image data volume, the rise of complex tasks like cross-modal image translation, and the urgent demands for security in image transmission have long posed insurmountable challenges to traditional image processing techniques. It is against this backdrop that AI, with its unparalleled ability to learn complex patterns, extract hierarchical features, and adapt to dynamic scenarios, has become the cornerstone of next-generation digital image processing technologies.

This book, *Artificial Intelligence in Digital Image Processing: Theories, Methods, and Applications*, is born from our collective aspiration to synthesize the rapidly evolving landscape of this interdisciplinary field. Over the past decade, we have witnessed breakthroughs that once seemed confined to science fiction: convolutional neural networks (CNNs) achieving precise classification of remote sensing images, generative artificial intelligence enabling seamless SAR-to-optical image translation, and reinforcement learning optimizing adaptive segmentation in cancer detection. However, alongside these advancements come new challenges—deepfake technologies disrupting media authenticity, adversarial attacks threatening computer vision systems, and the need for intelligent encryption to secure sensitive image data. Amid this flurry of innovation and disruption, there has been a lack of a comprehensive work that bridges theoretical fundamentals, cutting-edge methodological innovations, and diverse practical applications—one that guides readers from AI's mathematical underpinnings to hands-on deployment of algorithms in real-world scenarios.

We open with the theoretical bedrock: Chapter "Fundamentals of Deep Learning" lays the groundwork for all subsequent discussions, unpacking the core principles of deep neural networks that underpin modern image processing—

from basic architectures to optimization strategies, providing readers with the mathematical and conceptual tools needed to engage with advanced topics.

This book delves into specialized methodological innovations that address both opportunities and challenges in the field. We explore emerging technologies reshaping the landscape: Chapter "Deepfake in Image and Video Processing" dissects the AI-driven techniques behind synthetic media generation, while Chapter "Adversarial Attacks and Defenses in Computer Vision" confronts the vulnerabilities of AI systems and strategies to mitigate risks. The book also pioneers the intersection of AI and image security: Chapters "Adaptive Cryptographic System Orchestration via Intelligent Agents" and "Encryption/Decryption with Dynamic Algorithm-Based Optical Transform for Remote Sensing Images" bridge intelligent agents and optical transforms with cryptographic systems, addressing critical security needs in image transmission. Complementing these are chapters on enabling technologies: Chapter "SciData-Factory: An Automated Framework for Generating High-Quality Image-Text Datasets" addresses the foundational need for high-quality data, offering a solution to the data scarcity challenge that plagues many AI applications.

Finally, the book grounds these theories and methods in diverse real-world applications. We focus on high-impact domains: Chapters "Classification of Typical Remote Sensing Images Based on Deep Learning" and "SAR-to-Optical Image Translation Based on Generative Artificial Intelligence" demonstrate AI's transformative role in remote sensing, enabling enhanced Earth observation and cross-modal analysis. Chapter "Deep Learning Applications in Optical Cryptography" merges deep learning with optical systems for secure image transmission, while Chapter "Q-Learning Implementation for Decision Problems in Digital Image Processing: Application to Adaptive Image Segmentation in Histological Cancer Detection" showcases reinforcement learning's practical value in medical diagnostics—turning algorithmic innovation into life-saving tools. Each application chapter draws on the expertise of leading researchers, integrating their firsthand insights from academic research and industry collaborations.

Writing this book has been a journey of synthesis and collaboration, and we owe a debt of gratitude to many. Firstly, we thank the contributing authors—Prof. Bin Gao, Prof. Fei Tong, Dr. Gege Sun, Prof. Ravi Kumar, Prof. Lei Liu, and Prof. Camel Tanougast—whose specialized expertise has shaped the book's depth and breadth. Their groundbreaking work in deep learning foundations, media security, cryptographic systems, and medical imaging has been the backbone of this volume. We also thank the countless researchers whose open-source contributions and seminal papers have advanced the field, and our colleagues at Space Engineering University for their rigorous discussions that refined our ideas. Special thanks go to Mr. Thomas Ditzinger and Mr. Naveen Kumar at Springer Publishing, whose patience and expertise have made this book possible. Finally, we thank our families for their unwavering support during the long hours of writing and revision.

We hope this book serves as a valuable resource: for students, a textbook to build from AI fundamentals to specialized applications; for researchers, a reference to explore frontiers like deepfake detection, adaptive cryptography, and generative image translation; and for practitioners, a guide to translating academic innovation into solutions for remote sensing, healthcare, and image security. As AI continues to evolve, we anticipate the ideas herein will spark new breakthroughs. We welcome feedback from readers, as knowledge grows through dialogue and collaboration.

Beijing, China Hang Chen
Harbin, China Zhengjun Liu

Acknowledgments The editors would like to take this opportunity to express their sincere gratitude to the authors of the chapters for extending their wholehearted support in sharing some of their latest results. Without their significant contribution, this book volume could not have fulfilled its mission. The editor is very grateful to Editor-in-Chief Prof. Thomas Ditzinger and the editor Dr. Naveen Kumar for their help on our book. All the authors are indebted to the editorial staff at Springer, for their hard working and selfish attitude on helping our book plan.

Contents

Fundamentals of Deep Learning

Qi Li, Yutong Li, Shutian Liu, and Zhengjun Liu

Abstract This chapter offers a comprehensive exposition of the fundamental theories and key techniques of deep learning, encompassing its core concepts, foundational models, learning objectives, optimization strategies, and representative architectures. It begins by introducing the historical context and the central issue of contribution allocation, using artificial neural networks to illustrate the information flow and their advantages in representation learning over traditional methods. The training pipeline is then examined in detail, including model formulation, loss function design, gradient-based optimization, regularization, and early stopping, providing a unified view of deep learning from both theoretical and practical perspectives. Structurally, the chapter highlights the principles of local connectivity, weight sharing, and pooling in convolutional neural networks, with classic architectures used to illustrate their evolution. Finally, this chapter introduces variational autoencoders and generative adversarial networks, discussing their modeling frameworks and application scenarios.

Keywords Deep learning · Neural networks · Optimization algorithms · Convolutional neural networks · Generative models

1 Introduction

The models used in deep learning are generally quite complex, referring to the data flow from the original input of the samples to the output targets passing through multiple linear or nonlinear components. Since each component processes the information and subsequently affects the following components, it is unclear what the contribution of each component is when the final output result is obtained. This issue is called the contribution allocation problem. In deep learning, the contribution

Q. Li · Y. Li · S. Liu · Z. Liu (✉)
School of Physics, Harbin Institute of Technology, Harbin, China
e-mail: zjliu@hit.edu.cn

H. Chen, Z. Liu (eds.), *Artificial Intelligence in Digital Image Processing: Theories, Methods, and Applications*, Studies in Computational Intelligence 1246,
https://doi.org/10.1007/978-3-032-12891-1_1

allocation problem is a critical issue, as it relates to how to learn the parameters within each component [1].

A model that can better solve the problem of contribution distribution is the artificial neural network. Artificial neural networks, also known as neural networks, are mathematical models inspired by the way the human brain's nervous system works. Unlike the current structure of computers, the human brain's nervous system is a highly complex network composed of biological neurons, functioning as a parallel nonlinear information processing system. The human brain's nervous system can process signals such as sound and vision through multiple layers of encoding, continuously refining and abstracting from the most primitive low-level features to ultimately obtain a semantic representation of the original signal [2]. Similar to the human brain's neural network, artificial neural networks are composed of artificial neurons and the connections between them, with two special types of neurons: one type is used to receive external information, while the other type is used to output information. Thus, the neural network can be viewed as an information processing system that handles information from input to output. If we consider the neural network as a complex function controlled by a set of parameters, used to handle certain pattern recognition tasks (such as speech recognition, face recognition, etc.), the parameters of the neural network can be learned from data through machine learning. Because neural network models are generally quite complex, with relatively long paths for information transmission from input to output, the learning of complex neural networks can be regarded as a form of deep machine learning, that is, deep learning.

Neural networks and deep learning are not equivalent. Deep learning can use neural network models, but it can also employ other models (for example, deep belief networks are a type of probabilistic graphical model). However, because neural network models can relatively easily address the issue of contribution allocation, they have become the primary models used in deep learning [3]. Although deep learning was initially used to tackle representation learning problems in machine learning, due to its powerful capabilities, it is increasingly being used to solve some general artificial intelligence problems, such as reasoning and decision making.

To understand the significance or importance of deep learning, it is essential to analyze it from a broader perspective of machine learning or artificial Intelligence. In traditional machine learning, besides models and learning algorithms, features or representations are also crucial factors affecting the final learning outcomes, often more important than the algorithms in many tasks. Therefore, to develop a practical machine learning system, people often need to spend a lot of effort trying to design different features and feature combinations to enhance the final system's capabilities, which is known as the feature engineering problem. How to automatically learn effective data representations has become a key issue in machine learning. Early representation learning methods, such as feature extraction and feature selection, involved manually introducing some subjective assumptions for learning. This type of representation learning is not an end-to-end learning approach, and the obtained representations may not be effective for subsequent machine learning tasks. Deep learning, on the other hand, integrates representation learning and predictive model learning into an end-to-end learning process, requiring no manual intervention in between [4]. The problem that deep learning aims to solve is the contribution distribution problem, and neural networks happen to be an effective model for addressing this issue.

2 The Basic Elements of Deep Learning

Deep learning is about enabling computers to automatically learn from data and acquire certain knowledge (or patterns). As a discipline, deep learning typically refers to a class of problems and the methods for solving these problems, specifically how to find patterns from observed data (samples) and use the learned patterns (models) to predict unknown or unobservable data [5].

A collection of samples is called a dataset. Generally, a dataset is divided into two parts: the training set and the test set. The samples in the training set are used to train the model, also known as training samples, while the samples in the test set are used to evaluate the performance of the model, also known as test samples.

A sample is usually represented by a D-dimensional vector x, which is called the feature vector, consisting of all the features of the sample, where each dimension represents a feature. The label of the sample is usually represented by a scalar y.

The training set T consists of N samples, each of which is independently and identically distributed, meaning they are drawn independently from the same data distribution, denoted as

$$D = \{(x_1, y_1), (x_2, y_2), \cdots, (x_N, y_N)\} \tag{1}$$

The goal is to enable the computer to automatically find an optimal function $f(x)$ from a set of functions to approximate the true mapping relationship between each sample's feature vector x and label y.

How to find the function $f(x)$ is key to deep learning, and it generally needs to be accomplished through a learning algorithm A. This process of finding is usually referred to as the learning or training process, as Fig. 1.

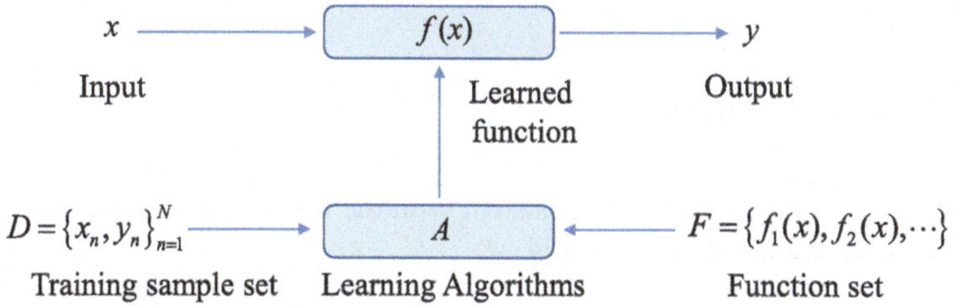

Fig. 1 Deep learning system example

Deep learning is the process of learning general patterns from limited observational data, and it can apply the summarized patterns to unobserved samples. Deep learning methods can be roughly divided into three basic elements: model, learning criteria, and optimization algorithm.

2.1 Model

For a machine learning task, the first step is to determine its input space and output space. The main difference between different deep learning tasks lies in the output space. Since the true mapping function $f(x)$ between x and y is unknown, we can only hypothesize a set of functions based on experience, known as the hypothesis space. We then select an ideal hypothesis by observing its characteristics on the training set. Common hypothesis spaces can be divided into linear and nonlinear types, corresponding models are referred to as linear models and nonlinear models, respectively.

The hypothesis space of linear models is a parameterized family of linear functions, represented as

$$f(x;\theta) = w^{\mathrm{T}}x + b \tag{2}$$

Among them, the parameter θ includes the weight vector w and the bias b.

Nonlinear models can be expressed as a linear combination of multiple non-linear basis functions $\phi(x)$, that is

$$f(x;\theta) = w^{\mathrm{T}}\phi(x) + b \tag{3}$$

where $\phi(x) = [\phi_1(x), \phi_2(x), \ldots, \phi_K(x)]$ is a vector composed of K nonlinear basis functions, which are learnable basis functions themselves, such as

$$\phi_k(x) = h\left(w_k^{\mathrm{T}}\phi(x) + b_k\right), \quad \forall 1 \le k \le K \tag{4}$$

where $h(x)$ is a nonlinear function, w_k and b_k are learnable parameters, then $f(x;\theta)$ is equivalent to a neural network model.

2.2 Learning Criteria

A good model $f(x;\theta)$ should be consistent with the true mapping function $y = g(x)$ for all possible values of (x, y), that is

$$|f(x,\theta) - y| < \varepsilon \tag{5}$$

where ε is a very small positive number. The quality of the model can be measured by the expected risk $R(\theta)$, which is defined as

$$R(\theta) = E_{(x,y)\sim p_r(x,y)}\left[L\Big(y, f(x;\theta)\Big)\right] \tag{6}$$

where $p(x, y)$ represents the true data distribution; $L(y,f(x;\theta))$ is the loss function used to quantify the difference between the two variables. The loss function is a non-negative real-valued function used to quantify the difference between model predictions and true labels. Below are several commonly used loss functions.

The most intuitive loss function is the error rate of the model on the training set, which is the 0-1 loss function, defined as

$$L(y, f(x;\theta)) = \begin{cases} 0, & \text{if } y = f(x;\theta) \\ 1, & \text{if } y \neq f(x;\theta) \end{cases} \tag{7}$$

Although the 0-1 loss function can objectively evaluate the quality of a model, its drawback is that it has poor mathematical properties: it is discontinuous and has a derivative of 0, making it difficult to optimize. Therefore, continuous and differentiable loss functions are often used as substitutes. The squared loss function is often used in tasks where the predicted label y is a real value, defined as

$$L(y, f(x;\theta)) = \frac{1}{2}[y - f(x;\theta)]^2 \tag{8}$$

The squared loss function is generally suitable for regression problems. The cross-entropy loss function is generally used for classification problems. Suppose the label y of the sample is a discrete category, and the model output is the conditional probability distribution of the category labels, that is

$$p(y = c|x;\theta) = f_c(x;\theta) \tag{9}$$

and meet

$$f_c(x;\theta) \in [0, 1], \quad \sum_{c}^{C} f_c(x;\theta) = 1 \tag{10}$$

The sample label can be represented by a C-dimensional one-hot vector y. Assuming that the label of the sample is k, then only the value of the kth dimension of the label vector y is 1, and the values of the remaining elements are 0. The label vector y can be regarded as the true conditional probability distribution of the sample label. For two probability distributions, cross entropy can generally be used to measure their difference. The cross entropy between the true distribution y of the labels and the predicted distribution is

$$\begin{aligned} L(y, f(x;\theta)) &= -y^{\mathrm{T}} \log f(x;\theta) \\ &= -\sum_{c}^{C} y_c \log f_c(x;\theta) \end{aligned} \tag{11}$$

A good model should have a relatively small expected error, but since the true data distribution and mapping function are unknown, it is actually impossible to calculate its expected risk. Given a training set T, what can be calculated is the empirical risk, which is the average loss on the training set

$$\theta^* = \arg\min_{\theta} R_D^{\mathrm{emp}}(\theta) \tag{12}$$

Therefore, a practical learning criterion is to find a set of parameters that minimizes the empirical risk, namely

$$\theta^* = \arg\min_{\theta} R_D^{\mathrm{emp}}(\theta) \tag{13}$$

This is the empirical risk minimization criterion. According to the large number law, when the size of the training set tends to infinity, the empirical risk tends to the expected risk. However, it is usually impossible to obtain unlimited training samples, and the training samples are often a very small subset of the real data or contain certain noise data, which cannot well reflect the true distribution of all da-ta. The empirical risk minimization can easily lead to a model with a low error rate on the training set but a high error rate on unknown data. This is called overfitting.

Overfitting problems are often caused by insufficient and noisy training data and strong model capabilities. In order to solve the overfitting problem, parameter regularization is generally introduced on the basis of empirical risk minimization to limit the model capacity so that it does not minimize the empirical risk excessively. This is the structural risk minimization criterion

$$\theta^* = \arg\min_{\theta} R_D^{\mathrm{struct}}(\theta) = \arg\min_{\theta} \frac{1}{N} \sum_{n=1}^{N} L(y_n, f(x_n;\theta)) + \frac{\lambda}{2} \|\theta\|^2 \tag{14}$$

where "||·||" is the regularization term of the $L2$ norm, which is used to reduce the parameter space and avoid overfitting; it is used to control the strength of regularization.

Other functions can also be used as regularization terms, such as the $L1$ norm. The introduction of the $L1$ norm usually makes the parameters sparse, so it is often used in many algorithms. From the perspective of Bayesian learning, regularization introduces a priori distribution of parameters so that they do not completely depend on training data.

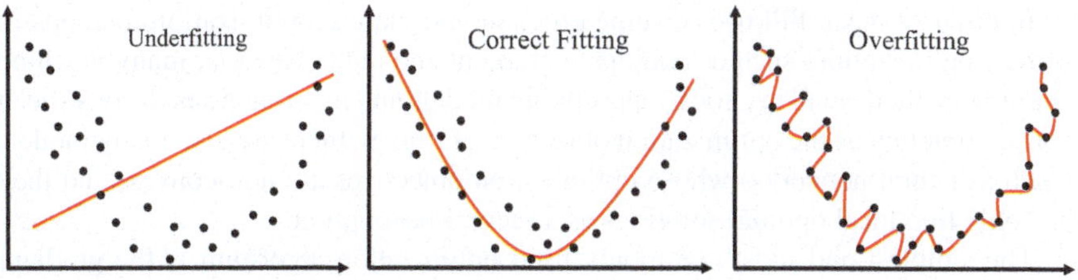

Fig. 2 Examples of underfitting and overfitting

The opposite of overfitting is underfitting, that is, the model cannot fit the training data well and the error rate on the training set is relatively high. Underfitting is generally caused by insufficient model capabilities, as Fig. 2.

The learning criterion in deep learning is not only to fit the data in the training set, but also to minimize the generalization error. Given a training set, the goal of deep learning is to find an "ideal" model with low generalization error from the hypothesis space in order to better predict unknown samples, especially samples that did not appear in the training set. Therefore, deep learning can be viewed as a generalization problem of obtaining more general rules from limited, high dimensional, noisy data.

2.3 Optimization Algorithm

After determining the training set, hypothesis space, and learning criteria, how to find the optimal model $f(x;\theta)$ becomes an optimization problem. The training process of deep learning is actually the process of solving optimization problems.

In deep learning, optimization can be divided into parameter optimization and hyperparameter optimization. The θ in model $f(x;\theta)$ is called the parameters of the neural network and can be learned through optimization algorithms. In addition to learnable parameters, there is another type of parameters that are used to define model structure or optimization strategy. These parameters are called hyperparameters.

Common hyperparameters include: the number of categories in clustering algorithms, the step size in gradient descent methods, the coefficient of regularization terms, the number of layers in neural networks, the kernel function in support vector machines, etc. The selection of hyperparameters is generally a combinatorial optimization problem, which is difficult to learn automatically through optimization algorithms. Therefore, hyperparameter optimization is a highly empirical technique in deep learning. It is usually set according to human experience, or through search methods to continuously adjust a set of hyperparameter combinations through trial and error.

In order to make full use of some efficient and mature optimization methods in convex optimization, such as conjugate gradient and quasi-Newton, many machine learning methods tend to choose appropriate models and loss functions to construct a convex function as the optimization objective. However, there are also many models (such as neural networks) whose optimization objectives are non-convex, and they can only find local optimal solutions as a second-best option.

The simplest and most commonly used optimization algorithm is the gradient descent method, which first initializes the parameters and then calculates the minimum value of the risk function on the training set T according to the following iterative formula

$$\theta_{t+1} = \theta_t - \alpha \frac{\partial R_D(\theta)}{\partial \theta} = \theta_t - \alpha \frac{1}{N} \sum_{n=1}^{N} \frac{\partial L(y_n, f(x_n; \theta))}{\partial \theta} \tag{15}$$

where θ_t is the parameter value at the tth iteration, and α is the learning rate. For the gradient descent optimization algorithm, in addition to adding regularization terms, overfitting can also be prevented by early stopping.

During the gradient descent training process, due to overfitting, the parameters that converge on the training samples are not necessarily optimal on the test set. Therefore, in addition to the training set and the test set, a validation set is sometimes used for model selection to test whether the model is optimal on the validation set. At each iteration, the newly obtained model is tested on the validation set and the error rate is calculated. If the error rate on the validation set does not decrease any more, stop iterating. This strategy is called early stopping, as Fig. 3.

In the gradient descent method, the objective function is the risk function over the entire training set, which is called batch gradient descent. The batch gradient

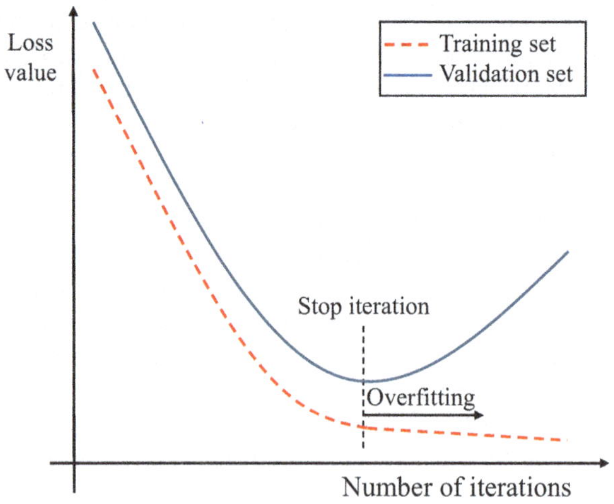

Fig. 3 Example of early termination

descent method needs to calculate the gradient of the loss function on each sample and sum it up at each iteration. When the number of samples in the training set N is large, the space complexity is relatively high and the computational overhead of each iteration is also large.

In deep learning, it is assumed that each sample is independently and identically distributed and randomly drawn from the true data distribution, and the real optimization goal is to minimize the expected risk. The batch gradient descent method is equivalent to collecting N samples from the true data distribution and using the gradient of the empirical risk calculated from them to approximate the gradient of the expected risk. In order to reduce the computational complexity of each iteration, it is also possible to collect only one sample in each iteration, calculate the gradient of the loss function of this sample and update the parameters, which is the stochastic gradient descent method. After a sufficient number of iterations, stochastic gradient descent can also converge to a local optimal solution.

The difference between batch gradient descent and stochastic gradient descent is whether the optimization objective at each iteration is the average loss function of all samples or the loss function of a single sample. Since stochastic gradient descent is simple to implement and converges very quickly, it is widely used. Stochastic gradient descent is equivalent to introducing random noise to the gradient of batch gradient descent. In non-convex optimization problems, stochastic gradient descent can more easily escape from local minima.

One disadvantage of the stochastic gradient descent method is that it cannot fully utilize the parallel computing power of computers. Mini batch gradient descent is a compromise between batch gradient descent and stochastic gradient descent. In each iteration, a small number of training samples are randomly selected to calculate the gradient and update the parameters. This can take into account the advantages of the stochastic gradient descent method and improve the training efficiency.

3 Neural Network Unit

Neural networks refer to a family of mathematical models inspired by biology and neuroscience. These models mainly simulate biological neural networks by abstracting the neural network of the human brain, constructing artificial neurons, and establishing connections between artificial neurons according to a certain topological structure. In the field of artificial intelligence, artificial neural networks are often referred to as neural networks or neural models.

A neural network can generally be viewed as a nonlinear model, whose basic components are neurons with nonlinear activation functions. Through the connection between a large number of neurons, the neural network becomes a highly nonlinear model. The connection weights between neurons are the parameters that need to be learned, which can be learned through the gradient descent method.

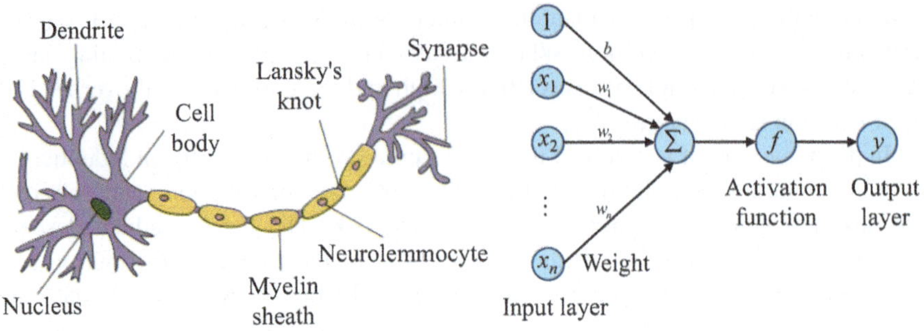

Fig. 4 Example of different types of neurons

3.1 Neurons

Artificial neurons, referred to as neurons, are the basic units that constitute neural networks. They mainly simulate the structure and characteristics of biological neurons, receive a set of input signals and generate outputs, as Fig. 4.

Suppose a neuron receives D inputs $x_1, x_2, \ldots, x_D$ and the net input z is used to represent the weighted sum of the input signal x obtained by a neuron. After passing through a nonlinear function f, the net input z obtains the output value a of the neuron, which can be expressed as

$$a = f\left(\sum_{d=1}^{D} w_b x_b + b\right) \tag{16}$$

where w is the D-dimensional weight vector, b is the bias term, and the nonlinear function f is called the activation function.

Activation functions are very important in neurons. In order to enhance the representation and learning capabilities of the network, the activation function needs to have the following properties

1. A nonlinear function that is continuous and differentiable (but not differentiable at a few points). Differentiable activation functions can directly use numerical optimization methods to learn network parameters.
2. The activation function and its derivative function should be as simple as possible, which is conducive to improving the network computing efficiency.
3. The value range of the derivative of the activation function should be within an appropriate range, neither too large nor too small, otherwise it will affect the efficiency and stability of training.

The following are several activation functions commonly used in neural networks. Sigmoid function refers to a type of S-shaped curve function, which is a saturation function at both ends. Commonly used Sigmoid functions include Logistic function and Tanh function. The logistic function is defined as

$$\sigma(x) = \frac{1}{1 + e^{-x}} \tag{17}$$

The logistic function can be regarded as a "squeezing" function, which "squeezes" an input in the real number domain to (0, 1). When the input value is near 0, the Sigmoid function is approximately a linear function; when the input value is close to both ends, the input is suppressed. The smaller the input, the closer it is to 0; the larger the input, the closer it is to 1. This characteristic is similar to that of biological neurons, which can produce excitement (output is 1) for some inputs and inhibition (output is 0) for other inputs. Compared with the step activation function used by the perceptron, the Logistic function is continuous and differentiable, and has better mathematical properties.

The tanh function is also a Sigmoid function, which is defined as

$$\tanh(x) = \frac{e^x - e^{-x}}{e^x + e^{-x}} \tag{18}$$

The tanh function can be viewed as an enlarged and translated logistic function, and its range is (−1, 1) as

$$\tanh(x) = 2\sigma(2x) - 1 \tag{19}$$

Figure 5 shows the shapes of the Logistic function and the Tanh function. The output of the Tanh function is zero-centered, while the output of the Logistic function is always greater than 0. The non-zero centered output will bias the input of the neurons in the next layer and further slow down the convergence of the gradient descent.

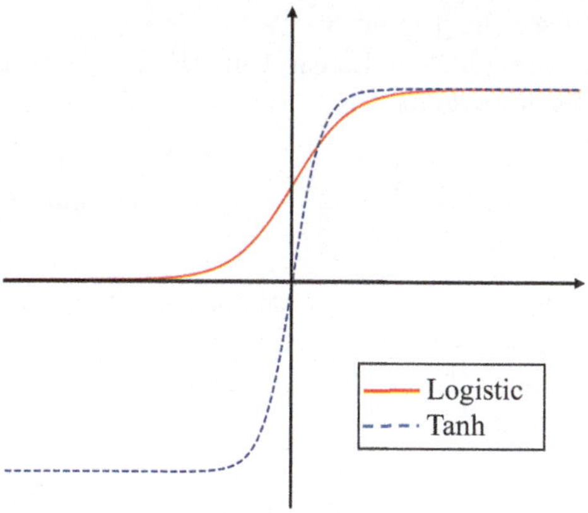

Fig. 5 Logistic function and Tanh function

Rectified Linear Unit (ReLU) is an activation function commonly used in deep neural networks, which is defined as

$$\mathrm{ReLU} = \begin{cases} x, & x \geq 0 \\ 0, & x < 0 \end{cases} = \max(0, x) \tag{20}$$

Neurons using ReLU only need to perform addition, multiplication and comparison operations, which is more efficient in calculation. The Sigmoid activation function will result in a non-sparse neural network, while ReLU has good sparsity, with about 50% of the neurons being activated. In terms of optimization, compared to the saturation at both ends of the Sigmoid function, the ReLU function is a left saturated function, and its derivative is 1 when $x > 0$, which alleviates the gradient vanishing problem of the neural network to a certain extent and accelerates the convergence speed of gradient descent.

Leaky ReLU maintains a very small gradient γ when the input $x < 0$. In this way, when the neuron is not activated, there is a non-zero gradient to update the parameters to avoid never being activated. It is defined as

$$\text{Leaky ReLU} = \begin{cases} x, & x \geq 0 \\ \gamma x, & x < 0 \end{cases} = \max(0, x) + \gamma \min(0, x) \tag{21}$$

Parametric ReLU (PReLU) introduces a learnable parameter, and different neurons can have different parameters. For the i-th neuron, its PReLU is defined as

$$\mathrm{PReLU} = \begin{cases} x, & x \geq 0 \\ \gamma_i x, & x < 0 \end{cases} = \max(0, x) + \gamma_i \min(0, x) \tag{22}$$

where γ_i is the slope of the function at $x < 0$. If $\gamma_i = 0$, then PReLU degenerates to ReLU. If γ_i is a small constant, then PReLU can be regarded as LReLU. PReLU can allow different neurons to have different parameters, or a group of neurons can share one parameter.

Exponential Linear Unit (ELU) is an approximately zero-centered nonlinear function defined as

$$\mathrm{ELU} = \begin{cases} x, & x \geq 0 \\ \gamma\left(e^x - 1\right), & x < 0 \end{cases} = \max(0, x) + \min\left(0, \gamma\left(e^x - 1\right)\right) \tag{23}$$

where $\gamma \geq 0$ is a hyperparameter that determines the saturation curve when $x \leq 0$ and adjusts the output mean to be around 0. Figure 6 gives examples of ReLU, Leaky ReLU, and ELU functions.

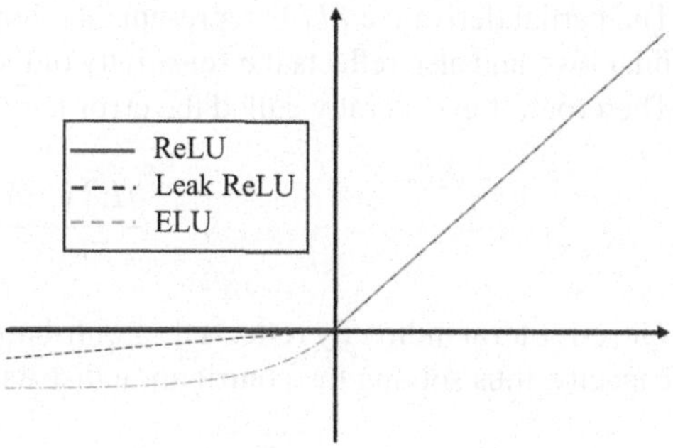

Fig. 6 ReLU, Leaky ReLU, and ELU functions

3.2 *Back Propagation Algorithm*

Using stochastic gradient descent for neural network parameter learning, given a data pair (x, y), it is input into the neural network model to obtain the network output. Assuming the loss function is L, to perform parameter learning, it is necessary to compute the derivative of the loss function with respect to each parameter [6].

Without loss of generality, calculate the partial derivatives of the parameters w and b in the l-th layer using the chain rule

$$\frac{\partial L\left(y, \hat{y}\right)}{\partial w_l} = \frac{\partial z_l}{\partial w_l}\frac{\partial L\left(y, \hat{y}\right)}{\partial z_l} \tag{24}$$

$$\frac{\partial L\left(y, \hat{y}\right)}{\partial b_l} = \frac{\partial z_l}{\partial b_l}\frac{\partial L\left(y, \hat{y}\right)}{\partial z_l} \tag{25}$$

The second term of both formulas is the partial derivative of the objective function with respect to the neuron l in the l layer, referred to as the error term. At this point, it is also necessary to calculate the partial derivatives of $\partial z/\partial w$, $\partial z/\partial b$, and $\partial L/\partial z$. Because $z = Wa + b$, the partial derivative is as follows

$$\frac{\partial z_l}{\partial w_l} = \left[\frac{\partial z_1}{\partial w_l}, \cdots, \frac{\partial z_i}{\partial w_l}, \cdots, \frac{\partial z_{M_l}}{\partial w_l}\right] = [0, \cdots, a_{l-1}, \cdots, 0] \tag{26}$$

According to the functional relationship between z and b, the partial derivative

$$\frac{\partial z_l}{\partial b_l} = I_{M_l} \in \mathbb{R}^{M_l \times M_l} \tag{27}$$

The partial derivative ∂*L*/∂*z* represents the impact of the *l*th layer of neurons on the final loss, and also reflects the sensitivity of the final loss to the *l*-th layer of neurons. Therefore, it is generally called the error term of the *l*th layer of neurons, that is

$$\delta_l = \frac{\partial L\left(y, \hat{y}\right)}{\partial z_l} \in \mathbb{R}^{M_l} \tag{28}$$

The error term indirectly reflects the contribution of different neurons to the network capacity, thus solving the contribution distribution problem better, because

$$\frac{\partial z_l}{\partial a_l} = W_{l+1}^{\mathrm{T}} \tag{29}$$

According to $a = f(z)$,where f is a bitwise function, thus there is

$$\frac{\partial a_l}{\partial z_l} = \frac{\partial f_l\left(z_l\right)}{\partial z_l} = \mathrm{diag}\left(f_l{}'\left(z_l\right)\right) \tag{30}$$

According to the chain rule, the error term at layer l is

$$\delta_l = \frac{\partial L\left(y, \hat{y}\right)}{\partial z_l} = \frac{\partial a_l}{\partial z_l}\frac{\partial z_{l+1}}{\partial a_l}\frac{\partial L\left(y, \hat{y}\right)}{\partial z_{l+1}} = \mathrm{diag}\left(f_l{}'\left(z_l\right)\right) W_{l+1}\delta_{l+1} \tag{31}$$

From the above equation, it can be seen that the error term of layer l can be calculated from the error term of layer $l + 1$, which is the essence of error backpropagation. The meaning of the backpropagation algorithm is that the error term (or sensitivity) of a neuron in layer l is the weighted sum of the error terms of all neurons in layer $l + 1$ that are connected to it. This is then multiplied by the gradient of the activation function of that neuron.

After calculating the error term for each layer, the gradient of the parameters for each layer can be obtained. Therefore, the training process of a feedforward neural network using the error backpropagation algorithm can be divided into the following three steps

1. Feedforward to compute the net input z and activation value a for each layer until the last layer.
2. Backpropagation calculates the error term for each layer.
3. Calculate the partial derivatives of the parameters at each layer and update the parameters.

In fact, the gradients of parameters can be automatically computed by the computer. Currently, mainstream deep learning frameworks include the functionality of automatic gradient computation [7], which means we can focus solely on the network structure and implement it with code, while the gradients can be calculated automatically without manual intervention, significantly improving development efficiency.

4 Convolutional Neural Network

Convolutional Neural Networks (CNNs) are generally composed of a feed forward neural network that is stacked with convolutional layers, pooling layers, and fully connected layers. CNNs have three structural characteristics: local connectivity, weight sharing, and pooling [8]. These characteristics provide CNNs with a certain degree of translation, scaling, and rotation invariance. Compared to feed-forward neural networks, CNNs have fewer parameters.

4.1 Convolution

Convolution is an important operation in analytical mathematics. It is often used in one-dimensional or two-dimensional convolution in signal processing or image processing.

One-dimensional convolution is often used in signal processing to calculate the delayed accumulation of a signal. Assume that a signal generator generates a signal x_t at every time t, and the decay rate of its information is w_k, that is, after $k - 1$ time steps, the information is w_k times the original. Then the signal y_t received at time t is the superposition of the information generated at the current time and the delayed information at the previous time as

$$y_t = w_1 x_t + w_1 x_{t-1} + \cdots + w_K x_{t-K+1} = \sum_{k=1}^{K} w_k x_{t-k+1} \tag{32}$$

where w is called the filter or convolution kernel, and K is the filter length. The convolution of the signal sequence $x_1, x_2, \ldots, x_K$ and the filter w is defined as

$$y = w * x \tag{33}$$

where $_*$ represents the convolution operation. Figure 7 gives an example of one-dimensional convolution of two filters. In general, the length of the filter K is much smaller than the length of the signal sequence x.

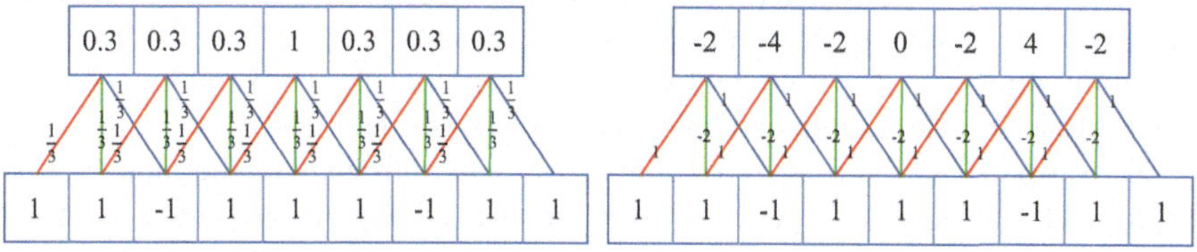

Fig. 7 One-dimensional convolution example

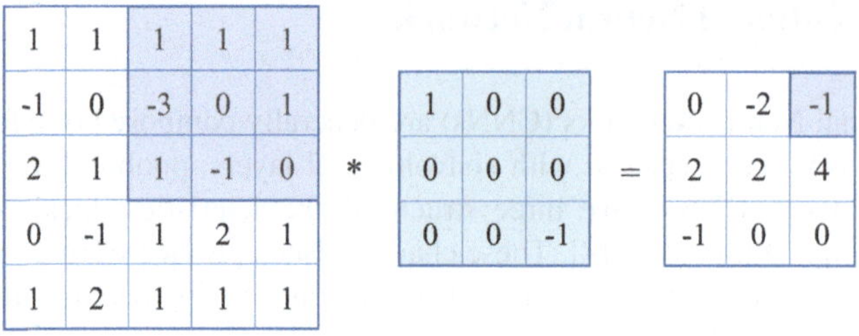

Fig. 8 Two-dimensional convolution example

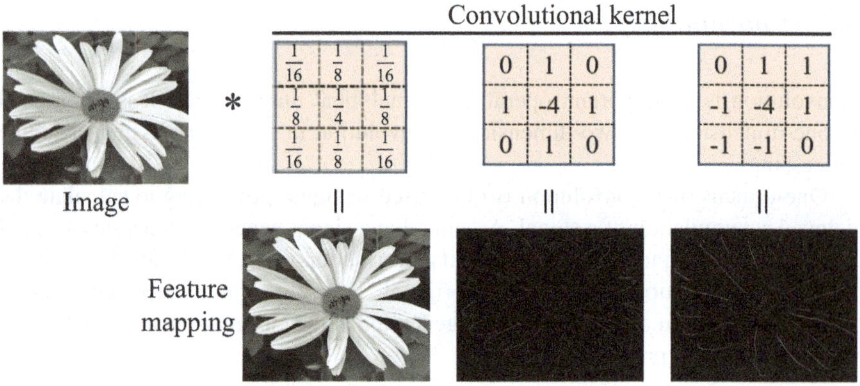

Fig. 9 Common filter examples

Convolution is also often used in image processing. Since the image is a two-dimensional structure, the one-dimensional convolution needs to be expanded, as Fig. 8. In image processing, convolution is often used as an effective method for feature extraction. The result obtained after applying the convolution operation to an image is called a feature map. Figure 9 presents several commonly used filters in image processing, along with their corresponding feature maps. The filter on the left side of the figure is a commonly used Gaussian filter, which can be used for smoothing and denoising images; the filters in the middle and on the right can be used to extract edge features [9].

Based on the standard definition of convolution, the sliding stride and zero padding of the convolution kernel can also be introduced to increase the diversity of convolution, allowing for more flexible feature extraction.

Stride refers to the time interval when the convolution kernel is sliding. Figure 10a shows an example of convolution with a stride of 2.

Zero padding is the process of adding zeros to both ends of the input vector. Figure 10b shows an example of convolution after adding one zero to each end of the input.

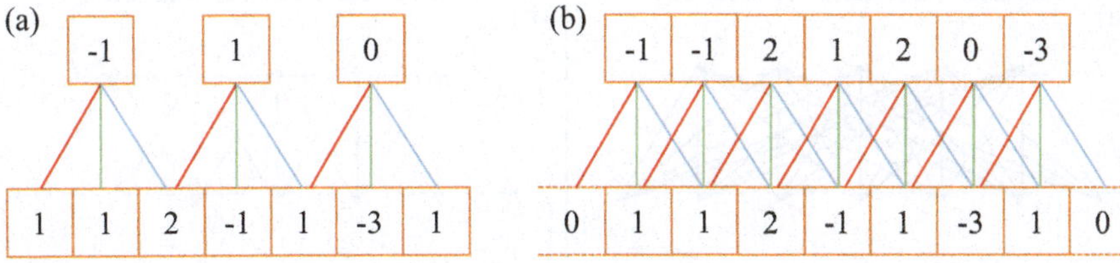

Fig. 10 Stride and padding. (**a**) Stride = 2; (**b**) Padding = 1

Assuming the number of input neurons in the convolutional layer is M, the convolution size is K, and the stride is S, with zero padding on both ends, the number of neurons in this convolutional layer is $(M - K + 2P)/S + 1$. Generally, there are three common types of convolutions: (1) Narrow convolution: stride $S = 1$, no zero padding on both ends, the output length after convolution is $M - K + 1$ (2). Wide convolution: stride $S = 1$, zero padding on both ends $P = K - 1$, the output length after convolution is $M + K - 1$ (3). Equal width convolution: stride $S = 1$, zero padding on both ends $P = (K - 1)/2$, the output length after convolution is M.

4.2 *Convolutional Network*

In a fully connected feedforward neural network, if the lth layer has M_l neurons, the l-1th layer has M_{l-1} neurons, and there are $M_l \times M_{l-1}$ connecting edges, that is, the weight matrix has $M_l \times M_{l-1}$ parameters. When M_l and M_{l-1} are both large, the weight matrix has many parameters and the training efficiency will be very low. If convolution is used instead of full connection, the net input z_l of the lth layer is the convolution of the l-1th layer activity value a_{l-1} and the convolution kernel w_l, that is

$$z_l = w_l * a_{l-1} + b_l \tag{34}$$

where the convolution kernel w is a learnable weight vector, and b is a learnable bias. According to the definition of convolution, the convolution layer has two very important properties.

1. Local Connection

   Each neuron in the convolutional layer is only connected to the neurons in a local window in the next layer, forming a local connection network. As Fig. 11, the number of connections between the convolutional layer and the next layer is greatly reduced, from the original $M_l \times M_{l-1}$ connections to $M_l \times K$ connections, where K is the convolution kernel size.

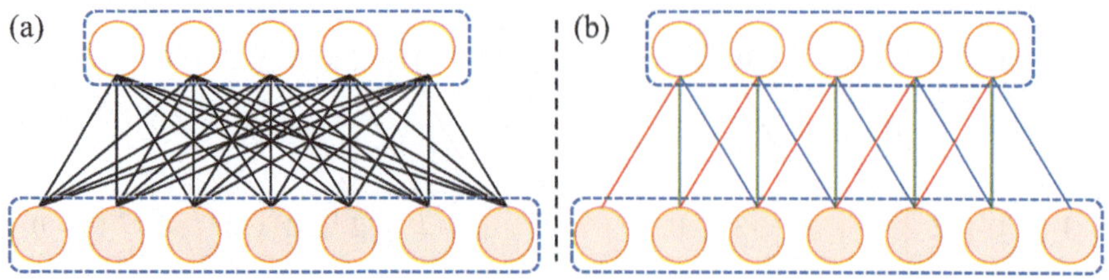

Fig. 11 Comparison of network layers. (**a**) Fully connected layer; (**b**) Convolutional layer

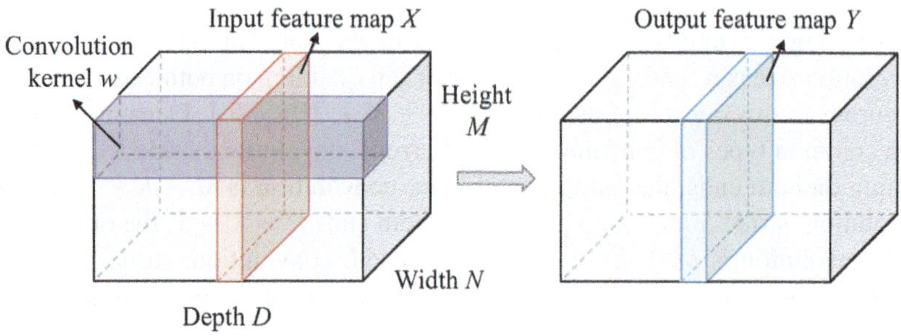

Fig. 12 Three-dimensional representation of convolutional layers

2. Weight Sharing

   It can be seen from the above formula that the convolution kernel w as a parameter is the same for all neurons in the l-th layer. As Fig. 11, the weights on all connections of the same color are the same. Weight sharing can be understood as a convolution kernel capturing only a specific local feature in the input data. Therefore, if you want to extract multiple features, you need to use multiple different convolution kernels.

Due to local connections and weight sharing, the parameters of the convolutional layer are only one K-dimensional weight w and one-dimensional bias b, for a total of $K + 1$ parameters. The number of parameters is independent of the number of neurons. In addition, the number of neurons in the lth layer is not chosen arbitrarily, but satisfies $M_l = M_{l-1} - K + 1$.

The function of the convolution layer is to extract the features of a local area. Different convolution kernels are equivalent to different feature extractors. Since convolutional networks are mainly used in image processing and images are two-dimensional structures, in order to make better use of the local information of the image, neurons are usually organized into three-dimensional neural layers with a size of height $M\times$ width $N\times$ depth D, consisting of D feature maps of size $M \times N$, as Fig. 12.

A feature map is a feature extracted from an image (or other feature map) after convolution. Each feature map can be used as a class of extracted image features. In order to improve the representation ability of the convolutional network, multiple

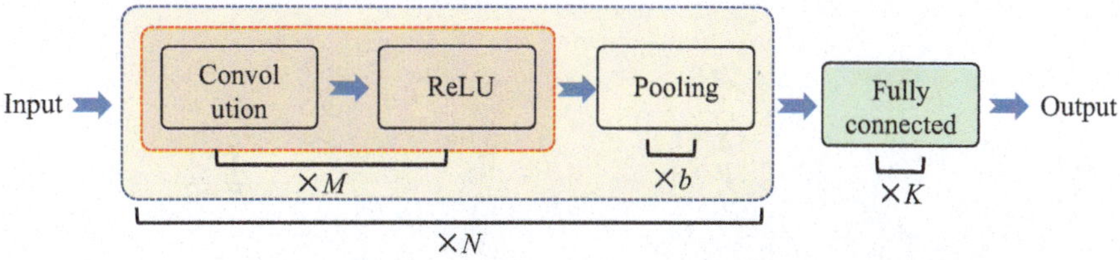

Fig. 13 Commonly used overall structure of convolutional networks

different feature maps can be used in each layer to better represent the characteristics of the image.

A complete convolutional network is composed of a cross-stack of convolutional layers, pooling layers, and fully connected layers [10]. The overall structure of the currently commonly used convolutional network is shown in Fig. 13. A convolutional block consists of consecutive M convolutional layers and b pooling layers. In a convolutional network, N consecutive convolutional blocks can be stacked, followed by K fully connected layers.

4.3 *Other Convolution Methods*

1. Transposed convolution

   The convolution operation can be used to convert high-dimensional features into low-dimensional features. If the step size is set greater than 1, the dimension of the output features can be further reduced. However, in some tasks, low-dimensional features need to be mapped to high-dimensional features, and it is still hoped to be achieved through convolution operations.

   A high-dimensional vector is x and a low-dimensional vector is z. Affine transformation can be used to achieve the mapping from high dimension to low dimension, that is $z = Wx$, where W is the transformation matrix. The reverse mapping from low dimension to high dimension can be easily achieved by transposing W, that is $x = W^{\mathrm{T}} z$.

   The convolution operation can also be written in the form of an affine transformation, that is

$$z_l = w * x = \begin{bmatrix} w_1 & w_2 & w_3 & 0 & 0 \\ 0 & w_1 & w_2 & w_3 & 0 \\ 0 & 0 & w_1 & w_2 & w_3 \end{bmatrix} x = Cx \tag{35}$$

   where C is a sparse matrix, and its non-zero elements come from the elements in the convolution kernel w. To achieve the mapping from the low-dimensional vec-tor z to the high-dimensional vector x, it can be implemented through the

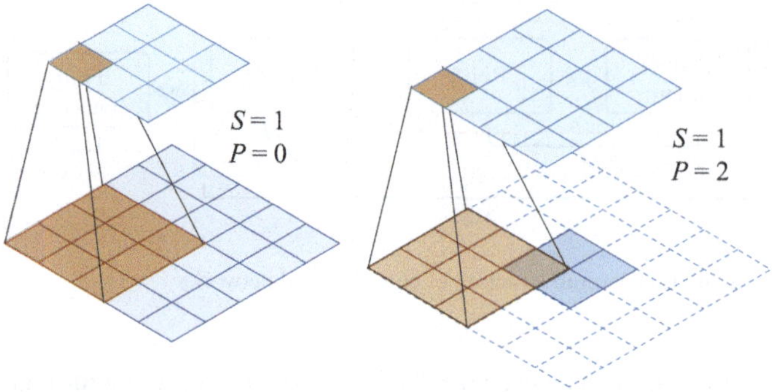

Fig. 14 The 2D convolution with stride $S = 1$ and no zero padding $P = 0$, and its corresponding transposed convolution

transpose of the affine matrix, that is

$$x = C^T x = \mathrm{rot180}(w) * z \tag{36}$$

where rot 180(·) denotes a 180° rotation. It can be seen that, from the perspective of affine transformations, the two convolution operations are also in a transposed relationship in form. Therefore, the convolution operation that maps low dimensional features to high-dimensional features can be called transposed convolution.

For an M-dimensional vector z and a convolution kernel of size K, if we want to map it to a higher-dimensional vector through convolution, we only need to pad the vector z with zeros on both ends by $P = K - 1$, and then perform the convolution, which will yield an $M + K - 1$ dimensional vector. Transposed convolution is also applicable to two-dimensional convolution, as Fig. 14.

2. Atrous convolution

For a convolutional layer, if you want to increase the receptive field of the output units, there are generally three ways to achieve this: increasing the size of the convolutional kernel; increasing the number of layers; or performing pooling operations before the convolution. The first two methods will increase the number of parameters, while the third method will lose some information [11].

Atrous convolution is a method that increases the receptive field of output units without increasing the number of parameters, as Fig. 15. Dilated convolution effectively increases the size of the convolution kernel by inserting 'dilations' between the elements of the kernel. If $D - 1$ dilations are inserted between every two elements of the convolution kernel, the effective size of the convolution kernel is

$$K' = K + (K - 1)(D - 1) \tag{37}$$

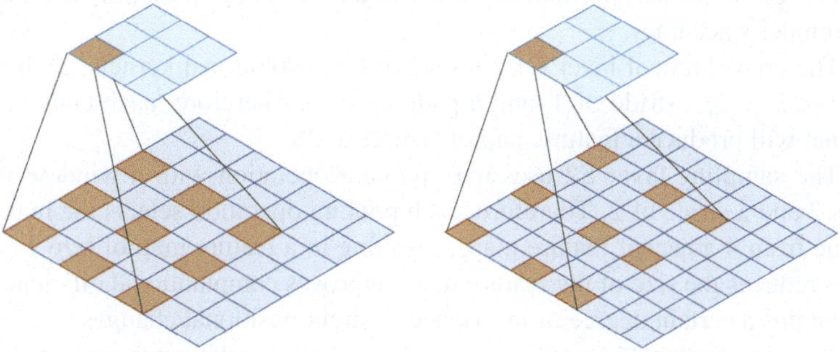

Fig. 15 Atrous convolution

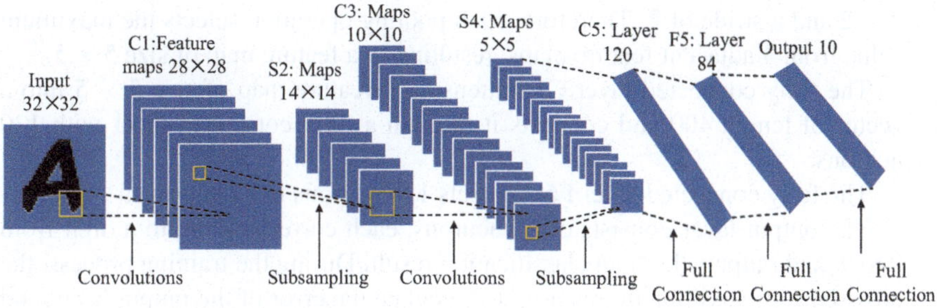

Fig. 16 LeNet-5 network architecture

where D is called the dilation rate. When $D = 1$, the convolution kernel is a standard convolution kernel.

4.4 *Classic Convolutional Neural Network*

1. LeNet-5

LeNet-5 was designed to address the problem of handwritten digit recognition and is considered one of the pioneering works in convolutional neural networks [12], as Fig. 16. This network was one of the first neural networks widely applied to digital image recognition and is also a milestone in the field of deep learning.

The basic structure of LeNet-5 includes a 7-layer network architecture (excluding the input layer), which consists of 2 convolutional layers, 2 subsampling layers (pooling layers), 2 fully connected layers, and an output layer.

The input layer receives handwritten digit images of size 32×32, which include grayscale values. In practical applications, input images are often preprocessed, such as normalizing pixel values, to speed up training and improve the model's accuracy.

The convolutional layer C1 consists of 6 convolutional kernels, each with a size of 5 × 5, a stride of 1, and a padding of 0. Therefore, each convolutional kernel will produce a feature map of size 28 × 28.

The sampling layer S2 uses max pooling operation with a window size of 2 × 2 and a stride of 2. Therefore, each pooling operation selects the maximum value from 4 adjacent feature maps, resulting in a feature map of size 14 × 14. This reduces the size of the feature map, improves computational efficiency, and maintains a certain degree of invariance to slight positional changes.

The convolutional layer C3 includes 16 convolutional kernels, each with a size of 2 × 2, a stride of 1, and a padding of 0. Therefore, each convolutional kernel will produce a feature map of size 5 × 5.

The sampling layer S4 uses max pooling operation with a window size of 2 × 2 and a stride of 2. Therefore, each pooling operation selects the maximum value from 4 adjacent feature maps, resulting in a feature map of size 5 × 5.

The fully connected layer C5 flattens each feature map of size 5 × 5 into a vector of length 400 and connects it through a fully connected layer with 120 neurons.

The fully connected layer F6 connects 120 neurons to 84 neurons.

The output layer consists of 10 neurons, each corresponding to a digit from 0 to 9, and outputs the final classification result. During the training process, the cross-entropy loss function is used to calculate the error of the output layer, and the weights of the convolutional kernels and fully connected layers are updated through the backpropagation algorithm.

2. AlexNet

AlexNet is the first modern deep convolutional network model, which was the first to use many techniques of modern deep convolutional networks, such as using GPUs for parallel training, adopting ReLU as the nonlinear activation function, using Dropout to prevent overfitting, and employing data augmentation to improve model accuracy [13]. The structure of AlexNet is shown in Fig. 17, which includes 5 convolutional layers, 3 pooling layers, and 3 fully connected layers (with the last layer being the output layer using the Softmax function).

Convolutional Layer C1: Filtering the input image of size 224 × 224 × 3 using 96 kernels, with a convolution kernel size of 11 × 11 × 3 and a stride of 4. A pair of feature maps of size 55 × 55 × 48 is passed through the ReLU activation function to generate activation maps. The activated images undergo max pooling with a size of 3 × 3 and a stride of 2, resulting in pooled feature maps of size 27 × 27 × 48.

Convolutional Layer C2: Using the output of Convolutional Layer C1 (after response normalization and pooling) as input, and applying 256 convolutional kernels with a kernel size of 5 × 5 × 48.

Convolutional Layer C3: It has 384 kernels, with a kernel size of 3 × 3 × 256, connected to the output of Convolutional Layer C2 (normalized and pooled).

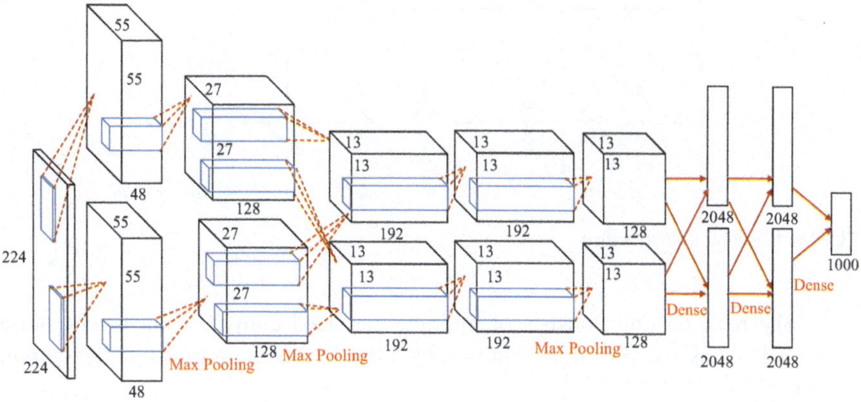

Fig. 17 AlexNet network architecture

Convolutional layer C4: There are 384 kernels with a kernel size of $3 \times 3 \times 192$.

Convolutional layer C5: It has 256 kernels, with a kernel size of $3 \times 3 \times 192$. Compared to layers C3 and C4, convolutional layer C5 includes a pooling layer, with the pooling kernel size also being 3×3 and a stride of 2.

The convolutional layers C3, C4, and C5 are interconnected, with no pooling or normalization layers in between.

Fully connected layer F6: As it is a fully connected layer, the convolution kernel size is $6 \times 6 \times 256$, with 4096 convolution kernels generating 4096 feature maps of size 1×1. Then it is passed through the ReLU function and Dropout processing. It is worth noting that AlexNet uses a Dropout layer to reduce the occurrence of overfitting.

Fully connected layer F7: Consistent with layer F6.

Fully connected layer F8: The output of the last fully connected layer is the input to a 1000-dimensional softmax, which will produce values for 1000 category predictions.

3. Residual network

Residual networks improve the efficiency of information propagation by adding direct connections to nonlinear convolutional layers [14]. In a deep network, it is expected that a nonlinear unit $f(x; \theta)$ approximates a target function $h(x)$. By splitting the target function into two parts: the identity function x and the residual function $h(x)$–x.

According to the universal approximation theorem, a nonlinear unit composed of neural networks has sufficient capacity to approximate the original target function or the residual function, but in practice, the latter is easier to learn. Therefore, the original optimization problem can be transformed into: allowing the nonlinear unit $f(x;\theta)$ to approximate the residual function $h(x)$–x, and using $f(x;\theta) + x$ to approximate $h(x)$.

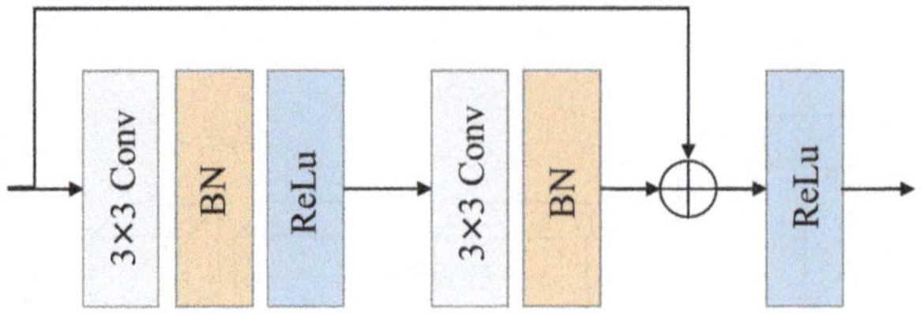

Fig. 18 Residual unit structure

The residual unit consists of multiple cascaded convolutional layers and a cross-layer skip connection, followed by a ReLU activation to obtain the output, as Fig. 18.

5 Deep Generative Models

Probabilistic generative model, referred to as generative model, is an important type of model in the field of probability statistics and machine learning. It refers to a series of models used to randomly generate observable data, that is, in a continuous or discrete high-dimensional space, there is a random vector that obeys an unknown data distribution. The generative model learns a parameterized model based on some observable samples to approximate the unknown distribution, and this model can be used to generate some samples so that the "generated" samples are as similar as possible to the "real" samples. Generative models usually contain two basic functions: probability density estimation and sample generation [15].

Generative models are widely used and can be used to model different data, such as images, text, sounds, etc. But for a complex distribution in a high dimensional space, density estimation and sample generation are usually not easy to achieve. Deep generative models use the ability of deep neural networks to approximate arbitrary functions to model a complex distribution or directly generate samples that conform to the distribution.

5.1 Variational Autoencoder

Assume that a generative model contains latent variables, meaning that some variables are unobservable. The observed variable X is a random vector in a high-dimensional space, while the latent variable Z is a random vector in a relatively low-dimensional space. The joint probability density function of this generative model can be decomposed as

$$p(x, z; \theta) = p(x | z; \theta)\, p(z; \theta) \tag{38}$$

where $p(z;\theta)$ is the probability density function of the prior distribution of the la-tent variable z, $p(x|z;\theta)$ is the conditional probability density function of the ob-served variable x when z is known, and θ represents the parameters of the two density functions. In general, it can be assumed that $p(z;\theta)$ and $p(x|z;\theta)$ belong to a certain parametric family of distributions, such as the normal distribution. The forms of these distributions are known, but the parameters θ are unknown, and estimation is performed by maximizing the likelihood.

Given a sample x, its log marginal likelihood log $p(z;\theta)$ can be decomposed as

$$\log p\left(x;\theta\right)=ELBO\left(q,x;\theta,\phi\right)+KL\left(q\left(z;\phi\right),p\left(z\left|x\right.;\theta\right)\right) \tag{39}$$

where $q(z;\phi)$ s the additional introduced variational density function with parameters ϕ, and $ELBO(q,x;\theta,\phi)$ is the evidence lower bound, that is

$$ELBO\left(q,x;\theta,\phi\right)=E_{z\sim q(z;\phi)}\log\frac{p\left(x,z;\theta\right)}{q\left(z;\phi\right)} \tag{40}$$

In each iteration of the solving algorithm, the theoretically optimal $q(z;\phi)$ is the posterior probability density function $p(z|x;\theta)$ of the latent variables, that is

$$q\left(z\left|x;\theta\right.\right)=\frac{p\left(x\left|z;\theta\right.\right)p\left(z;\theta\right)}{\int_{z}p\left(x\left|z;\theta\right.\right)p\left(z;\theta\right)\mathrm{d}z} \tag{41}$$

The calculation of the posterior probability density function $p(z|x;\theta)$ is a sta-tistical inference problem that involves integration. When the latent variable z is a finite one-dimensional discrete variable, the computation is relatively straightforward. However, in general cases, this posterior probability density function is difficult to compute and usually requires variational inference for approximate estimation. In variational inference, to reduce complexity, simpler distributions $q(z;\phi)$ are typically chosen to approximate the inference of $p(z|x;\theta)$. When $p(z|x;\theta)$ is quite complex, the approximation may not perform well. Additionally, the probability density function $p(x|z;\theta)$ is generally also quite complex, making it difficult to model directly using known families of distributions.

Variational autoencoders are a type of deep generative model that utilize neural networks to separately model two complex conditional probability density functions.

Using neural networks to estimate the variational distribution $q(z;\phi)$ is called the inference network. Theoretically, $q(z;\phi)$ can be independent of x. How-ever, since the goal of $q(z;\phi)$ is to approximate the posterior distribution $p(z|x;\theta)$, which is related to x, the variational density function is generally written as $q(z|x;\phi)$. The input of the inference network is x, and the output is the variational distribution $q(z|x;\phi)$.

Using neural networks to estimate the probability distribution $p(x|z;\theta)$ is called a generative network. The input of the generative network is z, and the output is the probability distribution $p(x|z;\theta)$.

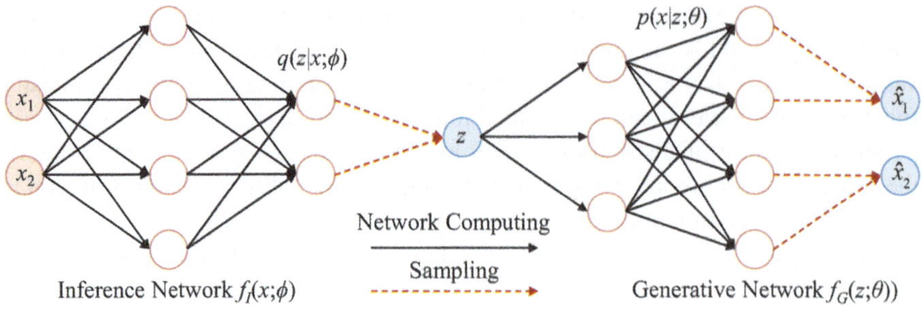

Fig. 19 The network structure of variational autoencoders

By combining the inference network and the generative network, we obtain the entire network structure of the variational autoencoder, as shown in Fig. 19, where solid lines represent network computation operations and dashed lines represent sampling operations.

The name of the variational autoencoder comes from its overall network structure, which is similar to that of an autoencoder. The inference network can be viewed as the "encoder", mapping observable variables to latent variables; the generative network can be seen as the "decoder", mapping latent variables back to observable variables. However, the principles behind variational autoencoders are completely different from those of autoencoders. In variational autoencoders, the outputs of the encoder and decoder are distributions rather than deterministic encodings.

The joint distribution $p(x|z;\theta)$ of the generative model can be decomposed into two parts: the prior distribution $p(z;\theta)$ of the latent variable z and the conditional probability distribution $p(x|z;\theta)$. For simplicity, it is generally assumed that the prior distribution of the latent variable z is an isotropic standard Gaussian distribution. Each dimension of the latent variable z is independent of the others. The conditional probability distribution $p(x|z;\theta)$ can be modeled using a generative network, which also represents the conditional probability distribution $p(x|z;\theta)$ with a parametric family of distributions, the parameters of which can be computed using the generative network. Depending on the type of variable x, it can be assumed that $p(x|z;\theta)$ follows different families of distributions.

If x is a binary vector of dimension D, we can assume that $p(x|z;\theta)$ follows a multivariate Bernoulli distribution, that is

$$p(x|z;\theta) = \prod_{d=1}^{D} p(x_d|z;\theta) = \prod_{d=1}^{D} \gamma_d^{x_d}(1-\gamma_d)^{1-x_d} \tag{42}$$

where γ is the parameter of the d-dimensional distribution. The parameter γ of the distribution can be predicted by the generative network.

If x is a continuous vector in D-dimensions, we can assume that $p(x|z;\theta)$ follows a Gaussian distribution with diagonal covariance, that is

$$p(x|z;\theta) = \mathrm{N}\left(x; \mu_G, \sigma_G^2 I\right) \tag{43}$$

Both μ_G and σ_G can also be predicted using a generative network. The goal of the generative network is to find a set of network parameters θ that maximizes the evidence lower bound, that is

$$\theta^* = \arg\max_{\theta} ELBO(q, x; \theta, \phi) \tag{44}$$

5.2 Generative Adversarial Network

Variational autoencoders explicitly construct the density function of samples and solve for parameters through maximum likelihood estimation, known as explicit density models. For example, the density function of variational autoencoders is $p(x,z;\theta) = p(x|z;\theta)p(z;\theta)$. Although a neural network is used to estimate $p(x|z;\theta)$, it still assumes that $p(x|z;\theta)$ belongs to a family of parametric distributions, and the neural network is merely used to predict the parameters of this parametric distribution family. This somewhat limits the capabilities of the neural network [16].

If you just want a model to generate samples that conform to the data distribution, you can estimate the density function of the data distribution without showing it. Assume that there is a simple and easy to sample distribution $p(z)$ in the low dimensional space, $p(z)$ is usually a standard multivariate normal distribution. Using a neural network to construct a mapping function $G(z)$ is called a generating network. By utilizing the powerful fitting ability of neural networks, $G(z)$ is made to obey the data distribution $p_r(x)$. This kind of model is called implicit density model. An implicit model does not explicitly model $p_r(x)$, but rather models the generation process. Figure 20 shows the process of generating samples by implicit model.

A key aspect of implicit density models is how to ensure that the samples generated by the generative network conform to the true data distribution. Since an explicit density function is not constructed, methods such as maximum likelihood estimation are employed for training. Generative adversarial networks (GANs)

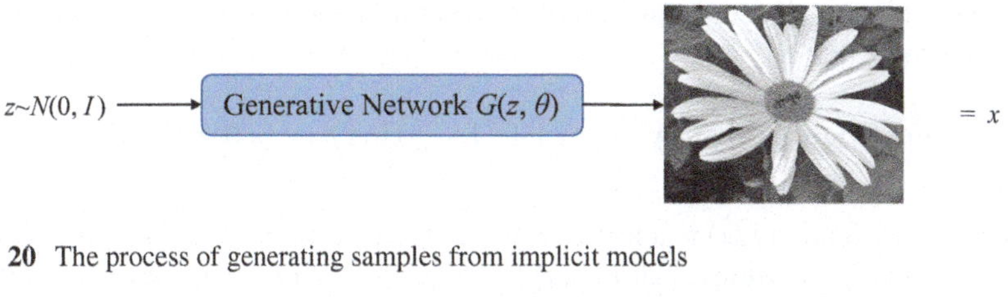

Fig. 20 The process of generating samples from implicit models

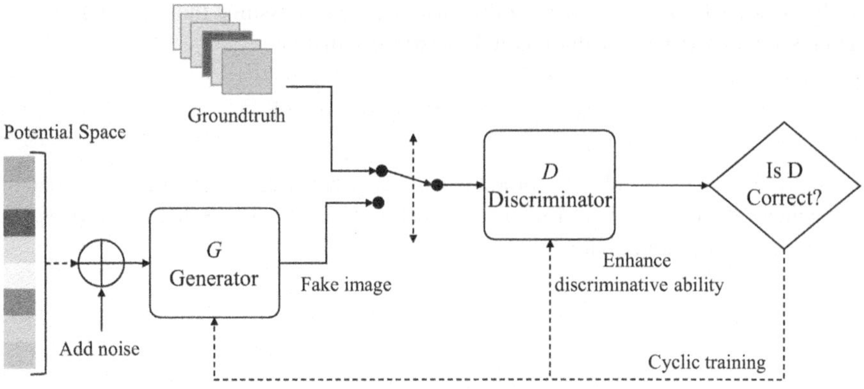

Fig. 21 Generative adversarial network architecture diagram

utilize adversarial training to ensure that the samples produced by the generative network conform to the true data distribution. In GANs, there are two networks engaged in adversarial training. One is the discriminator network, which aims to accurately determine whether a sample comes from real data or is generated by the generative network; the other is the generative network, which aims to produce samples that the discriminator cannot distinguish from real data. These two networks, with opposing objectives, are trained alternately [17]. When convergence is achieved, if the discriminator can no longer determine the source of a sample, it implies that the generative network can produce samples that conform to the true data distribution. The flowchart of the generative adversarial network is shown in Fig. 21.

The goal of the discriminator network D is to distinguish whether a sample x comes from the real distribution or from the generative model. Therefore, the discriminator network is essentially a binary classifier. A label $y = 1$ indicates that the sample comes from the real distribution, while $y = 0$ indicates that the sample comes from the generative model. The output of the discriminator network D is the probability that x belongs to the real data distribution, that is

$$p\left(y = 1\left|x\right.\right) = D\left(x;\phi\right) \tag{45}$$

The probability that the sample comes from the generative model is $p(y = 0|x) = 1\text{-}D(x;\phi)$. Given a sample (x,y), where $y = \{1, 0\}$ indicates whether it comes from $p_r(x)$ or $p_\theta(x)$, the objective function of the discriminator network is to minimize the cross-entropy, that is

$$\min_{\phi} -\left(E_x\left[y\log p\left(y = 1\left|x\right.\right) + (1-y)\log p\left(y = 0\left|x\right.\right)\right]\right) \tag{46}$$

when the distribution $p(x)$ is a mixture of the distribution $p_r(x)$ and the distribution $p_\theta(x)$ in equal proportions, that is $p(x) = [p_r(x) + p_\theta(x)]/2$. The above formula is equivalent to

$$\max_{\phi} E_{x\sim p_r(x)}\left[\log D(x;\phi) + E_{z\sim p(z)} \log\left(1 - D\left(G(z;\theta);\phi\right)\right)\right] \tag{47}$$

where θ and ϕ are the parameters of the generator network and the discriminator network respectively. The goal of the generative network is exactly the opposite of that of the discriminative network, that is, to let the discriminative network identify the samples it generates as real samples, as follows

$$\max_{\theta}\left(E_{z\sim p(z)}\left[\log D\left(G(z;\theta);\phi\right)\right]\right) = \min_{\theta}\left(E_{z\sim p(z)}\left[\log\left(1 - D\left(G(z;\theta);\phi\right)\right)\right]\right) \tag{48}$$

The two objective functions in the above equation are equivalent. However, in practical training, the former is generally used because it has better gradient properties. The gradient of the function log(x) is much smaller when x is close to 1 compared to when it is close to 0, approaching the "saturation" region. Thus, when the discriminator network D believes with high probability that the samples generated by the generator network G are "fake" samples, the gradient of the objective function with respect to θ becomes very small, which is detrimental to optimization.

Compared to single-objective optimization tasks, training generative adversarial networks (GANs) is relatively difficult and often unstable. Generally, it is necessary to balance the capabilities of the two networks. For the discriminator network, its initial discriminative ability should not be too strong, otherwise it will be difficult to enhance the capabilities of the generator network [18]. However, the discriminative ability of the discriminator network also cannot be too weak, or the generator network trained against it will not perform well either. During training, some techniques need to be employed to ensure that in each iteration, the discriminator network is slightly stronger than the generator network, but not too much.

6 General Conclusions

This chapter systematically elaborates on the fundamental theories and key technologies of deep learning. Starting from the essence of the problem, it deeply analyzes the core position of the contribution distribution problem in deep learning and clarifies the unique advantages of neural networks in feature representation and information processing. On this basis, it provides a detailed introduction to the basic components of deep learning, including model construction, learning criteria, and optimization algorithms, revealing the essential process of parameter learning through gradient descent and error backpropagation mechanisms. Subsequently, the chapter further analyzes the structural principles and typical architectures of convolutional neural networks, demonstrating their powerful capabilities and efficiency in image modeling. Finally, by introducing variational autoencoders and generative adversarial networks, it explores the theoretical framework and practical applications of deep generative models in probabilistic modeling and sample generation, highlighting the transition of deep learning from discriminative modeling to generative modeling. This chapter not only lays a solid foundation for readers in the theories and methods of deep learning but also provides a systematic knowledge map for understanding its wide applications in intelligent perception, pattern recognition, and data modeling.

Acknowledgments This work was supported by the National Natural Science Foundation of China (12374271, 62375065, 62405079); Fundamental Research Funds for the Central Universities (HIT.OCEF.2024055); the Fund of National Laboratory on Adaptive Optics, China (FNLAO-24-MS-O07); Natural Science Foundation of Heilongjiang Province of China (LH2024A009); Key Laboratory of Image Processing and Intelligent Control (Huazhong University of Science and Technology), Ministry of Education, Wuhan, China (IPIC2025-05); Key Laboratory of Advanced Medical Imaging and Intelligent Computing of Guizhou Province (AMIIC250204); China Postdoctoral Science Foundation-CCTEG Joint Support Program (2025T011ZGMK); State Key Laboratory of Ultrafast Optical Science and Technology, Xi'an Institute of Optics and Precision Mechanics, CAS (2025SKL-uFASTKF04).

Competing Interests The authors have no conflicts of interest to declare that are relevant to the content of this chapter.

References

1. Dong XP, Shen JB, Wang WG, Shao L, Ling HB, Porikli F. Dynamical hyperparameter optimization via deep reinforcement learning in tracking. IEEE Trans Pattern Anal Mach Intell. 2021;43(5):1515–29.
2. Wang X, Wang S, Liang XX, Zhao DW, Huang JC, Xu X, Dai B, Miao QG. Deep reinforcement learning: a survey. IEEE Trans Neural Netw Learn Syst. 2024;35(4):5064–78.
3. Li XH, Xiong HY, Li XJ, Wu XY, Zhang X, Liu J, Bian J, Dou DJ. Interpretable deep learning: interpretation, interpretability, trustworthiness, and beyond. Knowl Inf Syst. 2022;64(12):3197–234.
4. Zhang XC. Deep learning-based multi-focus image fusion: a survey and a comparative study. IEEE Trans Pattern Anal Mach Intell. 2022;44(9):4819–38.
5. Matsuo Y, LeCun Y, Sahani M, Precup D, Silver D, Sugiyama M, Uchibe E, Morimoto J. Deep learning, reinforcement learning, and world models. Neural Netw. 2022;152:267–75.
6. Wright LG, Onodera T, Stein MM, Wang TY, Schachter DT, Hu Z, McMahon PL. Deep physical neural networks trained with backpropagation. Nature. 2022;601(7894):549–55.
7. Mohammed A, Kora R. A comprehensive review on ensemble deep learning: opportunities and challenges. J King Saud Univ Com. 2023;35(2):757–74.
8. Shlezinger N, Whang J, Eldar YC, Dimakis AG. Model-based deep learning. Proc IEEE. 2023;111(5):465–99.
9. Chen YB, Mancini M, Zhu XT, Akata Z. Semi-supervised and unsupervised deep visual learning: a survey. IEEE Trans Pattern Anal Mach Intell. 2024;46(3):1327–47.
10. Xiao ZW, Xing HL, Zhao BW, Qu R, Luo SX, Dai PL, Li K, Zhu ZH. Deep contrastive representation learning with self-distillation. IEEE Trans Emerg Top Comput Intell. 2024;8(1):3–15.

11. Berahmand K, Daneshfar F, Salehi ES, Li YF, Xu Y. Autoencoders and their applications in machine learning: a survey. Artif Intell Rev. 2024;57(2):28.
12. Wu WC, Liu SJ, Xia YL, Zhang YA. Dual residual attention network for image denoising. Pattern Recogn. 2024;149:110291.
13. Kaveh M, Mesgari MS. Application of meta-heuristic algorithms for training neural networks and deep learning architectures: a comprehensive review. Neural Process Lett. 2023;55(4):4519–622.
14. Tian SS, Li LS, Li WJ, Ran H, Ning X, Tiwari P. A survey on few-shot class-incremental learning. Neural Netw. 2024;169:307–24.
15. Dubey SR, Singh SK, Chaudhuri BB. Activation functions in deep learning: a comprehensive survey and benchmark. Neurocomputing. 2022;503:92–108.
16. Zhao X, Wang LM, Zhang YF, Han XM, Deveci M, Parmar M. A review of convolutional neural networks in computer vision. Artif Intell Rev. 2024;57(4):99.
17. Gui J, Sun ZN, Wen YG, Tao DC, Ye JP. A review on generative adversarial networks: algorithms, theory, and applications. IEEE Trans Knowl Data Eng. 2023;35(4):3313–32.
18. Reyad M, Sarhan AM, Arafa M. A modified Adam algorithm for deep neural network optimization. Neural Comput Appl. 2023;35(23):17095–112.

Deepfake in Image and Video Processing

Bin Gao, Shuangcheng Wang, Xinhui Li, Wenshuo Zhang, Yuming Cui, and Chang Qi

Abstract Deepfake technology represents a major revolution in synthetic media, enabling highly realistic image and video processing. This chapter explores the core techniques behind deepfake generation, including autoencoders, generative adversarial networks (GANs), and diffusion models, each offering distinct advantages in quality and computational demands. It also examines cutting-edge detection methods, such as forgery analysis, noise detection, and temporal consistency verification, which are critical for identifying synthetic content.The development of deepfake technology hinges on robust datasets, efficient algorithms, and scalable evaluation metrics. By analyzing these components, this work aims to bridge the gap between synthetic media generation and detection, fostering innovation while addressing ethical challenges.

1 Introduction

Advances in artificial intelligence and image processing are transforming the way people interact with digital images and videos, posing unprecedented challenges to the authenticity and credibility of digital content. Deepfake is a technology based on deep learning that generates highly realistic multimedia content by replacing or synthesizing facial features, voices, or other visual elements in images or videos. Early deepfake technology focused on facial replacement, but it has since expanded to include emotion manipulation, voice synthesis, and even the creation of fully virtual avatars, becoming an important research direction in multimodal artificial intelligence. However, as the technical barriers have lowered, deepfake has increasingly been misused in areas such as fake news, identity forgery, and political manipulation, sparking widespread concerns about information security. In this

B. Gao (✉) · S. Wang · X. Li · W. Zhang · Y. Cui, · C. Qi
School of Computer and Big Data, Heilongjiang University, Harbin, China
e-mail: gaobin@hlju.edu.cn

H. Chen, Z. Liu (eds.), *Artificial Intelligence in Digital Image Processing: Theories, Methods, and Applications*, Studies in Computational Intelligence 1246,
https://doi.org/10.1007/978-3-032-12891-1_2

context, understanding the technical principles of Deepfake and researching reliable detection methods have become urgent needs for both academia and industry.

In 2014, Google researcher Ian Goodfellow proposed generative adversarial networks in his paper "Generative Adversarial Nets" [1], laying the core algorithmic foundation for deepfake. In 2018, NVIDIA launched the StyleGAN model, which significantly improved the quality of generated images. The model's face resolution reached 1024×1024 for the first time, making it difficult to distinguish between real and fake images. In 2021, OpenAI released DALL-E, based on diffusion models [2] and CLIP, demonstrating powerful multimodal generation capabilities. Deepfake technology expanded from face swapping to full-scene synthesis. This marks a new stage in the development of deepfake technology, with the introduction of diffusion models further enhancing the realism of the generated content. Simultaneously, deepfake technology emerged.

Achieving high-quality deepfakes relies on multiple technologies. This chapter primarily introduces deepfake implementations based on autoencoders [3], generative adversarial network models, and diffusion models [4]. Autoencoders excel at facial identity replacement through latent space operations, generative adversarial networks use a generator-discriminator adversarial training paradigm to generate high-fidelity image outputs, and diffusion models can further improve image quality through iterative denoising processes. In addition, we provide examples of feasible datasets used for training in the preparatory work and present an intuitive comparison based on the three different generation methods, offering a feasible analysis and demonstration for the implementation of deepfakes.

This chapter also systematically outlines multi-dimensional technical approaches for detecting deepfakes. Image forgery and noise detection analyzes physical traces and noise patterns in images, utilizing frequency domain transformations and deep learning models to identify inherent defects in generation algorithms. Local feature recognition and detection propose attention mechanism layers and local feature discovery techniques, enhancing the model's sensitivity and interpretability to subtle forgery traces by locating facial tampering regions and amplifying texture difference features. Dual-branch frame-level and temporal-level compressed video detection combines MesoNet and ResNet18 to capture forgery clues from the static features of key frames and the dynamic temporal consistency of residual frames. A soft maximum fusion strategy is used to optimize detection robustness. These methods collectively drive the development of detection technology toward multi-modal collaboration, fine-grained analysis, and anti-interference capabilities.

To promote the development of deepfake detection technology, researchers have also created a series of standardized datasets to facilitate model training, evaluation, and comparative analysis. These datasets cover a variety of forgery methods, image qualities, and application scenarios, enabling comprehensive evaluation of the algorithms' generality and robustness. Meanwhile, as real-world needs become increasingly apparent, efficiency-oriented metrics have become a supplement to the comprehensive performance evaluation of deepfake detection systems.

2 Implementation of Deepfake

Based on the development of artificial intelligence, deepfakes can typically be created using autoencoder technology, adversarial generative network technology, and diffusion models. We can choose the technology that best suits our needs based on the characteristics of these three different technologies. It is worth noting that before selecting a model, we typically need to carry out some necessary preparatory work, including data collection and preprocessing.

2.1 Related Works

2.1.1 Data Collection and Preprocessing

The quality of the output is constrained by the training dataset. The core objective of this stage is to obtain high-quality, diverse datasets of the target person and the replacement person, typically including image data and video data. For video data, it should cover as many angles, lighting conditions, and facial expressions as possible, with a duration typically ranging from 10 to 60 minutes, to ensure that the facial area is at least 200×200 pixels. Image data can be used as a necessary supplement to fill in missing angles or expressions in the video.

There are many common facial datasets for achieving deepfake, including CelebA [5] with 200,000 facial images and 40 annotated attributes, FFHQ [6] with 70,000 high-resolution facial images, and VGGFace2 [7] with 3.31 million facial images from 9131 individuals. Additionally, there are datasets specifically for video face swapping, such as FaceForensics++ [8], which includes 1000 original videos and four forgery methods.

In addition to using publicly available datasets, images or videos can be obtained from platforms like YouTube, TikTok, and X, provided that platform policies are adhered to.

The goal of data preprocessing is to transform raw data into standardized inputs sui for model training through operations such as cleaning and standardization. Specifically, this includes: unifying the format and size of the data to meet the model's input requirements, eliminating noise as much as possible to improve data purity, and thereby improving the quality of the model's output.

To unify the size of the data, methods such as forced scaling or a combination of scaling and cropping while maintaining the aspect ratio are often used. Additionally, for images that need to preserve their original aspect ratio, edge padding can be applied. These operations can be implemented using the resize function in OpenCV and the resize function in PyTorch. To standardize data format, it is necessary to ensure consistency in color space, data range, and file format. These functions can be implemented using the cvtColor function in OpenCV and the Normalize function in PyTorch.

To eliminate noise, it is necessary to identify the type of noise through quantitative analysis and locate the noise distribution area, such as using local variance maps to identify areas with concentrated high-frequency noise. Then, select an appropriate denoising method based on the characteristics of the noise: traditional filtering is suitable for mild noise and is computationally efficient, deep learning models can handle complex noise but require GPU acceleration, while hybrid strategies combine the advantages of both, using different parameters for local enhancement in key areas of the face. Finally, the denoising effectiveness is validated using objective metrics to ensure that the data quality meets the training requirements of the generative model.

After collecting and preprocessing the data, it is divided into training and validation sets in a 9:1 ratio.

2.1.2 Model Selection

Readers can select a model based on the information provided in the Table 1 to implement deepfake.

Of course, the above models can be used in combination to solve more complex deepfake generation problems.

2.2 Deepfake Implementation Based on Autoencoders

An autoencoder is a neural network that simulates the human brain's ability to compress and reconstruct information. It is primarily used for dimensionality reduction, feature extraction, and reconstruction. Its core architecture consists of two symmetrical neural networks: an encoder and a decoder. The encoder is responsible for extracting key features from the input image, compressing the original image from thousands of pixels to hundreds of pixels. It focuses on key facial features, eliminating noise and unimportant parts of the image, indicating that the image is a compressed version, which ultimately helps in remembering the basic features of the image. The decoder is responsible for decompressing information in the latent space to reconstruct the appearance of the original image. Since the autoencoder employs

Table 1 Multi-dimensional model comparison

Dimension	Autoencoder	GANs	Diffusion model
Training stability	High	Low	Medium
Generation quality	Low	High	Very high
Compute requirement	Consumer GPUs	High-end GPUs	Multi-GPU
Training speed	Hours	Days	Weeks
Data requirements	Thousands	Tens of thousands	Massive

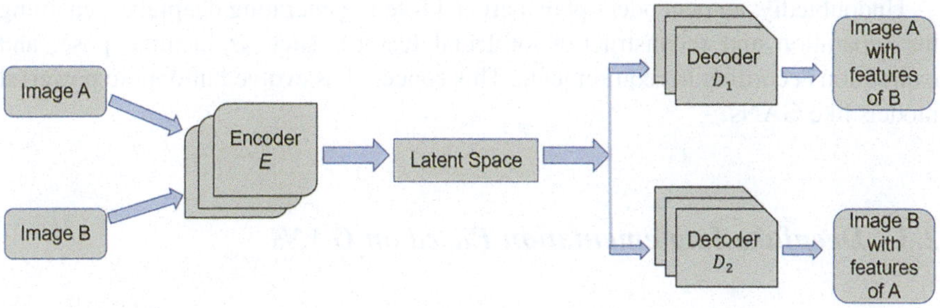

Fig. 1 The architecture of autoencoder

an unsupervised method, it compares the output image with the input image and makes adjustments between layers until it produces an image similar to or nearly a mirror image of the input.

The mathematical expression in the encoding stage can be expressed as

$$z = f_\theta(x) = \sigma(W_e x + b_e) \tag{1}$$

Among them, x is the input data, f_θ is the encoder network, z is the low-dimensional latent variable of the output, σ is the activation parameter, and W_e and b_e are the encoder parameters.

And the mathematical expression in the decoding stage can be expressed as

$$\hat{x} = g_\phi(z) = \sigma(W_d z + b_d) \tag{2}$$

Among them, $\hat{x}$ is the reconstructed output data, g_ϕ is the decoder network, σ is the activation parameter and W_d and b_d are the decoder parameters.

The ingenuity of the autoencoder structure lies in the latent space. It reduces the dimension of latent variables to be much smaller than that of the input data, forcing the network to discard redundant details in the input photo and retain only the most discriminative features. The purpose is to enable the model to recreate the image rather than simply output the same image. The performance of an autoencoder can be represented by the similarity between the input and output images. The higher the similarity between the input and output images, the better the encoder performance.

In the process of creating deepfakes, autoencoders typically adopt a single encoder-dual decoder architecture. As Fig. 1, in this architecture, two images—image A and image B—are input. An encoder E processes the two different images to retain their key features in the latent space, while two decoders D_1 and D_2 are trained separately to learn the common information between the two images. After training, the decoders aim to produce image A with the features of image B and image B with the features of image A. This is a face-swapping technique that is not limited to image synthesis. After making certain modifications to the algorithm, it is also feasible to achieve video synthesis [9].

Undoubtedly, autoencoders play a crucial role in generating deepfakes, enabling the separation and reconstruction of facial features such as identity, pose, and expression according to requirements. This concept has evolved into more powerful models like GANs.

2.3 Deepfake Implementation Based on GANs

GANs are the most widely used type of deep generative model for creating deepfakes, primarily employed for data augmentation, image generation, and synthesis. Compared to auto-encoders, GANs can generate more realistic images. Similarly, their core architecture consists of two neural networks: a generator and a discriminator. The generator is responsible for learning the distribution of real data and generating samples that are as realistic as possible, while the discriminator's task is to distinguish between real data and generated fake data. Obviously, in the early stages, the data generated by the generator is easily detected by the discriminator. However, since the discriminator continuously identifies flaws in the generated data, the generator makes targeted improvements and continuously optimizes its generation strategy, while the discriminator enhances its ability to distinguish generated data through iterative training. This iterative process is commonly referred to as the adversarial game between the generator and the discriminator to enhance the realism of the generated data.

In the process of competition between the generator G and the discriminator D, their ultimate goal is to generate the most realistic image which can be mathematically expressed as

$$\min_{G}\max_{D} V(D, G) = \mathbb{E}_{x\sim p_{\text{data}}}[\log D(x)] + \mathbb{E}_{z\sim p_z}[\log(1 - D(G(z)))] \tag{3}$$

Among them, $x \sim p_{\text{data}}$ represents the true data, $z \sim p_z$ represents the noise distribution, $D(x)$ represents the output probability of the discriminator for the true data, and $D(G(z))$ represents the output probability of the discriminator for the generated data.

As Fig. 2, the discriminator D receives two inputs simultaneously: one is a real face frame $x \sim p_{\text{data}}$, and the other is a face replacement frame $G(z)$ generated by the generator. It analyzes the local details and global features of the image layer by layer through a multilayer convolutional neural network, and finally outputs a probability value between 0 and 1, indicating the confidence in the authenticity of the image. When this probability value is equal to 0.5, it indicates that the model has reached the global optimal solution. As training progresses, the discriminator D becomes increasingly sensitive, and changes in its parameters influence the parameters of the generator G. The generator G takes as input a random noise vector $z \sim p_z$. The generator G maps the noise to a new data space, thereby generating realistic fake face data $G(z)$. Technically, the generator G employs

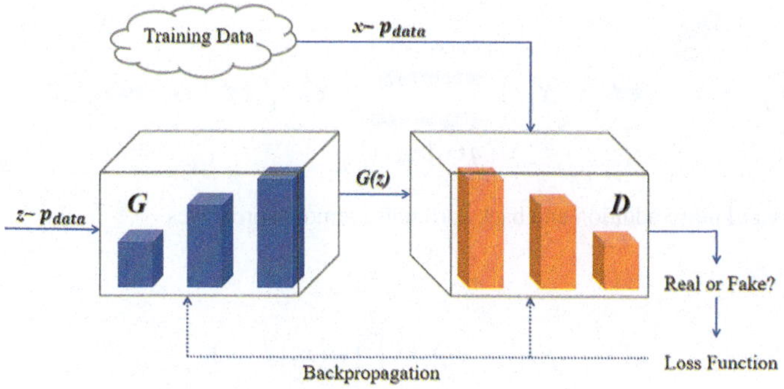

Fig. 2 Schematic description of a GANs learning framework

an inverted convolutional structure, enabling it to progressively assemble low-dimensional noise into a complete image, thereby generating a fake face image [10].

GANs also play a crucial role in the creation of deepfakes, but due to improper network structure design, GANs suffer from issues such as pattern collapse, non-convergence, and instability during training. The Big Data Analytics Research Center at the Hong Kong Polytechnic University has researched several solutions for better designing and optimizing GANs, including new objective functions and alternative optimization algorithm [11].

2.4 Deepfake Implementation Based on Diffusion Model

In 2020, the Denoising Diffusion Probabilistic Model (DDPM) was proposed, referred to as a diffusion model, further complementing autoencoder and GANs. The diffusion model framework comprises two distinct stages: the forward process and the backward process, with the forward process also known as the diffusion process. In the forward stage, original data such as images are gradually degraded into random noise by progressively adding Gaussian noise. In the backward stage, the neural network learns to gradually reconstruct the data from the noise. This process is similar to the generator in GANs, except that the generator undergoes dimensionality changes, while the DDPM backward process does not. Both the forward and backward processes are parameterized Markov chains, and their synergistic interaction enables the model to generate highly ordered samples from disordered noise, demonstrating significant mathematical rigor.

As Fig. 3, the forward process from x_0 to x_T can be referred to as a noise addition process, which gradually adds noise to the original image until it becomes a set of pure noise. In this process, the image x_t is only related to the previous moment x_{t-1} and can be regarded as a Markov process, satisfying

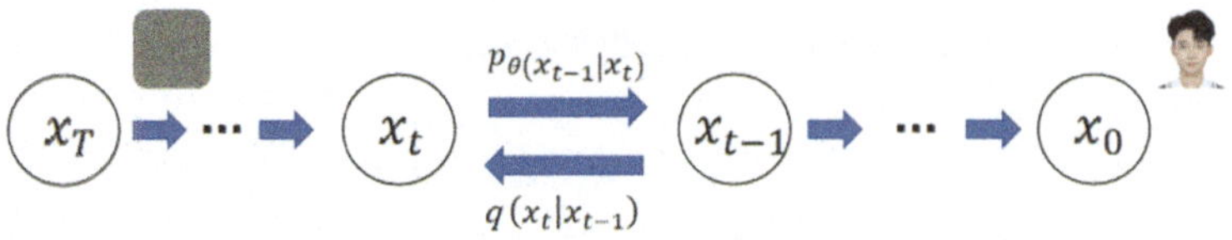

Fig. 3 Forward noise addition and backward noise removal processes

$$q(\boldsymbol{x}_{1:T}|\boldsymbol{x}_0) = \prod_{t=1}^{T} q(\boldsymbol{x}_t|\boldsymbol{x}_{t-1}) \tag{4}$$

And each step of the diffusion process is defined as

$$q(\boldsymbol{x}_t|\boldsymbol{x}_{t-1}) = \mathcal{N}\left(\boldsymbol{x}_t; \sqrt{1-\beta_t}\boldsymbol{x}_{t-1}, \beta_t\mathbf{I}\right) \tag{5}$$

Among them, $q(x_{1:T}|x_0)$ represents the probability distribution of obtaining the noisy image x_t after t steps of noise addition starting from the original image x_0; $q(x_t|x_{t-1})$ denotes the conditional probability distribution of obtaining the current image x_t given the previous image x_{t-1} at step t; This is a Gaussian distribution with a mean of $\sqrt{1-\beta_t}x_{t-1}$ and a variance of β_t, where β_t is a predefined noise variance coefficient, and I is the identity matrix.

According to the formula, x_t can be obtained by reparameterized sampling. $\epsilon \sim N(0, I)$, $\alpha_t = 1 - \beta_t$, then define $\overline{\alpha_t} = \prod_{i=1}^{T} \alpha_i$, so that the relationship between x_t and x_0 can be derived as follows.

$$q(x_t|x_0) = N(x_t; \sqrt{\overline{\alpha_t}}x_0, (1-\overline{\alpha_t})I) \tag{6}$$

The reverse process from x_t to x_0 is the process of restoring a set of random noise to the input, which can also be referred to as the denoising process. If the inverse process $q(x_{t-1}|x_t)$ is obtained, an image can be gradually reconstructed from the random noise x_t. Specifically, the diffusion model uses a neural network $p_\theta(x_{t-1}|x_t)$ to fit the reverse process $q(x_{t-1}|x_t)$, thereby deriving the reverse conditional distribution fitted by the neural network

$$p_\theta(x_{t-1}|x_t) = \mathcal{N}(x_{t-1}; \mu_\theta(x_t, t), \Sigma_\theta(x_t, t)) \tag{7}$$

Given t and x_t, $\mu_\theta = \frac{1}{\sqrt{\alpha_t}}\left(x_t - \frac{1-\alpha_t}{\sqrt{1-\overline{\alpha}_t}}\epsilon_\theta(x_t, t)\right)$, at this point, we only need to fit $\epsilon_\theta(x_t, t)$ using a neural network.

In the process of creating a deepfake, there are typically two stages: training and inference. During the training stage, image data of the target person is first collected and preprocessed. Then, a diffusion model is trained to learn how to reconstruct the target face from noise, while combining keypoint control to generate details. In the inference stage, conditional information such as expressions and poses is extracted from the original image or video frame. The trained model is then used to gradually denoise from random noise, generating a target face consistent with the pose of the photo or video frame. Finally, post-processing is applied to embed the generated face into the original data. This entire process ensures the realism and temporal consistency of the generated images [12].

The advantage of diffusion models over autoencoders and GANs lies in their higher generation quality, enabling the output of highly detailed and realistic data. For example, images generated by the popular image generation tool DALL-E not only exhibit high clarity and texture detail but also maintain semantic consistency. The generative mechanism of diffusion models effectively avoids the common mode collapse issues associated with GANs, making the generation of diverse data a reality. Additionally, diffusion models use direct loss functions during training and do not require complex adversarial training, resulting in a more stable training process.

3 Deepfake Detection Based on Image Forgery and Noise Detection

Deepfake detection technology is designed to identify false audiovisual content generated by artificial intelligence, such as images, videos, and audio. Its core principle is to distinguish between genuine and fake content by capturing inherent flaws left behind by the generation algorithm during the synthesis process. These flaws primarily manifest as abnormalities in biometric features, such as unnatural blink frequencies, synchronization discrepancies between lip movements and speech, distorted facial microexpressions, violations of physical laws, and traces of algorithmic techniques. Current mainstream methods rely on deep learning models to automatically extract these multi-dimensional features and are gradually evolving toward multi-modal joint analysis and forgery tracing to address continuously advancing generation technologies, ultimately providing a critical defense mechanism for the credibility of digital content.

3.1 Image Forgery

The core principle of image forgery detection lies in identifying subtle traces in images that violate real-world physical laws, imaging principles, or statistical characteristics.

The primary technical approaches include: analyzing underlying physical traces, such as detecting block artifacts or quantization inconsistencies caused by JPEG compression, analyzing local noise patterns, and capturing specific fingerprints left by generative models or tampering operations through frequency domain transformations. Verifying physical consistency is another key aspect, focusing on whether lighting direction, shadow projection, object reflection content, and perspective relationships are logically consistent across all elements in the image. Identifying biometric anomalies focuses on unnatural blinking, lip-sync discrepancies, distorted facial microexpressions, abnormal tooth/skin textures, and facial geometric proportions. For AI-generated content, the most powerful detection method currently available is to use deep learning models, which are trained on massive amounts of real and fake data to automatically learn and extract complex, multi-dimensional traces inherent to generative models or tampering artifacts that are difficult for the human eye to detect. Additionally, detecting edge/seam anomalies is an effective method for identifying traditional tampering techniques like copy-paste. Modern detection systems often integrate multiple principles and technologies, particularly relying on deep learning models for end-to-end feature learning and judgment. However, this field is in a state of continuous adversarial development, and detection technologies face significant challenges such as insufficient model generalization, difficulty in obtaining high-quality training data, post-processing interference, and the rapid evolution of generation techniques.

3.1.1 Detection Principle

Deepfake detection technology for identifying deepfake facial image forgery algorithms represents a refined evolution in the field of deepfake detection.

Not only does it allow for binary classification of facial images as genuine or forged, but it also provides specific traceability information about the forgery methods used. By constructing a large-scale facial video/image dataset, we utilized three classic models—Xception, GramNet, and F3-Net—to complete detection tasks on the ForgeryNet dataset, which included both genuine samples and forgery methods, significantly advancing the development of fine-grained deepfake detection. By combining the ForgeryNet and FF++ datasets to build a multi-branch system, we integrated binary and multi-class detection mechanisms. Using a dynamically expanding incremental learning framework, we improved detection accuracy while effectively reducing resource consumption and time costs. We conducted five-class detection research targeting diffusion models and GAN forgery techniques (real samples + 4 types of forgery methods), systematically evaluating the feasibility of Visual Transformers (ViTs) in multi-class deepfake detection, and addressing rapidly evolving forgery techniques through comparative analysis with traditional CNN models. Based on 9 GAN architectures and 4 diffusion models, we constructed a dataset and innovatively proposed a three-tier hierarchical detection framework: The first level distinguishes between real and AI-generated images, the second level separates GAN-generated images from diffusion model-generated images, and the third level identifies the specific generation architecture, significantly improving the system's accuracy and robustness. A momentum difficulty enhancement strategy was designed to target the forgery characteristics of diffusion models, dynamically allocating sample weights to enhance the model's adaptability to the heterogeneity of training data. Currently, this technology has broken through the limitations of basic authenticity discrimination and is continuing to evolve toward more granular and multi-category detection based on more comprehensive datasets.

3.1.2 Testing Methods

Airspace-based detection method Early deepfake detection methods based on general image classification models (such as Xception and EfficientNet) [13, 14] demonstrated basic performance but exhibited significant deficiencies in generalization and robustness when faced with unknown forgery techniques and image compression perturbations.

As Fig. 4, to capture increasingly realistic forgery traces, Zhao et al. [15] proposed a triple texture enhancement mechanism: the texture enhancement module strengthens texture feature extraction in shallow networks, the attention generation module focuses on subtle artifacts in multiple regions, and the bilinear attention pooling module achieves complementary coverage of attention regions, thus comprehensively extracting image detail textures. However, such methods are prone to overfitting specific forgery traces, leading to a sharp decline in detection performance for unknown forgery techniques. To address this, emerging research has shifted toward self-generated forgery paradigms: Li et al. [16] Face X-Ray model synthesizes face fusion boundary grayscale images and concurrently performs image authenticity classification and boundary pixel detection, achieving cross-domain generalization for unknown forgery methods; Zhao et al.'s [17] Pairwise Self-Consistency Learning (PCL) decomposes feature maps into local blocks, generates consistency maps via inter-block cosine similarity, and trains using an adversarial inconsistency image generator; Self-Blended Image (SBI) generates forgery samples based on single-face keypoint transformations, avoiding cross-face dependencies. While these methods significantly enhance generalization capabilities, they remain limited in full-face synthesis scenarios. Concurrently developed auxiliary enhancement techniques include: the reconstruction-guided attention mechanism proposed, which uses the difference map between the original image and the reconstructed image to locate tampered regions, combined with reconstruction classification learning (RECCE) to identify intrinsic differences; and the identity feature decoupling strategy implemented by Dong et al. [18], which separates identity-related/unrelated features using source-target dual encoders to enhance model robustness. These approaches collectively drive a strategic shift in detection paradigms from reliance on specific traces to generalized feature learning.

Frequency-domain-based detection methods: After being transmitted multiple times through network streaming media, videos and images often undergo multiple compression processes, making it difficult to discern forgery artifacts in low-quality images with the naked eye. Unlike traditional spatial-domain-based detection methods, researchers have begun to explore detection clues from the frequency-domain perspective.

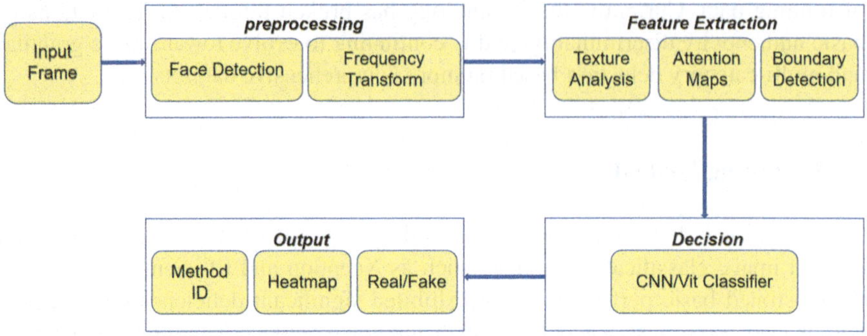

Fig. 4 Deepfake image detection flowchart

For example, Qian et al. [19] found that the frequency domain can effectively uncover detailed artifacts left by forgery operations. They designed a frequency-aware decomposition module to adaptively capture forgery clues in images and constructed a local frequency statistics module to statistically analyze frequency information in each local region of the image, then reorganized it into multi-channel feature maps as supplementary frequency-domain information. This method relies on frequency-domain information detection and maintains excellent performance even when processing highly compressed forged images. To more comprehensively capture artifacts in the frequency domain, Li et al. [20] proposed an adaptive frequency feature generation module to extract distinctive features from different frequency bands in a learnable manner, while introducing single center loss (SCL) to enhance the intra-class compactness of real faces while expanding the inter-class differences between them and forged faces. Addressing the multiscale characteristics of artifacts in images, Wang et al. [21] designed a multimodal multiscale Transformer model (M2TR) to detect local artifact details at different spatial levels, incorporating frequency-domain features as auxiliary information to ensure the model's detection capability for highly compressed images. Although the aforementioned frequency domain detection methods demonstrate strong detection capabilities for highly compressed forged images, their performance significantly degrades when faced with unknown forgery methods.

Time-domain-based detection methods: In video forgery detection research, to uncover dynamic inconsistencies between frames, Masi et al. [22] proposed a dual stream branch network, with one branch focusing on extracting dynamic temporal inconsistencies in consecutive video frames and the other using the LoG operator to amplify artifact details. Given the relevance between forgery detection and anomaly detection tasks, they introduced the Deep SVDD [23] loss function to enhance compactness within the real face class and discriminative power between real and fake face classes, thereby improving the model's generalization ability. Furthermore, Zheng et al. [24] found that setting the size of the temporal convolution kernel in the 3D convolution kernel to 1 can strengthen the network's representation of temporal information, enabling effective capture of temporal inconsistencies in forged videos

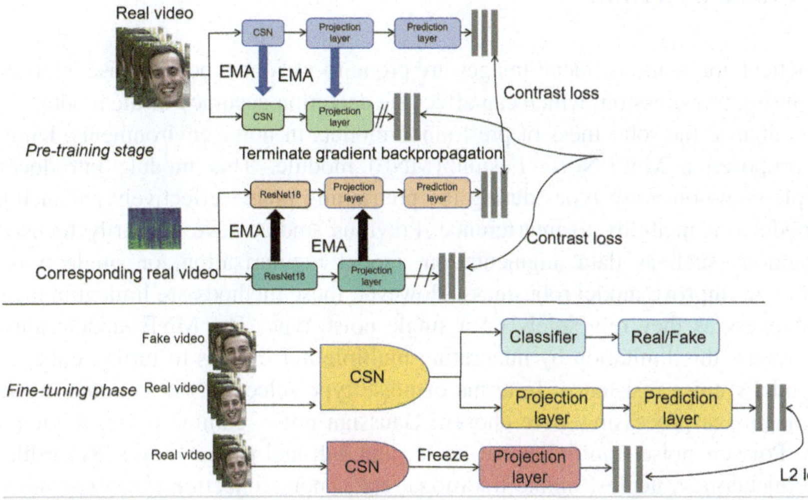

Fig. 5 Framework of the CLIP-based supplementary learning module for video forgery

and performing well when dealing with unknown forgery methods. However, since temporal inconsistencies are easily disrupted by noise, compression, and other factors, leading to damage to detection clues, the robustness of such methods is suboptimal.

Multi-modal detection methods: In the field of video forgery detection, multimodal detection methods focus on the principle of synchrony between speech and lip movements. The LipForensics model proposed by Haliassos et al. [25] is based on the fundamental assumption that forged videos disrupt the inter-frame continuity of lip movements. This model first undergoes lip reading pretraining in the LRW dataset [26] to learn the characteristics of natural lip movement, then fine-tunes it in a forgery detection dataset. Not only does it perform exceptionally well in cross-forgery method testing, but it also demonstrates significantly superior robustness compared to other methods because of the insensitivity of lip semantics to common perturbations. However, this model focuses solely on the mouth region and ignores information from other facial areas. To address this limitation, Haliassos et al. [27] further proposed the RealForensics model (as Fig. 5). This model uses a residual network (ResNet) [28] to extract audio features and a channel-separated convolutional network (CSN) [29] to extract video features, learning the natural correspondence between facial expressions and audio changes in real videos through self-supervised learning. After capturing the natural association between faces and audio, the model is fine-tuned using a forgery detection dataset. While this method offers advantages in terms of robustness and generalization, it relies on large-scale pretraining datasets and has high computational requirements, objectively setting a technical barrier for subsequent research.

3.2 Noise Detection

In practical applications, facial images are prone to various types of noise interference during transmission, which can affect the detection accuracy of the model.

To enhance the robustness of pre-trained models in noisy environments, Ding [30] proposed a Multi-Noise Fusion (MNF) module. This module introduces multiple common noise types during the pretraining phase, effectively enhancing the model's adaptability to interference. Previous studies have primarily focused on methods such as data augmentation, model regularization, or single noise injection to improve model robustness. However, these methods are limited in their effectiveness as they rely solely on a single noise type. The MNF module aims to overcome this limitation by integrating multiple noise types to further enhance the model's noise resistance. In terms of noise type selection, six common noise types in image processing were chosen: Gaussian noise, gamma noise, Rayleigh noise, Poisson noise, multiplicative noise, and salt-and-pepper noise. Regarding noise injection strategies, unlike traditional static noise injection (i.e. fixed noise type and intensity), a dynamic noise injection method based on linear combination is proposed. This method dynamically adjusts the noise type and intensity based on the model's training progress. During the early stages of pre-training, low-intensity single noise is introduced to help the model initially adapt to the noisy environment. As training progresses, the noise intensity is gradually increased, and a weighted combination of multiple noise types is introduced, enabling the model to learn more complex features and enhance its adaptability to diverse disturbances. Additionally, for different noise types, linear growth formulas are defined to ensure a smooth and targeted noise injection process. For the n different types of noise added, the linear growth of each noise i is as follows:

$$n_i(t) = n_{i_0} + (n_{i_{\max}} - n_{i_0})\frac{t}{T} \tag{8}$$

where $n_i(t)$ is the intensity of the i type of noise in the t round of training. n_{i_0} is the initial intensity of the i type of noise, $n_{i_{\max}}$ is the maximum intensity of the i type of noise, t is the current training round number, T is the total number of training laps. Then, combine the n types of noise according to the weight ω_i of each type of noise to form the total intensity $N(t)$ of the combined noise:

$$N(t) = \sum_{i=1}^{n} \omega_i \cdot n_i(t) \tag{9}$$

The weights are assigned based on the variance of each noise source. Noise sources with smaller variance have larger weights, as specified in the formula:

$$\omega_i = \frac{1/\sigma_i^2}{\sum_{j=1}^{n} 1/\sigma_j^2} \tag{10}$$

where σ_i^2 denotes the variance of the i noise source, and σ_j^2 denotes the variance of the j noise source.

Through this dynamic adjustment mechanism, the model not only performs well on clean data but also maintains high accuracy and strong robustness in the processing of noisy images in real-world scenarios. This design ensures that the model can fully leverage the advantages of high-quality training data while effectively addressing common image noise interference in practical applications, thereby maintaining excellent recognition performance and interference resistance in complex and changing real-world environments. The pretrained test model integrates a contrastive learning mechanism based on the UDenseNet architecture, significantly improving image classification performance. By pre-training PR-MAE to extract deep features from images and introducing the CLIP contrastive learning framework to construct positive and negative sample pairs, the model's ability to distinguish feature similarities and differences is enhanced, as Fig. 6. To optimize feature processing, the base model adds a convolutional neural network module: two convolutional-pooling operations are performed to refine feature processing. After feature map flattening and linear mapping, a ReLU activation function is used to construct a classifier tailored to the task requirements.

Fully leveraging the plug-and-play characteristics of the CLIP model [31, 32] to achieve cross-modal knowledge transfer: first, image features are extracted using Auto Processor and CLIP Vision Model With Projection, and the input size is unified to 224×224 via bilinear interpolation; second, inheriting the zero-shot inference capability of CLIP pretraining, it demonstrates excellent transferability and generalization in new tasks without the need for fine-tuning. In terms of loss function design, a similarity matrix is constructed from normalized feature vectors, and the average log-likelihood of positive samples is calculated, with its negative value used as the sample loss. The mean anchor loss is then aggregated to obtain the comparison loss value. During training, traditional classification loss and comparison loss are weighted and fused to form a joint loss function that guides the weight optimization of the classifier.

This flowchart illustrates an advanced image processing and learning framework that combines a pre-trained CLIP model with a newly designed UDenseNet

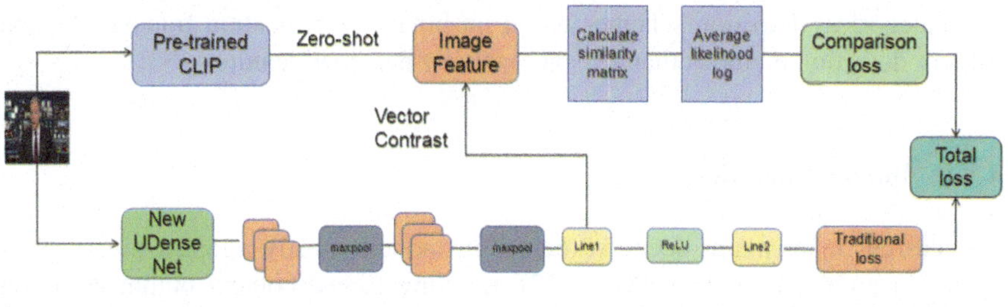

Fig. 6 Architecture of the hybrid UDenseNet and pre-trained CLIP model for image classification

network. First, the input image is fed into the pre-trained CLIP model, which can be directly applied to new tasks through zero-shot learning without requiring fine-tuning for specific tasks. The CLIP model extracts image features and maps these features to text features in a shared embedding space for image-text matching and retrieval.

Simultaneously, the input image is also fed into the newly designed UDenseNet network. UDenseNet is an improved dense connection network that enhances feature discriminative power through vector comparison, thereby extracting richer image features. In UDenseNet, feature maps undergo max pooling operations to reduce their size and extract the most prominent features. The pooled feature maps are then flattened into one-dimensional vectors for subsequent processing. The flattened feature vectors are passed through a series of fully connected layers, which may include activation functions such as ReLU, for further feature processing.

After the fully connected layers, a traditional loss is calculated, which may refer to a loss function related to a specific task, such as classification loss. Additionally, the image features extracted by the CLIP model are compared with text features to calculate similarity, generating a similarity matrix. The similarity matrix is processed to calculate the average likelihood ratio, and based on this, a comparison loss is calculated, which may refer to the loss function in contrastive learning.

Finally, the comparison loss and traditional loss are combined to form the total loss, which is used for model optimization. This framework optimizes model performance by combining the CLIP model and UDenseNet network, leveraging zero-shot learning and vector comparison, along with the calculation of traditional loss, thereby achieving better results in image processing and learning tasks.

In summary, this is a brief introduction to common methods for deepfake detection based on image forgery and noise detection.

4 Recognition and Detection of Deepfake Local Features

4.1 Related Works

The mainstream methods for detecting deepfake faces can be roughly divided into two categories: detection schemes based on binary classification frameworks and forgery detection methods that rely on manually designed features [33].

4.1.1 Binary Classifier

The current mainstream deepfake detection models mostly adopt a binary classification architecture [34, 35, 36, 37], marking forged content output as 0 and original frames as 1. Classification modules are typically built on deep learning architectures such as Convolutional Neural Networks (CNN), Long Short Term

Memory Networks (LSTM), or Visual Transformers (ViT), and their specific implementation logic varies depending on the type of deep neural network (DNN) used. For example, the blink features captured in the feature extraction stage can be analyzed in time sequence through the LSTM unit in the classification module, and whether they are counterfeit content can be determined by identifying temporal consistency anomalies in blink patterns between frames.

From the perspective of model structure, deepfake detectors often use fully connected layers as decision layers at the end. The high-level semantic features output by the convolutional layer are flattened and integrated into the layer, and the final classification result is generated through mapping calculation. As shown in Fig. 7, the typical deepfake image generation process is to replace the face area of the source image with the target image, which may result in the loss of identity representation information; the identity features of the forged image cannot be completely equivalent to the source image, and differences exist between these features and those of the target image. Especially when facial restoration is performed on forged images, the original identity features of the target image will undergo significant changes.

In the deepfake detection model, there are hidden decision boundaries in the identity space between genuine and fake images. During the training process, the binary classifier may misjudge specific identity groups as "real" or "fake" due to data bias. This biased feature representation can lead to misjudgments when evaluating across datasets: the classifier may make decisions based on prior biases of image facial features rather than fake traces, resulting in a decrease in detection accuracy.

4.1.2 Manual Forgery Detector

In existing research, many scholars have attempted to achieve forgery detection through manually designed features in modeling facial tampering techniques. In the past, Li et al. [38] noted that human physiological characteristics have features that are difficult to replicate by Manipulation technology, and proposed that video authenticity can be determined by analyzing the regularity of blink frequency. Pu et al. [39] utilized the unique fingerprint patterns formed by Generative Adversarial Networks (GANs) in the image noise space to detect GAN-like forged images, but the effectiveness of this method was significantly reduced for forged samples generated by non-GAN techniques. Manually designed deepfake detectors typically guide models to capture specific tampering traces, but such methods have obvious limitations: when forged samples contain novel tampering features that are not covered by the training data, the recognition performance of the detector will experience a significant decline.

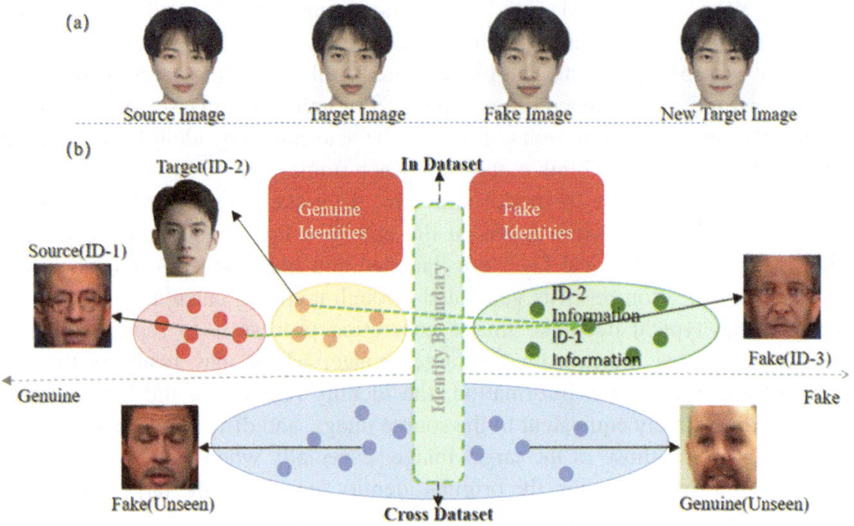

Fig. 7 The process of facial forgery technology. From the source identity to the target identity and then to the forgery result. The classification system for constructing the verification scenario in the figure includes the combination of same dataset and cross dataset, real and forged situations. (**a**) Illustrates the deepfake generation pipeline, which starts from a source identity, transfers facial attributes to a target identity, and produces a forged output. (**b**) Depicts the classification system designed for verification scenarios, showing the arrangement of genuine and fake identities within the same dataset as well as across different datasets, along with their separation by an identity decision boundary

4.2 *A Deepfake Face Detection Method Focusing on Forgery Areas*

This section describes a deepfake face detection framework that focuses on local tampering features. Embedding an attention mechanism module in the backbone network, generating region attention maps through separable convolution operations, this mechanism can accurately locate forged regions in facial images and enhance the model's perception of tampered details.

4.2.1 Attention Based Layers

As Fig. 8, Attention-based Layers (AbL) [40] can be applied to the feature maps of some classification models and focus attention on the regions that affect classification. And as Fig. 9, the input to AbL is a convolutional feature map $\mathbf{F} \epsilon \mathbb{R}^{N \times C \times H \times W}$ on several layers of the backbone network, where *N, C, H* and *W* represent the batch size, number of channels, height, and width, respectively. For the input feature map, an attention map $\mathbf{M} \epsilon \mathbb{R}^{N \times H \times W}$ is first calculated. The mapping

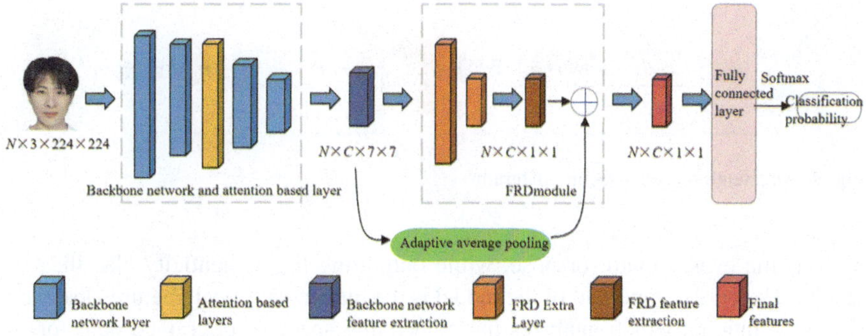

Fig. 8 A network architecture diagram

from$\boldsymbol{F}$ to $\boldsymbol{M}$ is expressed as:

$$\boldsymbol{M} = f(\boldsymbol{F}) \tag{11}$$

where $f(\cdot)$ is an operator. The implementation of the operator uses direct regression, that is, the attention map is computed through a convolution operation $f : \boldsymbol{F} \longrightarrow \boldsymbol{M}$ f can consist of a single or multiple convolutional layers. In this paper, f is implemented as a custom separable convolution layer, which includes two convolutional layers: Conv 1 and Pointwise. Conv 1 has the same number of input and output channels and a kernel size of 3×3. Pointwise has an input channel number of, an output channel number of 1, and a kernel size of 1×1. Input images first undergo feature extraction through some layers of the backbone networks and then generate a single-channel feature map (that is, the attention map $\boldsymbol{M}$) through the Conv 1 and Pointwise layers.

Afterwards, based on the attention map $\boldsymbol{M}$, $\boldsymbol{F}$ is transformed into an improved feature map, which is the final output of the layer:

$$\boldsymbol{F}' = \boldsymbol{F} \circ \sigma(\boldsymbol{M}) \tag{12}$$

4.3 Local Feature Discovery (LFD) Technique

With the increasing complexity of deepfake technology, it is becoming easier to generate highly realistic fake videos, which poses a serious challenge to the robustness of detection models. Traditional detection methods often suffer from decision black box problems, with insufficient ability to locate tampered evidence and interpretability of decision logic. To overcome the above bottlenecks, a locally interpretable self-mixing (LESB) deep forgery detection model is described. This model can locate the local facial regions that contribute the most to decision-making,

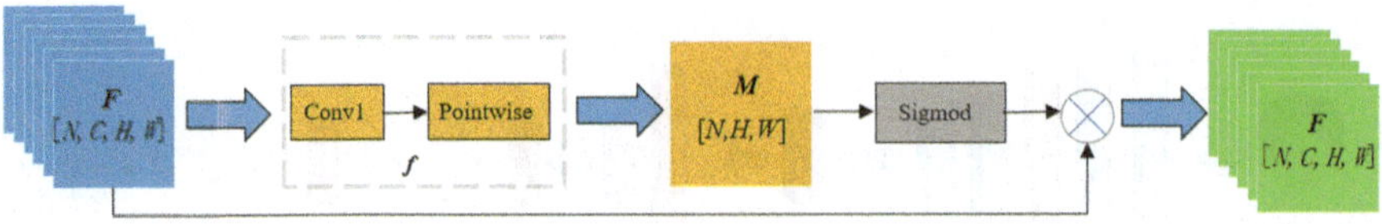

Fig. 9 An attention layer structure diagram

such as the eyes, mouth, or nose, while outputting the authenticity classification results. The core component of the model is the innovative Local Feature Discovery (LFD) technique, which analyzes the embedding space of pretrained detectors to discover key feature directions related to changes in the appearance of local facial features. This technology can explicitly recognize the subtle traces left by tampering operations in specific facial areas, providing traceable local evidence support for model decision-making.

4.3.1 The Specific Implementation Process of LFD

LFD technique and its core objective are to locate subspaces related to local facial component feature changes in the global facial representation embedding space of the pre-trained deepfake detector [41]. Specifically, this technology induces visual changes in local features by implementing spatially localized image transformation and enhancement operations on semantically significant facial regions such as the eyes, nose, and mouth. Taking the nasal region as an example, after generating various types of local variation samples, a dedicated PCA model is independently trained for each facial component to construct a subspace representing local feature variations in the global embedding space. This method can separate the local feature components embedded in the space of the pre-trained model, and then quantitatively analyze the contribution weight of feature changes in local areas such as the eyes and nose to the authenticity classification decision. By explicitly modeling the mapping direction of local features in the embedding space, LFD technology provides a decision-making basis for deepfake detection that can be traced back to specific facial components. The local feature discovery method includes several steps, including: (i) local mask generation, (ii) data augmentation, and (iii) PCA modeling.

The first step in the LFD process is to create a local mask corresponding to the prominent facial area. Specifically, the RetinaFace detector is first used to detect the facial regions of all input images [42], and then feature points are extracted from the face using the Dlib face alignment model with an integrated custom shape predictor. Based on the detected feature points, further construct binary masks for the three different facial regions of eyes, nose, and mouth, and enhance each binary mask through various shape transformations.

In the second step, researchers adopt a diversified data augmentation strategy to inject rich appearance variations into facial images. Define x as the facial image

Fig. 10 Local facial mask generation. Researchers employ the Dlib facial landmarking model to generate an 85-point annotation of the facial region, as Fig. 10 [43]. Leveraging these detected landmarks, we define local binary masks corresponding to the regions of interest on the face. The right-hand 4 × 4 grid demonstrates this process: each row features the base nose mask extracted from the input image (first column), with subsequent columns displaying shape, dimension, and positional variations of the mask

from the target dataset, and m as the binary mask shown in Fig. 9 (smoothed by a Gaussian filter), ψrepresents the set of operations selected from three types of enhancement techniques: color transformation, frequency transformation, and geometric transformation. They apply the above three types of transformations to the input image in a random combination to generate a globally distorted result $\psi(x)$, and then fuse the processed region with the original image $\boldsymbol{x}$ through the following method:

$$x_a = (1 - m) \odot x + m \odot x_d, \tag{13}$$

Figure 10 shows an example of enhancing the nose region. As can be seen, most of the pixel values in the enhanced image x remain unchanged, with only local changes occurring around the target facial area due to blending processing. This enhancement method can generate facial images with local appearance changes, making it easier to explore the embedding space of deep learning models and model local changes through principal component analysis (PCA). When using PCA to construct local subspaces, researchers first enhance N facial images in the dataset $\mathcal{D} = \{x^{(i)}\}_{i=1}^{N}$(each image is enhanced M times). Obtain the enhanced dataset $\mathcal{D}_a$: $\mathcal{D}_a = \{e_a^{(i,j)} \mid_{j=1:M}^{i=1:N}\}$, project the enhanced images onto the embedding space through pretrained CNN model ϕ, and obtain the d-dimensional embedding set $\varepsilon = \{e_a^{(i,j)} \mid_{j=1:M}^{i=1:N}\} \in \mathbb{R}^d$. These embeddings correspond to input images with similar pixel values except for locally enhanced regions. To model these local changes, they first remove information related to global image features. Specifically, it is necessary to calculate the conditional mean of the images embedded in the space:

$$\mu_j = \frac{1}{M}\sum_{k=1}^{M} e_a^{(i,k)}. \tag{14}$$

$$E = [e_a^{(i,j)} - \mu_j]|_{j=1:M}^{i=1:N} \tag{15}$$

Calculate the covariance matrix with E as the center:

$$\Delta = \frac{1}{N(N-1)}(E-\bar{E})(E-\bar{E})^T \tag{16}$$

and then solve the PCA-defined eigenproblem:

$$\Delta V = \Lambda V \tag{17}$$

Therefore, analyzing this subspace can achieve image inference of facial regions, which constitutes the core module of the proposed Local Feature Discovery (LFD) method.

4.4 Enhanced Local Differences

This section describes a deepfake detection method based on local texture difference feature enhancement, aimed at improving detection accuracy and generalization ability. This method identifies local texture differences as a common, inconsistent feature in fake facial images. By introducing a central difference convolution module, it enhances the model's ability to capture forged information, enabling it to learn more general local texture difference features, thereby improving the sensitivity of detecting subtle local differences. In addition, a convolutional block attention module is used to refine features: the spatial attention module locates the spatial positions of local texture differences, guiding the model to focus on key facial regions; The channel attention module filters and preserves differential features related to the detection task, while suppressing redundant information, thereby enhancing the feature representation and discriminative performance of the model [44].

4.4.1 Central Differential Convolution

Central Difference Convolution aims to enhance the ability to capture discriminative features of local textures by fusing pixel intensity features with spatial gradient information. This method first uses the 3 × 3 standard convolution structure of the classic Bottleneck module in the ResNet-50 network to preliminarily encode the pixel grayscale value distribution of the input image and obtain the basic inten-

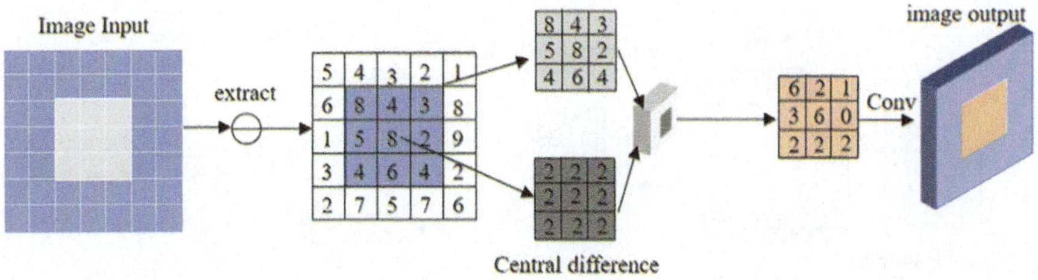

Fig. 11 The process of central differential convolution. The central differential convolution first extracts the upper and lower subregions of the input feature map, obtains the intermediate features through differential operation, and then convolves to obtain the output

sity feature representation. This traditional convolution-based strength extraction method can effectively preserve the low-frequency texture information of the image, providing a stable underlying semantic foundation for the fusion of gradient features in Fig. 11.

On this basis, the central differential convolution further introduces the gradient computing mechanism, which captures the edge changes and texture direction information of the image through differential operations on pixel neighborhoods. This strategy, which combines intensity features with gradient features, can significantly enhance the sensitivity of the model to local subtle texture differences—for example, high-frequency details such as pore distribution of real facial skin and synthetic edge artifacts of fake areas can be effectively characterized through intensity gradient joint features. Compared to traditional convolution operations that rely solely on strength information, this method provides a more discriminative texture feature space for subsequent authenticity discrimination tasks by constructing multi-dimensional local feature descriptors. The central differential convolution inherits the advantages of this structure while introducing gradient information, which extends the traditional convolution feature extraction ability, especially suitable for deepfake detection scenarios that require refined local feature analysis.

4.4.2 Convolutional Block Attention Module

After obtaining the representation of texture anomaly features, considering that the effective extraction of local texture discrepancy features requires simultaneous attention to both discrepancy information and the spatial positions of abnormal features, an attention mechanism inspired by CBAM (Convolutional Block Attention Module) is introduced. This mechanism guides feature extraction by generating adaptive weight distributions, enabling the model to focus on critical regions and enhance discriminative feature representation [45]. Drawing from the sequential design in CBAM, the dual-layer attention module consists of a channel attention module and a spatial attention module in cascaded composition. First, the channel attention module emphasizes "what" features are informative by leveraging inter-

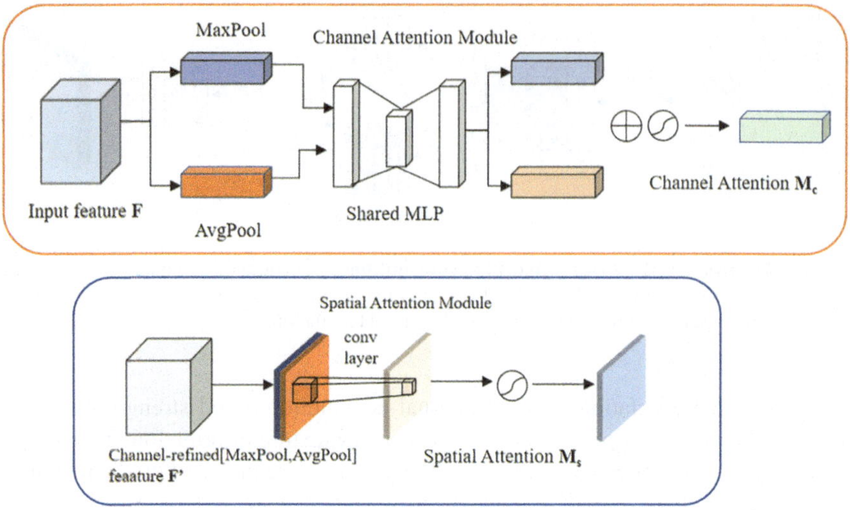

Fig. 12 The structure of a convolutional block attention module. The convolutional block attention module consists of spatial and channel attention modules.The channel attention module first applies spatial max-pooling and average-pooling to the input feature F, then feeds the outputs into a shared MLP, sums them, and activates to generate the channel attention map M_c. The spatial attention module processes the channel-refined feature F' : it performs channel-wise max-pooling and average-pooling, concatenates the results, and passes them through a convolution layer followed by activation to produce the spatial attention map M_s, thus adaptively enhancing key channel and spatial information

channel relationships. It processes the input feature map $\boldsymbol{F}\epsilon\mathbb{R}^{C\times H\times W}$ via average and max pooling along the spatial dimensions, generating two context descriptors. These descriptors are fed into a shared multi-layer perceptron (MLP) with a hidden layer (reduction ratio γ), and the outputs are summed and activated by a sigmoid to produce a channel-wise attention map $\boldsymbol{M}_c\epsilon\mathbb{R}^{C\times 1\times 1}$. Subsequently, the spatial attention module focuses on "where" the informative regions are. It applies average and max pooling along the channel axis, concatenates the results, and processes them through a 7×7 convolution layer to generate a spatial attention map $\boldsymbol{M}_s\epsilon\mathbb{R}^{1\times H\times W}$. The final refined feature map is obtained by element-wise multiplication of the input features with the channel attention map first, then with the spatial attention map:

$$F' = M_c(F) \otimes F \tag{18}$$

$$F'' = M_s(F') \otimes F' \tag{19}$$

where $\otimes$ denotes element-wise multiplication (Fig. 12) [44].

Firstly, channel attention captures inter-channel relationships by average-pooling and max-pooling along spatial dimensions, generating descriptors F^c_{avg} and F^c_{max}.

And then forwarded to a shared network to produce our channel attention map $M_c \epsilon \mathbb{R}^{C\times 1\times 1}$, the channel attention is computed as:

$$M_c(F) = \sigma(MLP(AvgPool(F)) + MLP(MaxPool(F))) \tag{20}$$

$$= \sigma(W_1(W_0(F^c_{avg})) + W_1(W_0(F^c_{max}))), \tag{21}$$

where σ denotes the sigmoid function, $W_0 \epsilon \mathbb{R}^{C/r\times C}$, and $W_1 \epsilon \mathbb{R}^{C\times C/r}$. Note that the MLP weights, W_0 and W_1, are shared for both inputs and the ReLU activation function is followed by W_0.

Secondly, they produce a spatial attention map by exploiting inter-spatial feature dependencies. Unlike channel attention, it focuses on informative locations, complementing channel attention. They compute it by applying avg and max pooling along the channel axis, then concatenating the results for feature description. The spatial attention is computed as:

$$M_s(F) = \sigma(f^{7\times 7}([AvgPool(F); MaxPool(F)]) \tag{22}$$

$$= \sigma(f^{7\times 7}([F^c_{avg}; F^c_{max}])), \tag{23}$$

where σ denotes the sigmoid function and $f^{7\times 7}$ represents a convolution operation with the filter size of 7×7.

5 Compressed Deepfake Video Detection Method Based on Dual-Branch Frame-Level and Time-Level Analysis

The compressed deepfake video detection method, based on dual branches of frame-level and time-level features, is a method for detecting compressed deepfake videos by learning frame-level and time-level features. On the one hand, compressed video adds a lot of redundant information to video frames, such as compression artifacts. The method in this chapter prunes redundant connections by designing a frame-level branch to prevent invalid connections from affecting the final detection. In addition, it should be noted that the method proposed in this chapter extracts key frames instead of the entire frames of the video, which greatly improves training efficiency. On the other hand, since fake videos ignore temporal consistency during the synthesis process, the method in this chapter designs a time-level branch to capture temporal features. These two branches will be trained independently, and the detection results will be fused at the end.

As Fig. 13, this method includes two branches: a frame-level branch and a time-level branch. The frame-level branch, which undergoes a convolutional neural network and a network pruning module, is used to classify whether frames are tampered with or authentic. First, extract keyframes from the video using FFmpeg [46], and use data augmentation to increase data diversity. At the same time, data cleaning

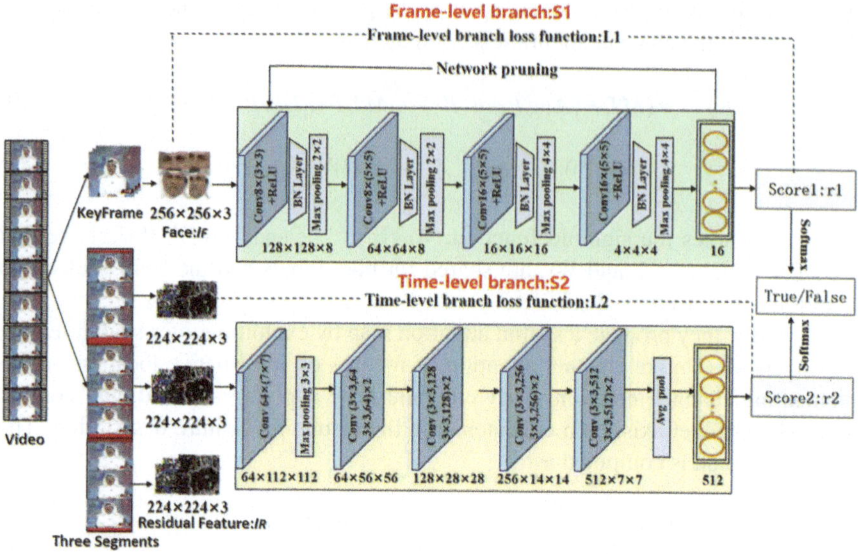

Fig. 13 A dual-branch method for detecting compressed Deepfake videos

is required to construct the input data. Then, the data is fed into the improved MesoNet [34], which extracts the mesoscopic features of the image. To avoid the model fitting the compression noise, the frame-level branch prunes MesoNet. The pruning process includes three steps: training network connections, pruning network connections, and continuing to train to adjust network weights. Finally, the probability value of the detection label is obtained. Since the compressed video not only contains keyframes but also predicted frames, where the keyframes have been used for the frame-level branch, the predicted frames are used for the time-level branch. The time-level branch is trained based on the temporal dependency features of the video, which are closely related to the predicted frames. Since inter-frame consistency is ignored during the generation of fake videos, the time-level branch first uses predicted frames to extract time-related features. Next, the time-level branch uses Resnet18 [28] to model the features. Finally, the model will output the predicted probability value. To combine these two branches, a series of strategies are used to obtain the final score. Finally, the softmax function is used to obtain the detection result.

5.1 Frame-Level Branch

For the frame-level branch, the first step is to extract frames from the compressed video. The Key frames in the compressed video retain complete information during the encoding process. In order to input as much information as possible into the network without affecting training efficiency, the frame-level branch no longer extracts images frame by frame, but only extracts keyframes and puts them into the network. To further expand the diversity of training samples, the frame-level branch also performs data augmentation by cropping the faces of keyframes at multiple scales. After multi-scale cropping, some images contain only faces, while others contain faces and surrounding areas. When cutting faces, a small amount of circular data is mistakenly recognized as faces, and this interfering data needs to be cleared to prevent it from affecting the results. Next, the faces are input into the improved MesoNet. Since the MesoNet network focuses on extracting mesoscopic features of the video, and mesoscopic features are between macroscopic and microscopic features, it is more conducive to detection. In fact, microscopic analysis cannot be applied to compressed video because the video loses a lot of information. Macroscopic analysis makes it difficult to distinguish forged videos because the videos do not leave obvious visual traces. In order to improve robustness, reduce overfitting, and improve the performance of detecting compressed deepfake videos, a kernel size of 4×4 is used in the third pooling layer instead of the original 2×2.

Finally, the modified network is pruned gradually. Quantization noise during video compression causes a specific visual error called compression artifacts. These artifacts result in a large amount of redundant noise in the compressed video, increasing the difficulty of detection. To avoid the impact of redundant noise on the network, the frame-level branch gradually prunes the modified MesoNet network. Pruning plays an important role in the robustness of the model. First, pruning unimportant connections can prevent invalid connections from affecting the final prediction. Second, pruning is an iterative process. Each iteration is a greedy search, and the classifier will select the most informative predictors, resulting in a smaller prediction variance. Third, the parameters of the pruned network become sparse, the complexity of the network is reduced, and overfitting can be reduced.

5.1.1 The Basic Principles of MesoNet

As shown in Fig. 14, the network comprises four convolutional layers and two fully connected layers. Each convolutional layer utilizes a ReLU (Rectified Linear Unit) activation function, followed by a batch normalization layer [47] and a max pooling layer. This regularizes the data, reduces its dimensionality, and prevents the vanishing gradient effect.

5.1.2 Network Pruning Technology

Network pruning techniques primarily reduce the computational cost of a model by removing unnecessary weights and connections, resulting in smaller, more efficient models while maintaining accuracy. Pruning can be performed on both Convolutional Neural Networks and fully connected networks, and has been proven

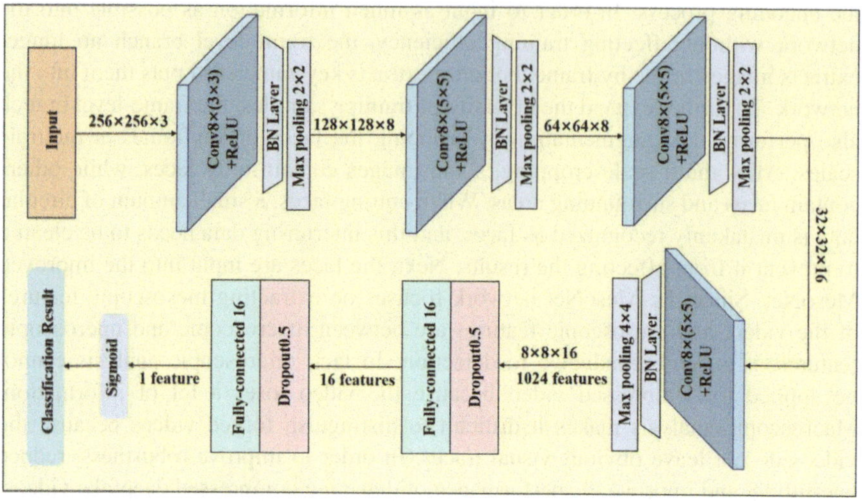

Fig. 14 MesoNet architecture

to effectively avoid overfitting, thus improving the robustness of the model. Network pruning mainly includes magnitude pruning and structured pruning [48].

Magnitude pruning is a technique for removing the smallest magnitude weights from a network, iteratively pruning weights by setting the smallest ones to zero. The network is then fine-tuned to recover its performance. Magnitude pruning can be applied to different layers of neural networks, including fully connected layers, convolutional layers, and recurrent layers. In fully connected layers, magnitude-based pruning removes connections with the smallest weights. In convolutional layers, magnitude-based pruning removes filters with the smallest weights. In recurrent layers, magnitude-based pruning removes connections between neurons with the smallest weights. All weights with absolute values below the pruning threshold are pruned, and the remaining weights are renormalized to maintain the network's original output. Magnitude pruning is typically performed iteratively, with multiple rounds of pruning and retraining. This iterative process allows the network to adapt to the reduced number of weights and maintain its performance. The pruning threshold is progressively lowered in each iteration, allowing more weights to be pruned in later rounds. One of the main advantages of magnitude pruning is that it can be applied to any type of neural network, including Convolutional Neural Networks and Recurrent Neural Networks.

Structured pruning involves removing entire filters or channels from a network, rather than individual weights. Structured pruning is a technique for reducing the size of deep neural networks by removing neurons or filters that contribute less to the network's accuracy. Unlike magnitude-based pruning techniques, structured pruning removes entire filters or layers while maintaining the original performance of the network. The basic idea of structured pruning is to identify and remove redundant or unnecessary components within a neural network. These components can include entire filters or layers that do not contribute significantly to the overall performance of the network. By removing these components, the network can become more compact and efficient, with fewer training parameters and faster inference time.

5.2 *Time-Level Branch*

For the time-level branch, firstly, the time-level branch uses 3 time segments to capture features. These three time segments refer to divide the entire video into three parts. Using time segments helps capture long-term dependencies. Each segment in the video generates its preliminary prediction for video detection. Then, the results of multiple segments are integrated as the overall video-level prediction. During the learning process, the score of the video-level prediction is averaged from the scores predicted by each time segment.

Then, use CoViAR [49] to extract the residual features of each video segment, and use the residual features to describe the strength of correlation between frames. Next, the residual features are input into Resnet18. By learning the extracted features, the inconsistency between frames of compressed deepfake videos is detected.

5.2.1 Video Frame Prediction

Video frame prediction is a challenging task, and even the most advanced models struggle to generate high-quality predictions for highly dynamic and complex video sequences [50]. The general procedure of video frame prediction includes the following steps:

(I) Build a prediction model. It takes a sequence of the past frames as input and outputs the prediction of the next frame in the sequence.
(II) Model training. The model used for video frame prediction can be any of the above types, and regardless of the choice, it must be trained on a large video frame dataset. To improve prediction accuracy, additional information, such as light flow or audio features, is usually also required.
(III) Optimize the model. The purpose of this step is to minimize the difference between the frames predicted by the model and the actual frames in the training dataset. This can be done using a Supervision Learning method. In the Supervision Learning method, the model is trained to base predictions on future frames in the training set. It can also be done using unsupervised or Reinforcement Learning methods. In this type of method, the model is trained to generate a sequence of frames that maintains time-series consistency with the input video data, without future frames for reference.

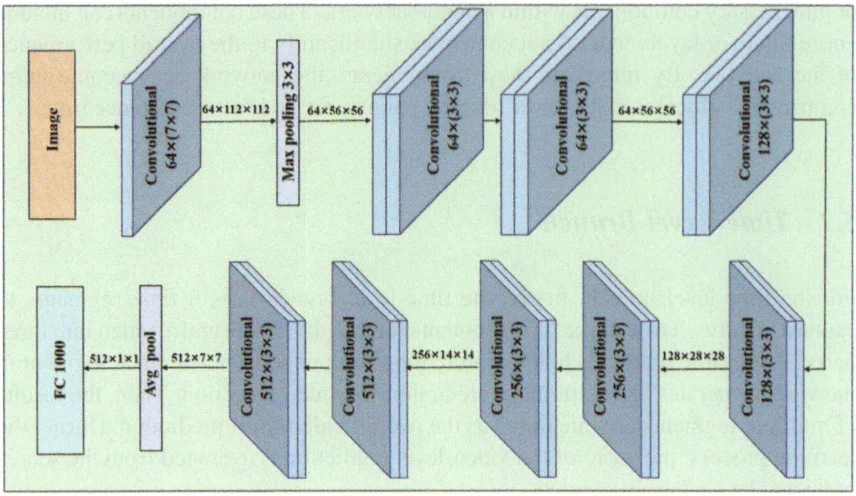

Fig. 15 ResNet18 architecture

5.2.2 The Basic Principles of ResNet18

The Residual Network (ResNet) alleviates the vanishing gradient problem by adding direct connections to nonlinear layers in neural network models, making it easier to train deep neural networks. As Fig. 15, the ResNet18 architecture includes one 7×7 convolutional layer, sixteen 3×3 convolutional layers (organized into 8 residual blocks, each block containing 2 convolutional layers), and one fully connected layer, resulting in smaller validation errors and faster convergence, thereby improving detection accuracy and efficiency.

5.3 Training and Loss

After keyframes are extracted from the video using a frame-level branch, the face I_F is input to train the network. The objective of the classification task is to minimize the L_1 loss. Additionally, frame-level branch applies pruning to the model. After feature extraction from the temporal branches, the residual features, I_R, are input to the network. The objective of the classification task is to minimize the L_2 loss. Both branches utilize Adam to optimize the model. Then, the training losses L_1 and L_2 are introduced, respectively. For the cross-entropy loss function, as the error increases, the gradient of the parameters increases, which is beneficial to the convergence of our detection model. Therefore, the cross-entropy loss is applied to train the frame-level branch. The input of the frame-level branch is keyframes, which are modeled by a convolutional neural network, MesoNet. The objective of the classification task is to minimize the loss L_1. L_1 can be expressed as:

$$L_1 = \frac{1}{N} \sum -[y_i \cdot \log(P_i) + (1 - y_i) \cdot \log(1 - P_i)] \tag{24}$$

where y_i represents the label of the frame, and p_i is the probability that the frame is predicted as a positive sample. The input of the time-level branch is the residual features of the video clip. Then, the Resnet18 in the time-level branch classifies the video using residual features. The training loss of the Resnet18 network can be expressed as:

$$L_2 = -\log \frac{\exp(v[l])}{\sum_j \exp(v[j])} \tag{25}$$

where v is the one-hot encoded prediction of the network, and l is the label of the video. The j in the denominator represents the index of the class. The task of social network compressed deepfake video detection has two classes, namely $j = 0, 1$. Therefore, the numerator is the probability of the target class, and the denominator is the probability of all classes. The greater the probability of the target class, the smaller the L_2.

5.4 *Two-Way Merge*

The final video detection results are obtained from these two branches. For each branch, the detection results are obtained from the last layer of the network output. The final score of the video is obtained through the following formula:

$$r = \text{softmax}(r_1) + \text{softmax}(r_2) \tag{26}$$

where r_1 represents the output of the frame-level branch, r_2 refers to the output of the time-level branch, and r is the final result. Since detection is a binary classification task, the model will output two values, representing the probability of a real video and the probability of a fake video, respectively. Both r_1 and r_2 are 1×2 vectors. softmax(r_1) is a 1×2 vector, containing two probability values, representing the probability of a real video and the probability of a fake video, respectively. The same applies to softmax(r_2). softmax(r_1) + softmax(r_2) represents the sum of the prediction results of the two branches. After addition, two values will be obtained, representing the probability of a real video and the probability of a fake video, respectively. If the probability score for a real video is higher than the probability score for a fake video, the video is predicted as real. Otherwise, the video will be predicted as fake.

6 Evaluation Metrics, Datasets, and Applications in Deepfake Detection

6.1 Common Evaluation Metrics

In the field of face artifact detection, it is very important to choose appropriate evaluation metrics. Evaluation metrics can objectively evaluate the performance of different algorithms or methods and help researchers fully understand the advantages and disadvantages of various techniques.

A combination of subjective and objective evaluation is usually used to evaluate the detection effect. Subjective evaluation relies entirely on the observer's visual judgment of the fake image, which makes the evaluation process highly subjective. In order to objectively evaluate the performance of the detectors, in addition to subjective observation of the fake images, objective evaluation metrics are introduced to comprehensively assess the performance of the detectors. Appropriate evaluation metrics can also help improve the efficiency of the deep fake face image detector.

Currently, some commonly used evaluation metrics are used to detect face forgery tasks to evaluate the performance of detectors. The evaluation metrics include accuracy (ACC), receiver operating characteristic (ROC), area under the curve (AUC), precision, recall, equal error rate (EER), and relative error reduction (RER).

(1) **Accuracy (ACC)**

Accuracy is the proportion of samples correctly classified by the classifier and is usually calculated according to the following formula:

$$ACC = \frac{TP + TN}{P + N} \tag{27}$$

In face artifact detection, TP (true positive) denotes the number of times a face is judged as real and judged correctly, TN (true negative) denotes the number of times a face is judged as a forgery and judged correctly, and P and N denote the number of real faces and forged faces, respectively.

ACC is the most used way to score and can be put to use for both two-choice and many-choice tasks. Often, when the ACC is high, the detector works well. But, if the data is not even, ACC might not be the best way to check since issues might happen even if the ACC is very high. So, at times, it helps to look at how the detector is doing with other measures, like precision, recall, ROC curves, and AUC values.

(2) **ROC and AUC**

As Fig. 16, in the ROC space, the vertical coordinate is the True Positive Rate (TPR) and the horizontal coordinate is the False Positive Rate (FPR). The ROC curve can visually characterize the difference in the performance of the detectors. The curve can visualize the performance of the classifier under

Fig. 16 ROC image

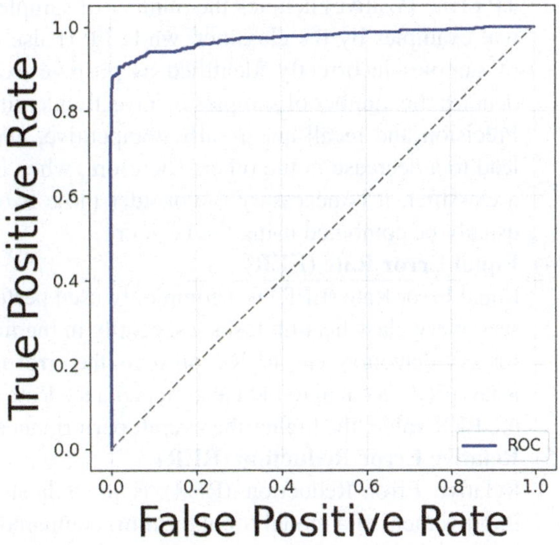

different thresholds. Most of the time, if the curve is near the top left corner, it means the classifier does a good job. AUC, which is a common way to check how well a detector works, looks at the space under the ROC curve. The AUC score goes from 0.5 to 1.0. The nearer the score is to 1.0, the better the classifier works. In most cases, if the AUC score is over 0.5, this shows that the classifier can guess well. With the ROC curve and AUC value, it is possible to visually compare the performance of different classifiers and select the best threshold or the best classifier. Especially when dealing with unbalanced datasets, ROC curves and AUC values can better reflect the overall performance of the classifiers, so they are also one of the commonly used evaluation metrics in deep forgery detection.

(3) **Precision and Recall**

Precision is used to quantify the number of true positive examples among the positive examples recognized by the classifier. The formula for accuracy is as follows:

$$\text{Precision} = \frac{TP}{TP + FP} \tag{28}$$

Recall is a metric that determines how many actual positive cases the classifier accurately identifies as positive. Recall is calculated using the following formula:

$$\text{Recall} = \frac{TP}{TP + FN} \tag{29}$$

TP (True Positive) denotes the number of samples correctly identified as positive examples by the classifier, while FP (False Positive) denotes the number of samples incorrectly identified as positive examples. FN (False Negative) denotes the number of samples incorrectly identified as negative samples.
Precision and recall are usually competitive, and improving one metric may lead to a decrease in the other. Therefore, when evaluating the performance of a classifier, it is necessary to consider these two metrics together, which can usually be combined using the F1 score.

(4) **Equal Error Rate (EER)**
Equal Error Rate (EER) is a commonly used performance evaluation metric for secondary classification tasks, especially in the fields of biometric recognition, forgery detection, etc. EER represents the error rate when the False Rejection Rate (FRR) is equal to the False Acceptance Rate (FAR). Therefore, the smaller the EER value, the higher the overall performance of the detector.

(5) **Relative Error Reduction (RER)**
Relative Error Reduction (RER) is an indicator to measure the error rate improvement of a model (or algorithm) compared with the baseline model, and is often used to evaluate the improvement effect. Set E_1 and E_2 as the error values between the two detectors ($E_1 > E_2$). RER is defined as follows:

$$RER = \frac{E_1 - E_2}{E_1} \tag{30}$$

6.2 *Common Datasets for Deepfake Detection*

Since 2018, the quick growth of Deepfake Technology (DFT) has pushed this area to make big steps in making real-looking fake face photos and videos. This shift makes it hard to use old ways to tell if images are true. This tech change has drawn a lot of eyes to the issue of spotting fake faces in both school and work areas. In order to effectively train and systematically evaluate various face forgery detection algorithms, many researchers and developers have constructed several representative datasets. These datasets usually cover face images and videos generated by different forgery methods, and also include real sample data, aiming to provide a training and testing basis for the algorithms with both diversity and complexity, so as to promote the continuous optimization and innovation of forgery detection technology in a data-driven way. Table 2 systematically organizes and summarizes the current mainstream face forensics datasets for subsequent research.

In the early days of deepfake techniques, researchers mainly used homegrown data, with undisclosed methods or small sample sizes, and a lack of uniform evaluation criteria. The research in this period was difficult to reproduce, limiting the comparison and development of algorithms. Starting from 2018, researchers gradually constructed some small-scale face forensics datasets, such as UADFV

Table 2 Summary of common Deepfake datasets

DataSet	Video(real)	Video(fake)	Number of people	Forgery method	Release time
UADFV	49	49	49	1	2018
DF-TIMIT	320	640	43	1	2018
FaceForensics++	1000	4000	–	4	2019
DFD	363	3068	28	1	2019
Celeb-DF	590	5639	59	1	2019
DFDC	23,654	104,500	960	8	2019
WildDeepfake	3805	3509	–	1	2020
DeeperForensics-1.0	50,000	10,000	100	1	2020
ForgeryNet	99,630	121,617	5400	15	2021

[51] and DF-TIMIT [52]. Among them, Yang et al. [51] collected 49 real face videos from the YouTube platform, and subsequently generated 49 fake face videos corresponding to them through the FakeAPP tool, which composed a face dataset of 98 videos in total, UADFV; while Korshunov et al. [52] collected 49 real face videos from the publicly available VidTIMIT dataset, manually screened 16 pairs of individuals with similar appearance, and based on these samples produced the DF-TIMIT dataset containing 640 face-swapped videos. However, with a small number of dataset samples and a single forgery technique, these datasets are not well-suited for deep learning tasks.

With the development of deepfake technology, the deepfake dataset has entered into a large-scale stage. Rössler et al. [53] first constructed the FaceForensics++ dataset containing 5000 real and fake face videos, which supports real application scenarios, has stable quality, and is suitable for model training and evaluation, significantly promoting the research progress of face forgery detection technology. In addition, researchers have successively released several large-scale face forensics datasets, such as DFD [54], Celeb-DF [55], and WildDeepfake [56], which play an important role in promoting the research of deepfake detection. However, these datasets still have certain limitations in terms of the diversity of forgery techniques, the authenticity of data, and the coverage of complex scenarios, making it difficult to fully simulate complex forgery situations in the real world. For this reason, some research teams have further constructed large-scale datasets with multiple forgery methods, more participants and more complex scenes, such as DeeperForensics-1.0 [57], DFDC [58] and ForgeryNet [59], to better support the performance evaluation and enhancement of deep forgery detection algorithms in real-world applications. Next, this paper describes some of the classical deepfake datasets.

(1) **FaceForensics++**
The FaceForensics++ dataset is one of the most representative and widely used resources in the field of face forgery detection. It was built by a research team from the Technical University of Munich, Germany, to advance the development of deepfake detection technology. The dataset contains 1000

real face video sequences, which are taken from high-quality news clips on YouTube, with natural expressions and various lighting conditions, providing a good sense of realism and diversity. Based on this, the researchers used four typical face forgery techniques: deepfake, Face2Face [60], FaceSwap [61], and NeuralTextures [62] to tamper with these videos to generate the corresponding forged versions. To accommodate different experimental needs, the dataset also provides three compression quality versions of the videos: original uncompressed, visual lossless high quality (quantization parameter of 23), and visual lossy low quality (quantization parameter of 40) to simulate the image degradation due to compression in reality. Examples of face images generated by the four forgery techniques are shown in Fig. 17, where the red boxes represent high-quality video images and the blue boxes represent low-quality video images. These videos cover multiple actors and different poses of facial expressions, making the dataset somewhat diverse and challenging. As a standard dataset for forgery detection research, FaceForensics++ has been widely used in the training and evaluation of various algorithms, which helps researchers to deeply understand the features of different forgery techniques and promote the advancement of face forgery detection technology.

(2) **DFD**

Researchers at Google have constructed a DeepFake Detection (DFD) dataset designed to advance face forgery recognition technology. To create the dataset, Google worked with 28 actors to record hundreds of real videos in a variety of scenarios and synthesized more than 3000 fake videos using publicly available deepfake generation methods. The DFD dataset contains a total of 363 real and 3068 fake videos, with forged content featuring diverse postures, expressions, and lighting conditions, and is widely used to train and evaluate deep forgery detection algorithms. Figure 18 shows some face images in the DFD dataset.

(3) **Celeb-DF**

Celeb-DF is a large-scale, high-quality deep forgery video dataset widely used in face forgery detection research. The dataset is based on real videos of 59 public figures of different genders, ages, and ethnic backgrounds on the YouTube platform, which cover a wide range of expressions, gestures, and speaking scenes. The researchers used the improved deepfake method to fake these videos, generating 5639 high-quality fake videos with a total frame count of more than 2 million. Compared with earlier datasets such as UADFV and DF-TIMIT, Celeb-DF significantly improves the quality of the forgeries, effectively mitigates common artifacts such as mouth distortion, skin color inconsistency, and poor edge blending, and makes the forged videos more natural and realistic, enhancing the applicability and challenge of practical detection tasks. This dataset has become one of the important benchmarks for evaluating the generalization ability of forgery detection algorithms in recent years. Figure 19 shows some face images in the Celeb-DF dataset.

(4) **DFDC**

The DFDC (DeepFake Detection Challenge) dataset is a large-scale deepfake detection dataset created and maintained by Facebook with multiple partners to

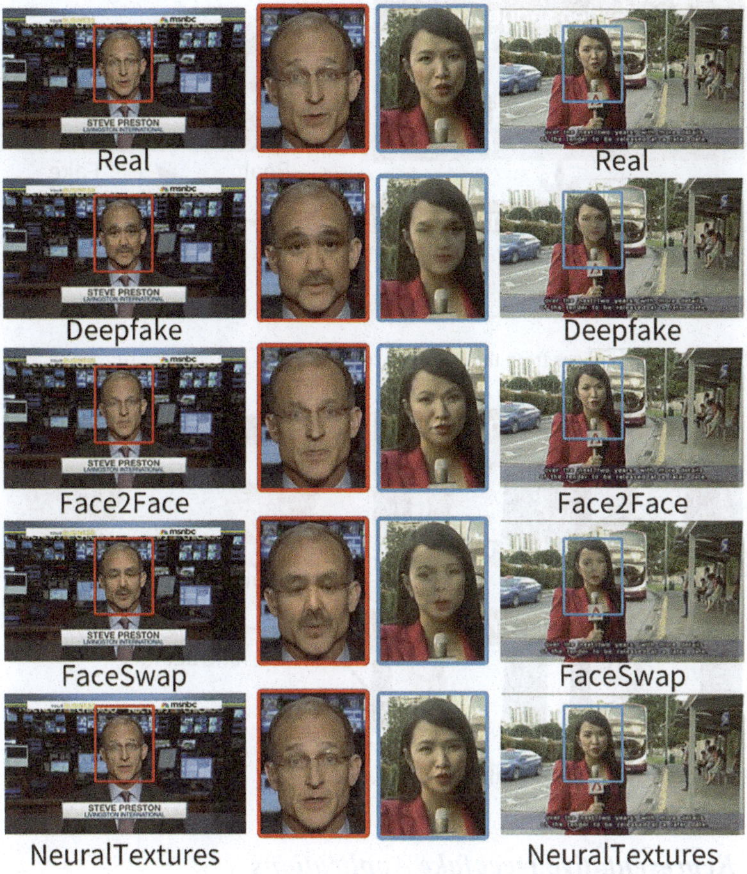

Fig. 17 Example of images from the FaceForensics++ dataset

advance the development of forgery video detection algorithms. The dataset contains more than 120,000 video clips, including real recorded videos and forged videos generated by multiple methods. All subjects participated voluntarily, possessed different genders, ages, and skin colors, and were recorded in diverse contexts, enhancing the visual diversity and realistic representation of the data. The DFDC dataset aims to be a standard benchmark for training, validating, and evaluating deepfake detection algorithms. It not only provides a challenging testbed for researchers, but also technically supports the response to the information distortion and privacy risks that can be associated with forged videos, making it one of the most representative research resources in the field. Figure 20 shows some face images in the DFDC dataset.

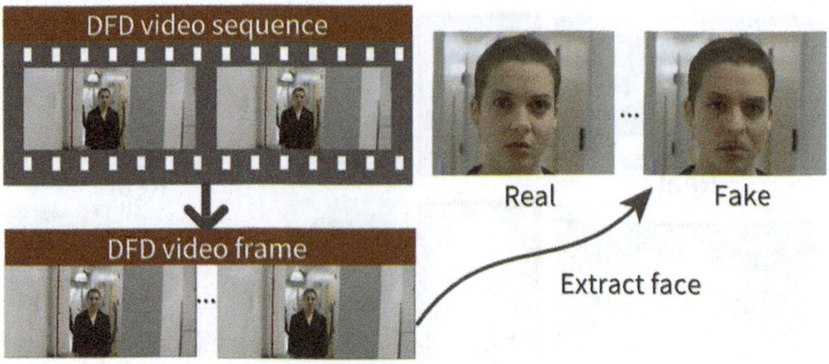

Fig. 18 Example of images from the DFD dataset

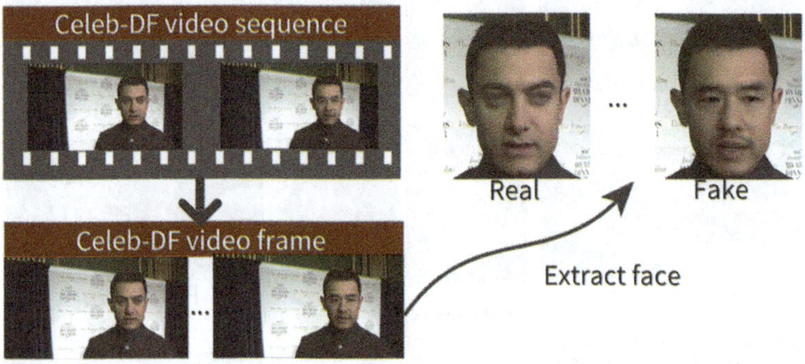

Fig. 19 Example of images from the Celeb-DF dataset

6.3 Representative Deepfake Applications

Deepfake technology, as an important branch of artificial intelligence, is continuing to promote the innovative development of various industries, giving rise to diverse application scenarios such as virtual image generation, film and television post-production, advertising, and marketing. At the same time, tools and platforms around deepfake technology are also emerging. As the Table 3, a variety of deepfake tools have been developed and publicly released, ranging from open source projects to commercial software. These tools are usually characterized by easy operation and rich functionality, which greatly reduces the threshold of deep forgeries and enables non-professionals to generate deep forged content with ease, thus accelerating the promotion and application of the technology in the public domain.

Launched in 2018, FaceApp [63] is a popular mobile app that users can use to manipulate and enhance facial features in photos. The app offers a variety of filters and effects, such as aging, gender transformation, and facial distortion, through artificial intelligence algorithms. With its easy-to-use interface and advanced technology, FaceApp has quickly become a global phenomenon, with millions of

Fig. 20 Example of images from the DFDC dataset

Table 3 Representative Deepfake applications

Application name	Year of release	Main functionality
FaceApp	2018	Modifying facial expressions and attributes
ZAO	2019	Face swapping with movie/TV celebrities
DeepFaceLab	2019	Face replacement in videos
Impersonator	2019	Character movement and posture transfer
Doublicat	2020	Face swapping in images or animated emoji packs
Reface	2020	Face swapping with celebrities in videos or GIFs
Wombo AI	2020	Controlling character movements and expressions
Avatarify	2021	Transferring facial expressions in real-time streams
DALL·E 2	2022	Generating images from text prompts
Stable Diffusion	2022	Text-to-image model training and generation

downloads and shares on the social media. The app is loved by users for its ability to generate realistic and interesting image transformations, and is often used as an entertainment and creative tool.

In 2019, Chinese company Momo launched ZAO [64], a deep-fake face-swapping app. The app utilizes deep-learning technology to superimpose a photo of a user's face onto a character in a preset short video clip, allowing users to select a movie or TV show scene to replace their own face into. Since its release, ZAO has become rapidly popular in China, with users sharing their face-swapping videos on social media.

DeepFaceLab [65] is a free and open source software that specializes in creating deep fake videos. It realizes face swapping in videos and images through deep learning algorithms to generate high-quality faked effects. Users can make use of different types of neural networks for face replacement and enhance the realism of the video through techniques such as mask drawing, overlaying and color correction. The software is equipped with a user-friendly interface, built-in pre-trained models and operation tutorials to help users efficiently create deep faked content. DeepFaceLab has a wide user base among content creators, researchers, and enthusiasts of deep faking technology due to its practicality and ease of use.

Impersonator [66] is a deep learning-based tool that helps users create high-quality and realistic deep faked videos. Developed by a team of researchers from China's Beijing Film Academy and Tsinghua University, the software uses Generative Adversarial Networks technology to learn and map one person's facial features to another's face to produce seamless and realistic video content. It also features the ability to adjust facial expressions and lip movements to ensure a high match to the target person's audio. Impersonator comes with a user-friendly interface that offers multiple customization options, including detail level adjustment and selection from multiple pretrained models. With its superior video generation results, the tool has gained widespread attention and adoption in creative fields such as film, advertising and art installations.

Doublicat [67] is a mobile app developed by Ukrainian company RefaceAI that utilizes deep forgery technology to enable users to replace faces in videos and motion graphics with their own or celebrity faces. The app uses artificial intelligence and machine learning algorithms to seamlessly blend new faces into videos, generating effects that are both realistic and funny. Users can choose from a wide variety of videos and emoji animations and share their creations on social media platforms such as Instagram and TikTok. Doublicat was released in 2020 and quickly became popular, with more than 10 million downloads in just two months.

Reface [68] is an app made by a company in Ukraine. It uses deep fake tech to let users swap their face with faces from videos, films, or emojis. You can make real and fun swapped face stuff fast and easy, then post it on social sites like TikTok and Instagram. This app works on both iOS and Android phones. It has grown to be one of the top face-swap tools around the world because it's simple to use and gives good, real looks.

Wombo [69] is a mobile app that uses deep learning technology to create funny fake singing videos. It allows users to select a song from a list of popular songs and then uses artificial intelligence to animate the user's face to make it look like they are singing the lyrics. Available for iOS and Android devices, Wombo has become popular for making humorous videos and sharing them on social media platforms. Despite the technology used being based on deep learning algorithms, the app has gained attention for its ability to create high-quality animations that look realistic and engaging. Users can share their videos on a variety of social media platforms including Instagram, TikTok and Twitter.

Avatarify [70] is a popular deep faking software tool that allows users to superimpose their own faces onto various characters and images in real time. The tool allows users to choose from a variety of pre-existing images, including celebrities and historical figures, as well as upload their own images for conversion. One of the unique features of Avatarify is its ability to work in real-time, allowing users to create real-time depth fakes during video calls or live streams. This makes it a popular tool for content creators and social media influencers to create engaging and interactive content.

DALL-E 2 [71] is an artificial intelligence program developed by OpenAI that generates images from textual descriptions.Released in 2022, DALL-E 2 is capable of generating complex image. DALL-E 2 uses a neural network architecture known as a transducer to process textual descriptions and generate the corresponding images. The program has been trained on a large image and text dataset, enabling it to recognize and reproduce a variety of objects, scenes, and concepts. Users can enter textual cues such as "tree made of clouds" or "cake that looks like a castle" and DALL-E 2 will generate an image that matches the description. DALL-E 2 has a wide range of potential applications, including creating visual aids for people with disabilities and generating images for marketing and advertising.

Stable Diffusion [72] is one of today's increasingly popular generative art tools based on the diffusion model that takes simple text prompts and creates incredible images seemingly out of nothing, generating character model art for role-playing games, art for those who can't afford to pay an artist's commission, and cool art to inspire writers and other creatives. The software page allows users to enter a text prompt for the image and adjust any settings they personally prefer, then select the Generate button to create an image with stable diffusion. Users can also adjust the resolution of the image using the width and height settings, or increase the sampling step for a higher quality image. This is an example of the growing capabilities of artificial intelligence and machine learning, technologies that have the potential to revolutionize the way we create and interact with visual content.

References

1. Goodfellow IJ, Pouget-Abadie J, Mirza M, Xu B, Warde-Farley D, Ozair S, Courville A, Bengio Y. Generative adversarial nets. Paper presented at the 28th International Conference on Neural Information Processing Systems; 2014. p. 2672–80.
2. Croitoru F-A, Hondru V, Ionescu RT, Shah M. Diffusion models in vision: a survey. IEEE Trans Pattern Anal Mach Intell. 2023;45(9):10850–69.
3. Li P, Pei Y, Li J. A comprehensive survey on design and application of autoencoder in deep learning. Appl Soft Comput. 2023;138:110176.
4. Karras T, Laine S, Aila T. A style-based generator architecture for generative adversarial networks. Paper presented at the IEEE/CVF Conference on Computer Vision and Pattern Recognition; 2019. p. 4401–10.
5. Liu Z, Luo P, Wang X, Tang X. Deep learning face attributes in the wild. Paper presented at the IEEE International Conference on Computer Vision; 2015. p. 3730–8.
6. Karras T, Laine S, Aittala M, Hellsten J, Lehtinen J, Aila T. Analyzing and improving the image quality of Stylegan. Paper presented at the IEEE/CVF Conference on Computer Vision and Pattern Recognition; 2020. p. 8110–9.
7. Cao Q, Shen L, Xie W, Parkhi OM, Zisserman A. Vggface2: a dataset for recognising faces across pose and age. Paper presented at the 13th IEEE International Conference on Automatic Face & Gesture Recognition; 2018. p. 67–74.

8. Rössler A, Cozzolino D, Verdoliva L, Riess C, Thies J, Nießner M. Faceforensics++: learning to detect manipulated facial images. Paper presented at the IEEE/CVF International Conference on Computer Vision; 2019. p. 1–11.
9. Patel Y, Tanwar S, Gupta R, Bhattacharya P, Davidson IE, Nyameko R, Aluvala S, Vimal V. Deepfake generation and detection: case study and challenges. IEEE Access. 2023;11:143296–323.
10. Guarnera L, Giudice O, Battiato S. Fighting deepfake by exposing the convolutional traces on images. IEEE Access. 2020;8:165085–98.
11. Pan Z, Yu W, Yi X, Khan A, Yuan F, Zheng Y. Recent progress on generative adversarial networks (GANs): a survey. IEEE Access. 2019;7:36322–33.
12. Ronneberger O, Fischer P, Brox T. U-net: convolutional networks for biomedical image segmentation. Paper presented at the 18th International Conference on Medical Image Computing and Computer-Assisted Intervention; 2015:234–41.
13. Chollet F. Xception: deep learning with depthwise separable convolutions. Paper presented at the IEEE Conference on Computer Vision and Pattern Recognition; 2017:1800–7.
14. Tan M, Le Q. EfficientNet: rethinking model scaling for convolutional neural networks. Paper presented at the 36th International Conference on Machine Learning; 2019:6105–14.
15. Zhao H, Zhou W, Chen D, Wei T, Zhang W, Yu N. Multi-attentional deepfake detection. Paper presented at the IEEE/CVF Conference on Computer Vision and Pattern Recognition; 2021. p. 2185–94.
16. Li L, Bao J, Zhang T, Yang H, Chen D, Wen F, Guo B. Face X-ray for more general face forgery detection. Paper presented at the IEEE/CVF Conference on Computer Vision and Pattern Recognition; 2020. p. 5000–9.
17. Zhao T, Xu X, Xu M, Ding H, Xiong Y, Xia W. Learning self-consistency for deepfake detection. Paper presented at the IEEE/CVF International Conference on Computer Vision; 2021. p. 15003–13.
18. Dong S, Wang J, Liang J, Fan H, Ji R. Explaining deepfake detection by analysing image matching. Paper presented at the European Conference on Computer Vision; 2022. p. 18–35.
19. Qian Y, Yin G, Sheng L, Chen Z, Shao J. Thinking in frequency: face forgery detection by mining frequency-aware clues. Paper presented at the European Conference on Computer Vision; 2020. p. 86–103.
20. Li J, Xie H, Li J, Wang Z, Zhang Y. Frequency-aware discriminative feature learning supervised by single-center loss for face forgery detection. Paper presented at the IEEE/CVF Conference on Computer Vision and Pattern Recognition; 2021. p. 6458–67.
21. Wang J, Wu Z, Ouyang W, Han X, Chen J, Jiang Y-G, Li S-N. M2tr: multi-modal multi-scale transformers for deepfake detection. Paper presented at the International Conference on Multimedia Retrieval; 2022. p. 615–23.
22. Masi I, Killekar A, Mascarenhas RM, Gurudatt SP, AbdAlmageed W. Two-branch recurrent network for isolating deepfakes in videos. Paper presented at the European Conference on Computer Vision; 2020. p. 667–84.
23. Ruff L, Vandermeulen R, Goernitz N, Deecke L, Siddiqui SA, Binder A, Müller E, Kloft M. Deep one-class classification. Paper presented at the International Conference on Machine Learning; 2018. p. 4393–402.
24. Zheng Y, Bao J, Chen D, Zeng M, Wen F. Exploring temporal coherence for more general video face forgery detection. Paper presented at the IEEE/CVF International Conference on Computer Vision; 2021. p. 15044–54.
25. Haliassos A, Vougioukas K, Petridis S, Pantic M. Lips don't lie: a generalisable and robust approach to face forgery detection. Paper presented at the IEEE/CVF Conference on Computer Vision and Pattern Recognition; 2021. p. 5037–47.
26. Chung JS, Senior A, Vinyals O, Zisserman A. Lip reading sentences in the wild. Paper presented at the IEEE Conference on Computer Vision and Pattern Recognition; 2017. p. 3444–53.
27. Haliassos A, Mira R, Petridis S, Pantic M. Leveraging real talking faces via self-supervision for robust forgery detection. Paper presented at the IEEE/CVF Conference on Computer Vision

and Pattern Recognition; 2022. p. 14950–62.
28. He K, Zhang X, Ren S, Sun J. Deep residual learning for image recognition. Paper presented at the IEEE Conference on Computer Vision and Pattern Recognition; 2016. p. 770–8.
29. Tran D, Wang H, Torresani L, Feiszli M. Video classification with channel-separated convolutional networks. Paper presented at the IEEE/CVF International Conference on Computer Vision; 2019. p. 5552–61.
30. Ding B, Lu T, Peng S, Wang L. A pre-training-based recognition model for face image forgery algorithms. J Front Comput Sci Technol. 2025:1–16.
31. Zheng JS, Hu XY, Chen C, Zhou YC, Gao DY, Tang ZM. A new deepfake detection model for responding to perception attacks in embodied artificial intelligence. Image Vision Comput. 2024;151.
32. Moon K-H, Ok S-Y, Lee S-H. SupCon-MPL-DP: supervised contrastive learning with meta pseudo labels for deepfake image detection. Appl Sci. 2024;14(8):3249.
33. Dong S, Wang J, Ji R, Liang J, Fan H, Ge Z. Implicit identity leakage: the stumbling block to improving deepfake detection generalization. Paper presented at the IEEE/CVF Conference on Computer Vision and Pattern Recognition; 2023. p. 3994–4004.
34. Afchar D, Nozick V, Yamagishi J, Echizen I. Mesonet: a compact facial video forgery detection network. Paper presented at the IEEE International Workshop on Information Forensics and Security; 2018. p. 1–7.
35. Bayar B, Stamm MC. A deep learning approach to universal image manipulation detection using a new convolutional layer. Paper presented at the ACM Workshop on Information Hiding and Multimedia Security; 2016. p. 5–10.
36. Cozzolino D, Poggi G, Verdoliva L. Recasting residual-based local descriptors as convolutional neural networks: an application to image forgery detection. Paper presented at the ACM Workshop on Information Hiding and Multimedia Security; 2017. p. 159–64.
37. Nguyen HH, Yamagishi J, Echizen I. Capsule-forensics: using capsule networks to detect forged images and videos. Paper presented at the IEEE International Conference on Acoustics, Speech and Signal Processing; 2019. p. 2307–11.
38. Liy CM, InIctuOculi L. Exposing AI created fake videos by detecting eye blinking. Paper presented at the IEEE International Workshop on Information Forensics and Security; 2018. p. 1–7.
39. Vudayagiri P, Sana R, Abdallah HA, Agarwal N, Agarwal S. Synthetic shadows: the interplay of forensic detection and anti-forensic techniques in GAN-generated images. Multimed Syst. 2025;31(4):1–26.
40. Dang H, Liu F, Stehouwer J, Liu X, Jain AK. On the detection of digital face manipulation. Paper presented at the IEEE/CVF Conference on Computer Vision and Pattern Recognition; 2020. p. 5781–90.
41. Soltandoost E, Plesh R, Schuckers S, Peer P, Struc V. Extracting local information from global representations for interpretable deepfake detection. Paper presented at the Winter Conference on Applications of Computer Vision; 2025. p. 1629–39.
42. Deng J, Guo J, Ververas E, Kotsia I, Zafeiriou S. Retinaface: single-shot multi-level face localisation in the wild. Paper presented at the IEEE/CVF Conference on Computer Vision and Pattern Recognition; 2020. p. 5203–12.
43. Sukno FM, Waddington JL, Whelan PF. 3-D facial landmark localization with asymmetry patterns and shape regression from incomplete local features. IEEE Trans Cybern. 2014;45(9):1717–30.
44. Woo S, Park J, Lee J-Y, Kweon IS. CBAM: convolutional block attention module. Paper presented at the European Conference on Computer Vision; 2018. p. 3–19.
45. Zeiler MD, Fergus R. Visualizing and understanding convolutional networks. Paper presented at the European Conference on Computer Vision; 2014. p. 818–33.
46. Lei X, Jiang X, Wang C. Design and implementation of a real-time video stream analysis system based on FFMPEG. Paper presented at the World Congress on Software Engineering; 2013. p. 212–6.

47. Ioffe S, Szegedy C. Batch normalization: accelerating deep network training by reducing internal covariate shift. Paper presented at the International Conference on Machine Learning; 2015. p. 448–56.
48. Han S, Pool J, Tran J, Dally W. Learning both weights and connections for efficient neural network. In: Advances in Neural Information Processing Systems; 2015. p. 28.
49. Wu C-Y, Zaheer M, Hu H, Manmatha R, Smola AJ, Krähenbühl P. Compressed video action recognition. Paper presented at the IEEE Conference on Computer Vision and Pattern Recognition; 2018. p. 6026–35.
50. Liu B, Chen Y, Liu S, Kim H-S. Deep learning in latent space for video prediction and compression. Paper presented at the IEEE/CVF Conference on Computer Vision and Pattern Recognition; 2021. p. 701–10.
51. Yang X, Li Y, Lyu S. Exposing deep fakes using inconsistent head poses. Paper presented at the IEEE International Conference on Acoustics, Speech and Signal Processing; 2019. p. 8261–5.
52. Korshunov P, Marcel S. Deepfakes: a new threat to face recognition? Assessment and detection. arXiv preprint arXiv:1812.08685; 2018.
53. Rössler A, Cozzolino D, Verdoliva L, Riess C, Thies J, Nießner M. FaceForensics: A large-scale video dataset for forgery detection in human faces. arXiv preprint arXiv:1803.09179; 2018
54. Dufour N, Gully A, Karlsson P, et al. Deepfakes detection dataset by Google and Jigsaw; 2019. Available at: https://research.google/blog/contributing-data-to-deepfake-detection-research/.
55. Li Y, Yang X, Sun P, Qi H, Lyu S. Celeb-DF: a large-scale challenging dataset for deepfake forensics. Paper presented at the IEEE/CVF Conference on Computer Vision and Pattern Recognition; 2020. p. 3207–16.
56. Zi B, Chang M, Chen J. WildDeepfake: a challenging real-world dataset for deepfake detection. Paper presented at the ACM International Conference on Multimedia; 2020. p. 2382–90.
57. Jiang L, Li R, Wu W, et al. DeeperForensics-1.0: a large-scale dataset for real-world face forgery detection. Paper presented at the IEEE/CVF Conference on Computer Vision and Pattern Recognition; 2020. p. 2889–98.
58. Dolhansky B, Bitton J, Pflaum B, et al. The Deepfake detection challenge Dataset. arXiv preprint arXiv:2006.07397; 2020.
59. He Y, Gan B, Chen S, et al. ForgeryNet: a versatile benchmark for comprehensive forgery analysis. Paper presented at the IEEE/CVF Conference on Computer Vision and Pattern Recognition; 2021. p. 4360–9.
60. Thies J, Zollhöfer M, Stamminger M, Theobalt C, Nießner M. Face2face: real-time face capture and reenactment of RGB videos. Paper presented at the IEEE Conference on Computer Vision and Pattern Recognition; 2016. p. 2387–95.
61. Korshunova I, Shi W, Dambre J, et al. Fast face-swap using convolutional neural networks. Paper presented at the IEEE International Conference on Computer Vision; 2017. p. 3677–85.
62. Thies J, Zollhöfer M, Nießner M. Deferred neural rendering: image synthesis using neural textures. ACM Trans Graphics. 2019;38(4):1–12.
63. FaceApp. FaceApp: Neural face transformation mobile application. Wireless Lab OOO; 2017. Available at: https://www.faceapp.com.
64. ZAO. ZAO: Real-time face swap application. Momo Inc.; 2019. Available at: https://zaodownload.com.
65. Iperov. DeepFaceLab: Open-source deepfake face-swapping tool; 2018. Available at: https://github.com/iperov/DeepFaceLab.
66. Liu W, Piao Z, Min J, et al. Liquid Warping GAN: a unified framework for human motion imitation, appearance transfer and novel view synthesis. Paper presented at the IEEE/CVF International Conference on Computer Vision; 2019. p. 5904–13.
67. Doublicat. Doublicat: Face swap video app (early version of Reface). NeoCortext Inc.; 2019. Available at: https://get.reface.app.
68. Reface. Reface: Face swap videos and memes with your photo. NeoCortext Inc.; 2020. Available at: https://get.reface.app.

69. Wombo AI. Wombo: AI-powered lip sync and face animation app. Wombo Studios Inc.; 2020. Available at: https://www.wombo.ai.
70. Avatarify. Avatarify: Real-time AI face animator. Open-source project; 2020. Available at: https://github.com/alievk/avatarify.
71. OpenAI. DALL·E 2: AI system for realistic image generation from text; 2022. Available at: https://openai.com/dall-e-2.
72. Stability AI. Stable Diffusion: High-resolution text-to-image synthesis model; 2022. Available at: https://stability.ai/blog/stable-diffusion-public-release.

Adversarial Attacks and Defenses in Computer Vision

Yu Ji, Wenzhi Wu, Hang Chen, and Zhengjun Liu

Abstract Computer vision systems have demonstrated excellent performance in tasks such as image classification, target detection, and face recognition. However, the deep neural networks that drive these systems are highly susceptible to interference from tiny, carefully designed perturbations, namely adversarial samples. These samples are virtually imperceptible to the human eye yet can lead to serious misjudgments in the model, revealing the fragility of its decision boundary and its fundamental difference from human perception. Conversely, defense strategies aim to improve robustness of models against adversarial attacks and ensure accurate outputs in the face of malicious attacks.

Keywords Computer vision · Adversarial attacks · Defenses strategy · Decision boundary

1 Introduction

The threats against samples are universal. They can migrate across models, affect models with different structures or training data, and manifest in the physical world. For instance, printed adversarial perturbation patterns can deceive camera systems, causing autonomous driving systems to misidentify "stop" signs as "speed limit" signs, medical AIs to misclassify benign tumors as malignant, and face-recognition access control to be illegally bypassed, as Fig. 1. Therefore, studying the generation

Y. Ji · W. Wu
School of Electronic Engineering, Heilongjiang University, Harbin, China
e-mail: yuji@hlju.edu.cn; wuwenzhi@hlju.edu.cn

H. Chen
School of Space Information, Space Engineering University, Beijing, China

Z. Liu (✉)
School of Physics, Harbin Institute of Technology, Harbin, China
e-mail: zjliu@hit.edu.cn

H. Chen, Z. Liu (eds.), *Artificial Intelligence in Digital Image Processing: Theories, Methods, and Applications*, Studies in Computational Intelligence 1246,
https://doi.org/10.1007/978-3-032-12891-1_3

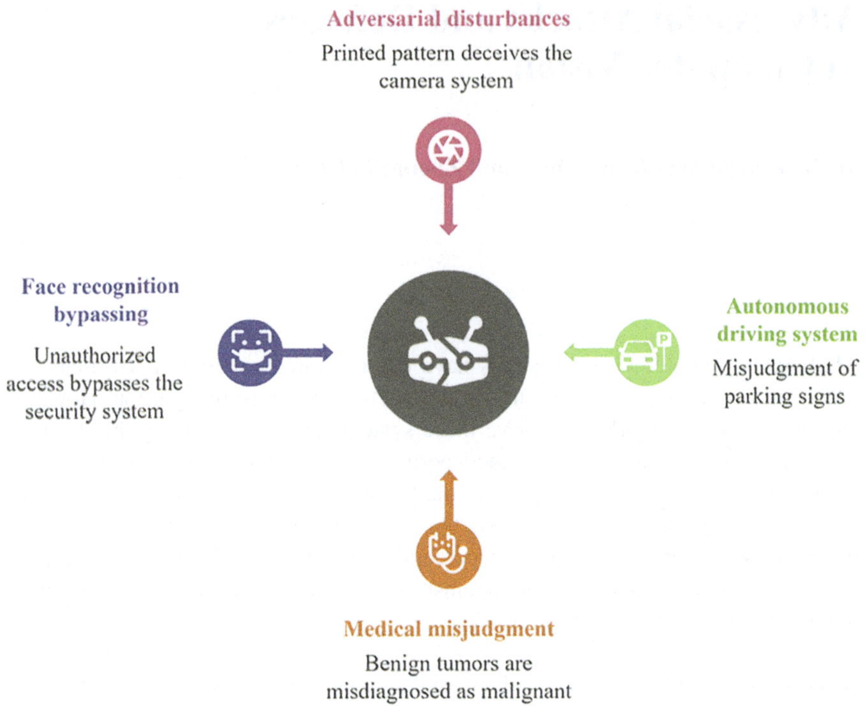

Fig. 1 Threat of adversarial samples

mechanism of counterattacks and developing effective defense strategies is essential to ensuring the safe, reliable, and trustworthy deployment of computer vision–based systems.

The motivation behind adversarial attacks stems from the observation that deep learning models, while powerful in many domains, can be easily misled by carefully crafted examples [1]. For instance, even a small, carefully designed noise added to an image can lead a model to misclassify it with high confidence. This phenomenon has been demonstrated in numerous studies [2], which introduced the concept of adversarial examples and highlighted their potential to deceive state-of-the-art models. The existence of such vulnerabilities poses significant risks in real-world applications, where the integrity and reliability of AI systems are paramount.

Defenses can be broadly categorized into three main approaches: modifying the training process, altering the input data, or using external modules to detect or mitigate adversarial examples [3]. One of the most prominent defense strategies is adversarial training, where models are trained on both clean and adversarial examples to improve their robustness [4]. Other methods include modifying loss functions, activation functions, or network architectures to enhance resilience against attacks. Additionally, some approaches focus on detecting adversarial perturbations before they reach the model, such as feature denoising or input normalization techniques [5].

Despite these advancements, the development of effective defenses remains an ongoing challenge. Adversarial attacks continue to evolve, with new methods emerging that exploit the limitations of existing defense strategies. For example, physical-world attacks, which involve manipulating the environment in which the model operates (adversarial lights or patches), present unique challenges that are not easily addressed by traditional digital defenses [6]. Similarly, black-box attacks, where the attacker has no direct access to parameters of the model, require different defense strategies compared to white-box attacks [7].

Adversarial attacks and defenses in computer vision are a dynamic and rapidly evolving field. While deep learning has revolutionized many aspects of computer vision, the vulnerability of these models to adversarial examples necessitates continued research into both attack and defense mechanisms. As the technology becomes more integrated into critical systems, ensuring the robustness and security of AI models against adversarial threats will be essential for their widespread and trustworthy deployment [8].

This chapter introduces research progress on adversarial attacks and defenses in computer vision. An in-depth analysis of the principles and evolution of classical algorithms and the main attack paradigms are provided, such as white-box, black-box, and physical-world attacks. It also examines the mechanisms and effects of mainstream defense strategies such as input preprocessing, robust model training, adversarial sample detection, and model integration. Finally, it explores current challenges and future research directions such as verifiable robustness, human-machine perceptual alignment, and the exploration of new architectures. The schematic diagram of adversarial attacks and defenses in computer vision is shown in Fig. 2.

2 Basic Principle of Adversarial Attacks

In computer vision, an adversarial attack is a technique that exploits vulnerabilities of a model to produce misleading results. Attackers add carefully designed perturbations to the input data, deviating output of the model from what is expected. These perturbations are virtually imperceptible to human observers. These attacks focus on exploiting the nonlinear properties and high-dimensional feature space of the model, allowing small input changes to lead to significant changes in output of the model. For instance, altering the values of specific pixels in an image can cause the model to misclassify a cat as a dog.

An adversarial attack can be broken down into three steps. First, the attacker chooses a target model, which may be a deep neural network or another machine learning algorithm. Next, the attacker generates an adversarial sample to mislead decisions of the model by adding perturbations to the original sample. This process involves optimizing the algorithm to minimize the perturbations while maximizing

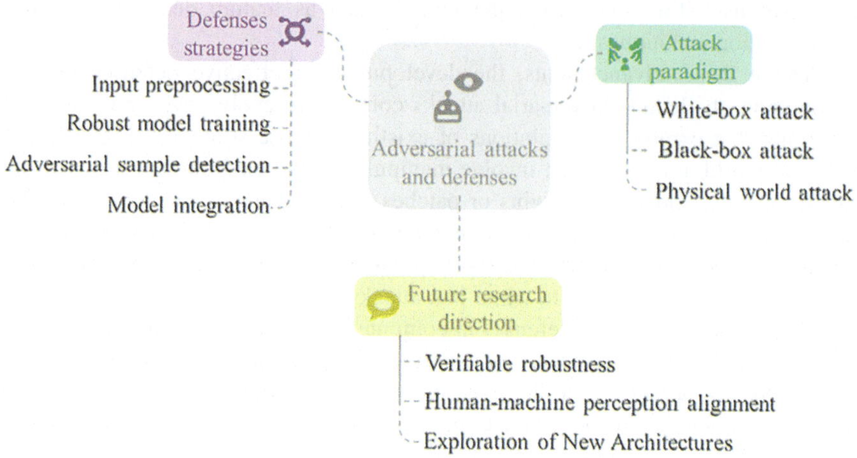

Fig. 2 Diagram of adversarial attacks and defenses in computer vision

classification error of the model. Finally, the attacker feeds the adversarial sample into the target model and observes whether output of the model is misled as expected.

The goal of the adversarial attack is to generate an adversarial sample $x' = x + \delta$ that is very similar to the original sample x, such that the target model f makes an incorrect output $f(x') \neq y_{\text{true}}$ or $f(x') = y_{\text{target}}$. Generally, the perturbation size δ is constrained by an L_p norm of $\|\delta\|_p \leq \epsilon$. The attack can be classified in multiple dimensions based on the amount of information the attacker has, the expected goal, and the application scenario.

2.1 Classification Based on Attacker Knowledge

Attacker knowledge can be classified into various types, mainly including white-box attacks and black-box attacks, as Fig. 3. This classification method not only reflects the attacker's understanding of the target system, but also influences the choice of attack strategies and the effectiveness of the attack.

2.1.1 White-Box Attacks

The attacker has complete knowledge of the target model, including its architecture, weights, training data, distributions, and gradient information. This is the most powerful and easiest attack scenario to execute, typically using gradient information to directly optimize the perturbation.

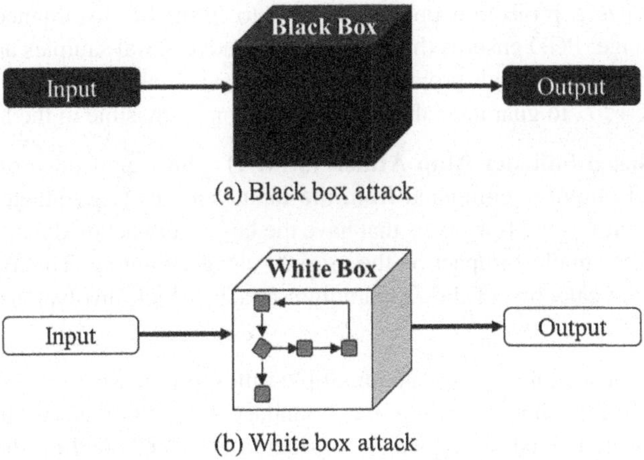

Fig. 3 White-box attacks and black-box attacks. (**a**) Black box attack. (**b**) White box attack

Fast Gradient Sign Method (FGSM) The core idea of this classic method is to add perturbations in the direction of the sign of the gradient of the loss function $J(\theta, x, y)$ with respect to the input $x' = x + \varepsilon \cdot \text{ sign } (\nabla_x J(\theta, x, y))$, where ε is a hyperparameter to be chosen. Although FGSM is computationally efficient, the generated perturbations are often large, resulting in a relatively low attack success rate [9].

Basic Iterative Method (BIM) To address the shortcomings of the fast gradient sign method, a simple approach was introduced. The fast method is to be applied multiple times with small step sizes [10]. Furthermore, the pixel values of the intermediate results must be clipped after each step to ensure they remain within the ε-neighborhood of the original image. This process can be mathematically expressed as

$$x'_{t+1} = \text{clip}_{x,\varepsilon}\left(x'_t + \alpha \cdot \text{sign}\left(\nabla_x J\left(\theta, x'_t, y\right)\right)\right) \tag{1}$$

where α is the step size, $\nabla_x J\left(\theta, x'_t, y\right)$ is the gradient of the model on the current sample, and x_t is the adversarial sample in the t-th iteration.

Projected Gradient Descent (PGD) Another optimization method is the projected gradient descent method, whose core idea is to gradually adjust the input samples through an iterative optimization process to maximize the predictive loss of the model under the premise of satisfying the perturbation constraints. This is currently the popular and well-studied attack criterion for evaluating the robustness of the model. The formula for PGD is

$$x'_{t+1} = \prod\nolimits_{\varepsilon}\left(x'_t + \alpha \cdot \text{sign}\left(\nabla_x J\left(\theta, x'_t, y\right)\right)\right) \tag{2}$$

where $\prod_{\varepsilon}(\cdot)$ is a projection operation, used to limit the disturbance within the allowable range. PGD ensures that the generated adversarial samples are within the perturbation range through projection operations, such as the L_{∞} norm constraint $\|x_{adv} - x\|_{\infty} \leq \varepsilon$, to guarantee that the perturbation is invisible to the human eye.

Jacobian-Based Saliency Map Attack (JSMA) The significance of the change in the model output is computed from the input features (e.g., image pixels) and iteratively modifies the few pixels that have the largest impact on the target category output and the smallest impact on the original category output. This type of attack falls under the category of the L_0-paradigm attack, which involves perturbing the sparse [2].

DeepFool The samples are iterated multiple times, progressing toward the nearest decision boundary until crossing the boundary leads to misclassification. The perturbations generated are typically considered a lower bound on the robustness of the classifier [11].

2.1.2 Black-Box Attacks

Attackers are only able to query the target model to obtain input-output pairs, such as prediction labels or confidence scores, without having access to the internal details of the model, such as structure, parameters, and gradients. This type of attack bears a closer resemblance to real-world attack scenarios.

Transfer-Based Attacks Attackers leverage their own surrogate model to generate adversarial samples and then utilize the relocatability of these samples to directly attack the target black-box model [12]. The ultimate efficacy of the concluding offensive maneuver is contingent upon the degree of resemblance between the surrogate model and the target model. The integration of surrogate models through model-based data enhancement has been demonstrated to enhance the efficacy of migration processes, thereby improving the success rate of such endeavors.

Query-Based Attacks

1. *Zeroth Order Optimization*

   Zero-order optimization (ZOO) is a method of optimization that never require the explicit computation of gradient information [13]. Rather, it approximates the optimal solution through the evaluation of function values. The fundamental principle underlying this approach involves the utilization of the function value of the objective function to estimate the direction of the gradient through an iterative process. This estimation is then employed to update the variables accordingly, thereby gradually approaching the extreme point of the objective function. The gradient can be estimated by numerical difference. If a small positive perturbation, designated as μ, exists in the vector $\boldsymbol{x}$ at the current optimization point, the gradient of the objective function $f(\cdot)$ can be expressed as follows:

$$\hat{\nabla} f(\mathbf{x}) = \frac{f(\mathbf{x} + \mu \mathbf{u}) - f(\mathbf{x} - \mu \mathbf{u})}{2\mu} \mathbf{u} \tag{3}$$

where **u** refers to a standard Gaussian vector. The gradient is estimated through the use of a finite difference technique, thereby facilitating optimization in the absence of explicit gradient information. The application of a white-box-like optimization method to generate adversarial samples necessitates a substantial number of numerical queries, which is a significant computational burden.

2. *Natural evolution strategies*

   The estimation of the gradient of the objective function through probability distribution sampling, leveraging the concept of evolutionary strategy, enables the attacker to access only the confidence score of the model output [14]. This allows for the approximation of the gradient direction through adaptive sampling direction, which is efficiently optimized while reducing the number of queries. In comparison with the finite difference method, the variance is reduced, and the high-dimensional space is more efficient. Nevertheless, the execution of this process necessitates the execution of hundreds to thousands of queries, with the outcome contingent upon the confidence score output.

3. *Boundary attack*

   The model requires solely model prediction labels or hard decision information [15]. It commences with initial adversarial samples or random noise from the target class. Then, it iteratively scales down the magnitude of the perturbation along the decision boundary. In this manner, it gradually reduces the perturbation while maintaining adversariality. This method requires only predictive labels and is widely applicable. However, convergence is slow, and the quality of the initial adversarial samples affects performance.

4. *Surrogate model-based attacks*

   The attacker gathers input-output data by querying the target model repeatedly, trains a local agent model, performs a white-box attack on the agent model to generate adversarial samples, and attacks the target model using mobility [16]. The iterative training of the agent model, also known as active learning, has been demonstrated to enhance efficiency, eliminate the necessity to query the target model in real time, and facilitate rapid generation. However, the efficacy of this migration is contingent upon the similarity between the models in question, rendering it ineffective for robust training methods, such as defense distillation.

2.2 Classification Based on Attack Targets

1. **Targeted Attacks**

A targeted attack is defined as an attempt by an adversary to force a model to misclassify an adversarial sample x' into a specific target category y_{target} that the attacker has specified, as Fig. 4. The implementation of targeted attacks allows for a high degree of control, with the potential to generate misleading results, leading to the misclassification of benign transactions as fraudulent ones. Furthermore, they have the capacity to generate more nuanced perturbations, which are frequently imperceptible to humans, with the objective of approximating the target category. Furthermore, targeted attacks facilitate directed disruption by implanting specific backdoors, such as identifying specific faces as authorized persons. However, such attacks are usually more difficult to execute than untargeted attacks, and the implementation process is complex, requiring both $f(x_{\text{adv}}) = y_{\text{target}}$ and perturbation constraints to be satisfied. Black-box scenarios necessitate a substantial amount of trial and error, target class decision boundaries, and elevated query costs. Perturbations designed for a particular model are challenging to generalize to other models and demonstrate

Fig. 4 Targeted attacks

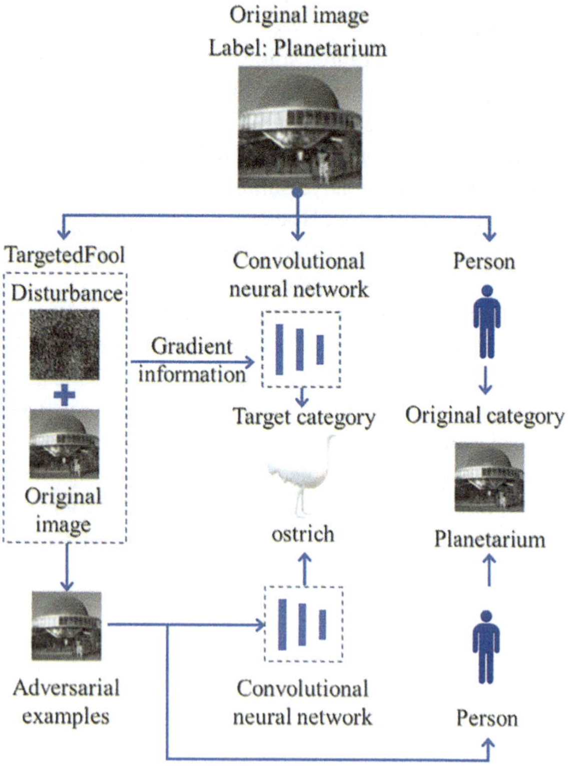

suboptimal migration. Slight deviations in perturbation can result in the failure of the attack, which carries a significant risk of failure.

2. **Untargeted Attacks**
   A targetless attack is defined as output of the adversarial sample $x^{'}$ not being equal to its true category y_{true} without specifying a particular error category. A targetless attack that deviates from the correct label, without constraining the error class, is more likely to succeed because it allows for a larger output space. The optimization objective is a single-peak function, with a clear direction of the gradient and faster convergence. The model output space requirements are minimal and universally applicable. Due to the flexibility inherent in the objective, smaller perturbations to the paradigm can be investigated enhance robustness. However, the consequences of a targetless attack are difficult to predict and may lead to the misclassification of non-critical categories. It has been demonstrated that this process may result in the generation of perturbations that are disproportionately large in size, owing to the presence of untargeted constraints. The utilization of this model is constrained, the attainment of precise misclassification is arduous, the capacity for destruction is restricted, and substantial perturbations may initiate anomaly detection mechanisms.

2.3 *Physical-World Attack*

Physical-World Attack is defined as attack behavior involving the direct manipulation of physical entities or the intrusion into physical space for the purpose of compromising the integrity of an information system, stealing sensitive data, or interfering with the system normal operation. Overcoming the complex transformations of distance, angle, illumination changes, camera noise, occlusion, etc. during the imaging process is a key challenge in attacking the actual system when applying the digital domain-generated adversarial perturbations to the physical-world objects.

1. **Expectation over Transformation**
   Expectation over transform (EOT) is a general framework, and the core idea of EOT is to ensure the robustness of the generated adversarial samples in real physical environments by optimizing the expected loss of the adversarial perturbations under multiple physical transformations [17]. It has been demonstrated that the adversarial performance against random rotations, translations, scaling, brightness adjustments, noise additions, and other perturbations can be maintained under the optimization of the expected transform distribution. When the original input image x has an adversarial perturbation δ to be optimized, the objective function of the target model $f(\cdot)$ can be defined as follows

$$\min_{\delta}\mathbb{E}_{t\sim T}\left[\mathcal{L}\left(f\left(t\left(x+\delta\right)\right),y_{\text{target}}\right)\right] \tag{4}$$

where $t \sim T$ is used to represent the physical transformation function that is sampled from the transformation distribution T. The symbol L is used to represent the loss function. The symbol y_{target} is used to denote the target error class. However, the high-dimensional transform space necessitates a substantial number of samples and imposes a considerable computational overhead, as Fig. 5.

2. **Adversarial Patches**

   The design of patches is typically informed by considerations of physical deployability and viewpoint invariance [18]. Without disturbing the whole image, deep learning models, especially visual models, are prompted to generate false predictions by employing physical patterns with adversarial perturbations that are pasted in the target object or scene. The fundamental objective is to migrate the adversarial sample technique from the digital domain to the physical world, thereby transcending the virtual boundaries of traditional cyber-attacks and directly impacting the real environment.

3. **Adversarial Texture/Camouflage**

   Adversarial texture/camouflage represents a sophisticated form of physical world attack, whereby continuous patterns with adversarial perturbations are generated and applied to modify the visual characteristics of a target object surface. This is intended to deceive deep learning models, particularly target detection and recognition systems. The focus of this research is the transformation of digital adversarial perturbations into physically deployable texture schemes that adhere to realistic constraints, thereby combining attack robustness and stealth. A combination of three-dimensional modeling and rendering techniques is required.

4. **Challenges**

   The design objectives for attacks in the physical world are robustness, stealth, and generalizability. However, at present, the cost and universality of these technologies remain constrained, and the high cost of chip-level protection hinders their widespread adoption for consumer-grade devices. Secondly, challenges arise in the domains of dynamic environment adaptability, sensor noise, target motion, and high real-time requirements. Furthermore, physical counterattacks must continuously cope with these emerging attack vectors. Furthermore, there is an absence of a uniform evaluation framework for physical countermeasures

Fig. 5 Sampling accuracy rate

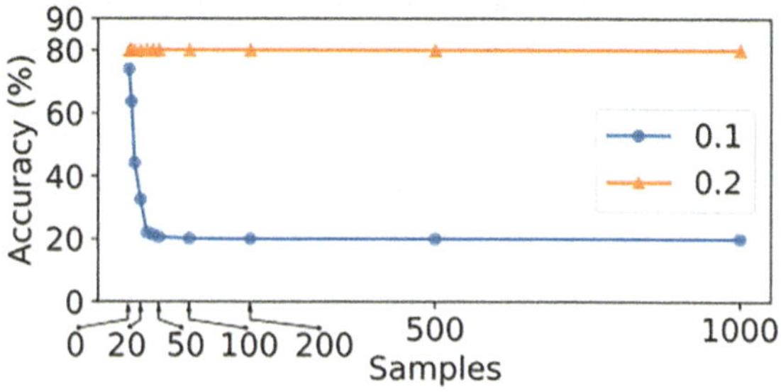

samples. It is imperative for the defense to incorporate a multifaceted approach encompassing physical security measures, hardware hardening protocols, robust model design, and dynamic detection mechanisms. Concurrently, it must address the challenges posed by cost considerations and the necessity for adaptability in a rapidly evolving technological landscape. The proliferation of the internet of things and autonomous driving has led to a heightened threat to critical infrastructure from cyberattacks. This threat will continue to escalate and require attention from multiple disciplines.

2.4 *Other Types of Attacks*

1. Universal adversarial perturbation

   In Universal Adversarial Perturbation, the generation of a single perturbation vector δ is undertaken to ensure its independence from the input image [19]. This perturbation vector is then employed to modify any given input image, thereby gradually deviating the current perturbation point from the original classification region. Through a systematic adjustment of the perturbation vector, the classifier is consequently led to misclassify $f(x + \delta) \neq f(x)$ with a high degree of probability, as Fig. 6. Such perturbations are remarkably generalizable, capable of migrating between disparate models and datasets. This poses a grave threat to the security of deep learning models, which are characterized by highly structured decision boundaries and sharing weaknesses.
2. Backdoor/trojan attacks

   Two significant categories of threats to the security of deep learning models are known as backdoor attacks and trojan attacks. These attacks involve the insertion of malicious triggers into the training phase of a model, resulting in effective performance of the model under typical inputs but aberrant behavior under specific trigger conditions. This is due to the fact that the aforementioned subjects are extremely stealthy and difficult to defend.

   In the context of model training, the term backdoor attack refers to the process by which an attacker surreptitiously incorporates a backdoor into the model through the use of malicious means, such as data poisoning. It is important to note that these backdoors will not be triggered when the model is operating in a normal manner. However, in the event that the input data contains specific triggers, the model will misclassify according to intention of the attacker. This category of cyberattack typically involves the exploitation of vulnerabilities inherent in the model training process. For instance, an attacker may gain access to training data or parameters of the model, and subsequently modify them. A distinguishing feature of backdoor attacks is their stealthy nature. In the case of models implanted with backdoors, these models function normally in the face of normal inputs. However, under specific conditions, they reveal their malicious behavior.

A trojan attack bears a resemblance to a backdoor attack, yet it is initiated in a distinct manner. In the context of information technology, Trojan attacks are defined as malicious cyber-attacks that exploit system vulnerabilities or malicious programs to tamper with model parameters located in system memory. The objective of such attacks is to take control of the model and implant a backdoor, thereby facilitating unauthorized access to critical systems and data. An adversary can inject a backdoor into a model by directly perturbing the model weight values. A hallmark is their ability to blend in, where the perpetrator typically disguises malicious code as a legitimate program, deceiving the user into installing and executing it.

The backdoor attacks and trojan attacks are often used interchangeably; however, their attack mechanisms and implementations differ. It has been demonstrated that backdoor attacks are principally executed through data poisoning or model parameter modification. In contrast, Trojan attacks are predominantly characterized by the exploitation of system vulnerabilities or malicious programs. However, it is important to note that both of these factors have the

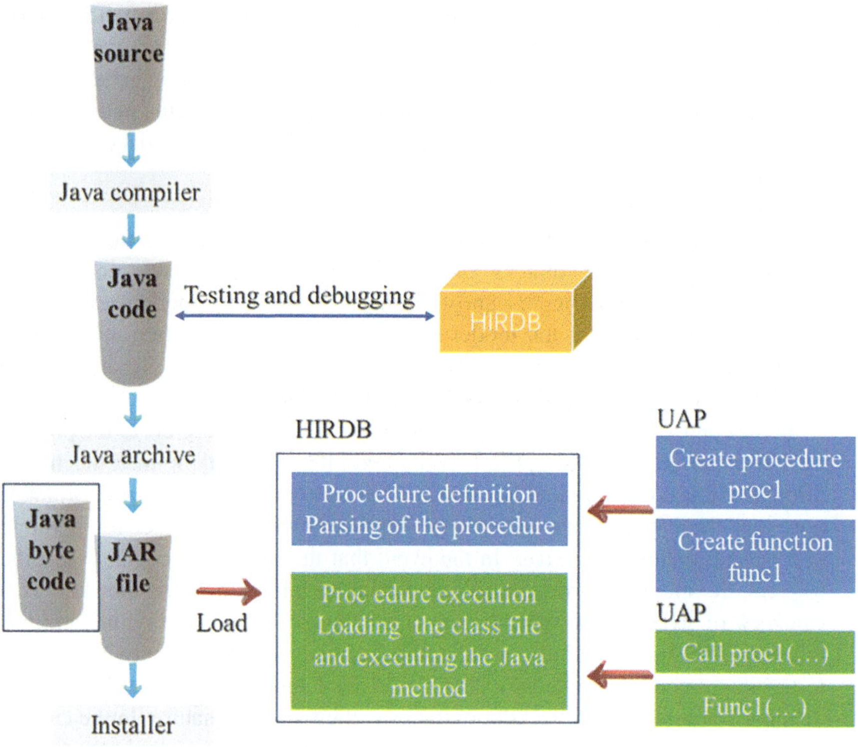

Fig. 6 Universal adversarial perturbation

potential to induce anomalous behavior in the model under specific conditions. This, in turn, can compromise the security and reliability of the system.

3. Adversarial reprogramming

   Adversarial reprogramming is a novel attack method that allows an attacker to reconfigure a neural network to perform different tasks by manipulating the inputs rather than modifying the model parameters. The objective of this manipulation is to reprogram the target model to perform the task specified by the attacker. Adversarial Reprogramming is a process that modifies an image classification model to function as a counting model, thereby demonstrating the model's capacity for modification in terms of its computational capabilities, as Fig. 7.

4. Generative attack

   Generative attacks are a type of cyberattack that leverage generative artificial intelligence techniques, such as generative adversarial networks and variable auto-encoders, to emulate or fabricate the behavior of legitimate users. This tactic enables the circumvention of conventional security safeguards, thereby evading traditional protective measures. The implementation of generative attacks usually relies on deep learning models such as generative adversarial networks. These models can generate high-quality synthetic data, such as text and images, to imitate the distribution and characteristics of real data.

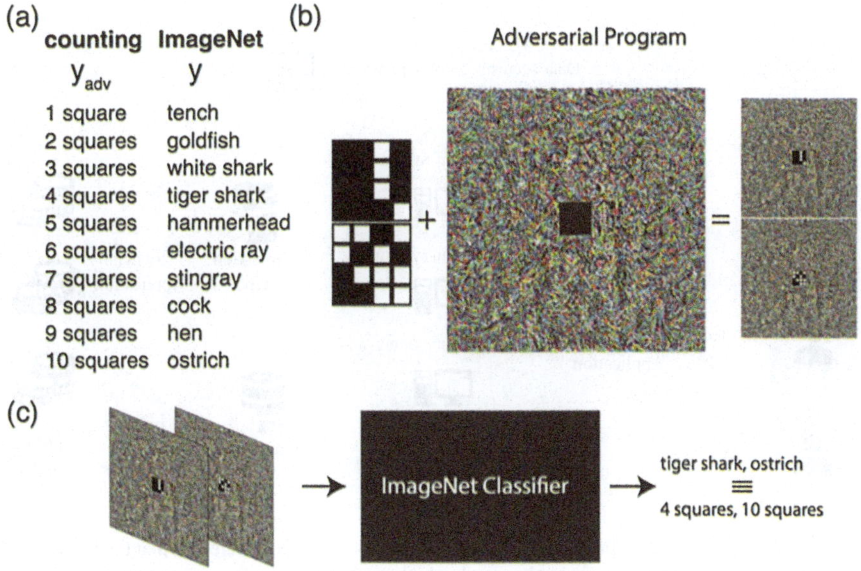

Fig. 7 Illustration of adversarial reprogramming. (**a**) Mapping ImageNet labels to adversarial task labels (the squares in the image). (**b**) An image from the adversarial task (left) is embedded at the center of the adversarial program (middle), resulting in an adversarial image (right). (**c**) Illustration of inference using the adversarial image

AI-generated attacks, particularly deep forgery and automated scripting attacks, possess the capability to generate realistic content due to their high emulation and robust stealth. This capacity renders traditional rule-based security defense mechanisms ineffective in their ability to identify and intercept such attacks. In order to address the aforementioned challenge, it is necessary to employ advanced data security tools, which generally possess the following characteristics, as Fig. 8.

1. Modern security tools leverage artificial intelligence to discern anomalous patterns or behaviors, a technique that is of particular importance for identifying AI-generated attacks.
2. Real-time monitoring has been shown to facilitate rapid identification and response to potential threats, thereby mitigating the damage caused by cyber-attacks.
3. The integration of threat intelligence into these tools enables the recognition of novel attack patterns in advance, facilitating the implementation of protective measures.

The adversarial attack method aims to evaluate and improve the robustness of the model by adding small perturbations to the input data to make the model produce erroneous outputs. The adversarial attack methods are compared from the perspectives of attack type, representative algorithm, attacker knowledge, target, disturbance constraint and characteristics, which can be summarized into Table 1.

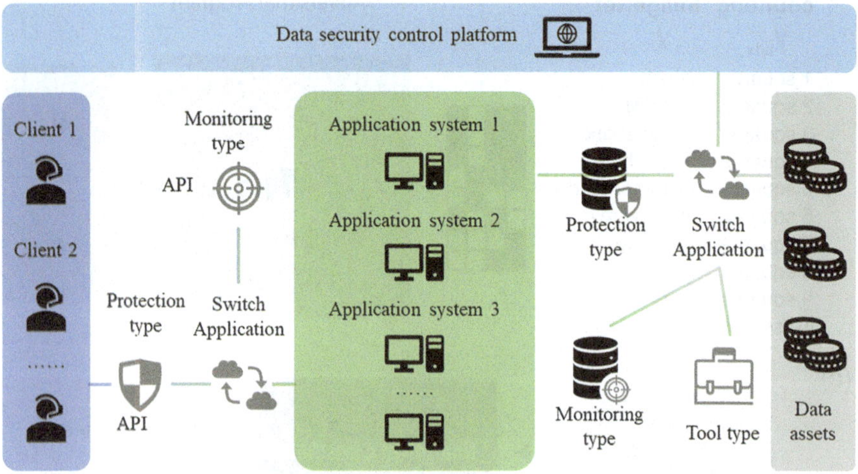

Fig. 8 Generative attack. The required data security tools. It has been detected that the behavior of a certain user has suddenly become abnormal, which might be due to the account being controlled by AI

Table 1 Summary of the main counter-attack methods

Attack type	Representative algorithm	Attacker knowledge	Target	Disturbance constraint	Characteristic
White-box attacks	FGSM	Completely	With a goal	L_∞	Fast, simple, with significant disturbance
	PGD/BIM	Completely	Without/with a goal	L_p	Strong attacks, the gold standard for robustness evaluation, high computational overhead
	JSMA	Completely	With a goal	L_0	Sparse perturbation, modification of a few pixels
	DeepFool	Completely	Without a goal	L_2	Approximate minimum perturbation
Black-box attacks (transfer-based attacks)	Proxy model	None (requires training)	Without/with a goal	L_p	Rely on model similarity, transferability
Black-box attacks (query-based attacks)	ZOO	Queried access	Without/with a goal	L_p	Estimated gradient, query overhead are huge
	NES	Queried access	Without/with a goal	L_p	Efficient search, relatively few query times
	Boundary attack	Access decision	Without/with a goal	L_2	Only labels are needed, starting from the target class/noise
Physical-world attack	EOT	Generally white box	Without/with a goal	L_p	Model physical transformations, enhance robustness
	Adversarial patches	Generally white box	Without/with a goal	Spatial locality	Physically printable, focusing on local areas, effective for target detection attacks
Other types of attacks	UAP	Black-box/white-box	Without a goal	L_2	A single perturbation deceives a large number of samples
	Backdoor/Trojan attacks	Implantation during the training stage	With a goal	Trigger specific	Highly concealed, difficult to detect, contaminated during the training stage
	Generative attack	Generally white box	Without/with a goal	–	High naturalness sample

3 Confrontational Defense: Strategies and Techniques

Adversarial defense strategies are designed to enhance resilience of the model to adversarial attacks and to guarantee that the model produces accurate outputs in the presence of malicious perturbations, as Fig. 9.

These defense strategies generally encompass methods such as adversarial training, gradient masking, and model integration. Adversarial training has been demonstrated to enhance the robustness of a model by introducing adversarial samples during the training process. This allows the model to learn a more robust feature representation. Conversely, gradient masking thwarts such exploitation by modifying the gradient information of the model. The incorporation of multiple models, through a process known as model integration, has been demonstrated to enhance the overall robustness of the system. This enhancement is attributed to the aggregation of predictions from multiple models, which compensates for the potential variations in the sensitivities of these models to the same adversarial sample. Defense strategies can be broadly classified into four categories, input preprocessing/transformation, robust model training, adversarial sample detection, and model integration and diversification.

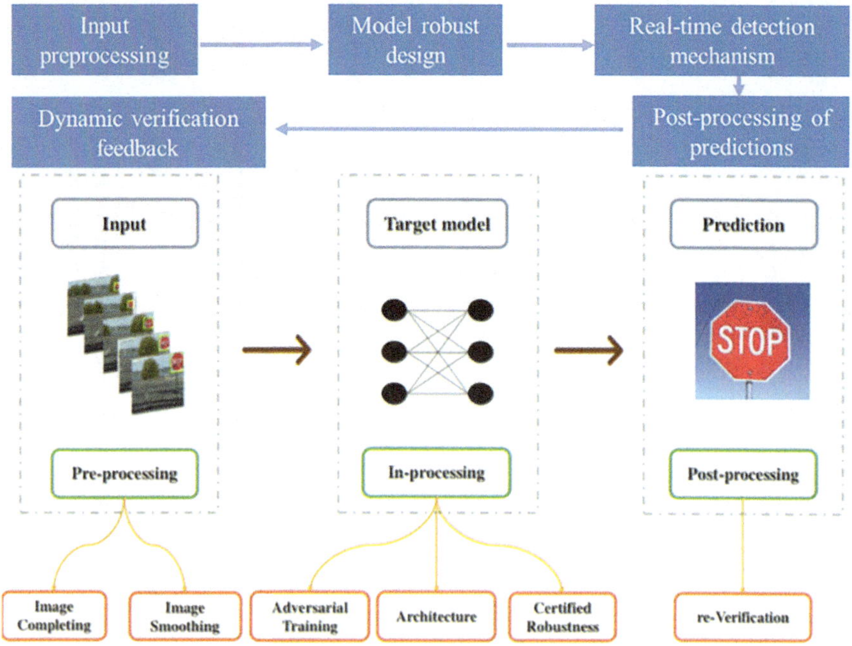

Fig. 9 Systematic defense framework. From input preprocessing, model architecture optimization, to post-processing verification, a closed loop is formed

3.1 *Input Preprocessing/Transformation*

Input preprocessing/transformation involves processing the input data to eliminate or reduce the effects of antagonistic perturbations before the model processes it. This approach does not alter the model itself, but rather improves its robustness by modifying the input data. Common image transformations include randomization, denoising, filtering, JPEG compression, and feature squeezing. Randomization can be defined as the incorporation of random noise (Gaussian or Bernoulli) into the input or the implementation of operations such as random resizing or random padding. The efficacy of the perturbation is counterbalanced by the introduction of randomness. The process of denoising or filtering involves the application of image processing techniques, including but not limited to Gaussian blur, median filter, total variation minimization, and deep learning-based image denoising methods, with the objective of smoothing out counteracting perturbations. However, this approach may concomitantly result in the degradation of the semantic information contained within the image. The efficacy of removing high-frequency adversarial perturbations by leveraging the lossy nature of JPEG compression is well-documented; however, its effectiveness is limited in the face of adaptive attacks. Feature squeezing is a process that involves the reduction of color depth or the implementation of spatial smoothing on the input features.

The advantages of this approach include its simplicity, low computational cost, and minimal relevance to the model. The implementation is relatively straightforward and does not necessitate extensive modifications to the model. The preprocessing step is typically characterized by a low computational demand, rendering it well-suited for real-time applications. This approach is not contingent upon a particular model structure and is applicable to a broad range of models. Potential disadvantages include degradation of model performance, limited effectiveness against specific attacks, and misclassification of normal data. It is imperative to note that the preprocessing step may introduce additional noise or distortion, which can adversely affect the model's normal performance. It has been demonstrated that certain complex adversarial attacks may not be effectively defended against by the preprocessing step. In certain instances, the preprocessing phase may inadvertently modify standard data, resulting in model misclassification. The capacity to attain robustness gains is constrained, frequently at the expense of maintaining the accuracy of clean samples. Moreover, these gains are ineffective in the defense against sophisticated white-box attacks.

3.2 *Robust Model Training*

Robust model training is a process that aims to enhance the resilience of a model to adversarial attacks. This is achieved by introducing adversarial samples or other robustness enhancement techniques during the training process. Common methods include adversarial training and regularization methods. Among these methods, Adversarial Training encompasses two distinct approaches: Standard adversarial training and Improved Adversarial Training methods.

The core idea of standard adversarial training is to incorporate adversarial samples into the training data to minimize data loss as

$$\min_\theta \mathbb{E}_{(x,y)\sim\mathcal{D}}\left[\max_{\delta\in\mathcal{D}}\ell\left(f_\theta\left(x+\delta\right),y\right)\right] \tag{5}$$

where θ represents the model parameters. (x, y) are the training data sampled from the data distribution $\mathcal{D}$, and $\ell(f_\theta(x+\delta), y)$ is the loss value calculated for the adversarial sample $x+\delta$. The projection gradient descent method is typically used to maximize the internal value and generate strong adversarial samples. The goal is to find the perturbation δ that can cause the model to make errors. The external minimization is achieved by training the model parameters θ to minimize the loss caused by the perturbation. Although it significantly improves the model's robustness against specific attack types L_∞, it faces many challenges. The computational cost is high, and adversarial samples need to be generated in each training step. Robust overfitting occurs (the robust test accuracy decreases in the later training stage), and robust accuracy of the model on clean samples is usually lower than that of the standard model. The types of attacks and perturbations used for defense during training are constrained by norms.

Improved adversarial training methods are tradeoff-inspired adversarial defense via surrogate-loss [20]. The loss must be decomposed into natural and bounded or adversarial risks. The balance between these two types of risks must be explicitly constrained with KL scatter. This will result in a better trade-off between robustness and accuracy. The misclassification aware adversarial training method pays special attention to samples misclassified by adversarial samples [21]. These samples are assigned higher weights. This emphasizes the learning of the correct boundaries of these difficult samples. Friendly adversarial training focuses on samples that are close to the decision boundary but not misclassified, contending that learning the boundaries of these samples is more important for improving robust generalization. Semi-supervised and unsupervised adversarial training leverage substantial volumes of unlabeled data to enhance the robustness of the system. The application of architecture and regularization methods involves the integration of specific network architectures or regularization terms, such as gradient regularization and Lipschitz continuity constraints, with the objective of enhancing the robustness of the system.

The objective of certified robust training is to develop a model whose robustness can be mathematically substantiated to ensure accurate classification within a specified perturbation range, thereby providing a lower bound guarantee. Interval bound propagation is a method that calculates the upper and lower bounds of the activation values of each layer of the network by symbol propagation and minimizes the upper bound loss that contains all possible inputs within the perturbation range. The process of training is arduous and necessitates a gradual increase in perturbations, initiated from minimal initial displacements. Convex relaxation methods are a class of algorithms that employ linear programming, semi-positive definite programming, or convex relaxation to compute robust bounds on the output of a network and to train based on these bounds. Randomized smoothing involves the training of a base classifier, followed by the construction of a smoothing classifier through the sampling of the input multiple times to vote under Gaussian noise perturbations [22]. This smoothed classifier is provably robust within a specific L_2 perturbation radius. The radius of provable robustness is contingent upon the accuracy and noise level of the underlying classifier when operating on noisy data. At present, it stands as the most pragmatic and verifiable defense method for scaling to large datasets.

Consequently, rigorous model training engenders significantly more robust defenses, particularly adversarial training and verifiable methods. These methods constitute the primary means of defending against sophisticated white-box attacks and can be seamlessly integrated into a broad spectrum of models and tasks, exhibiting excellent generalizability. However, it should be noted that adversarial training necessitates substantial computational resources and time, particularly in the context of large-scale datasets. During the training process, models may prioritize enhancing the robustness of their predictions over achieving absolute accuracy on clean data. It has been demonstrated that models that demonstrate strong performance on training data may not necessarily exhibit comparable performance on unseen attacks.

3.3 Adversarial Sample Detection

Adversarial sample detection is defined as the process of identifying and rejecting adversarial samples during the inference phase of a model. This identification and rejection is achieved through the use of a feature representation of the detector or validation input. This method typically relies on the discrepancy between the adversarial sample and the normal sample to make decisions. Adversarial sample detection encompasses three primary methodologies: inconsistency-based detection, confidence/uncertainty-based detection, and input gradient-based detection.

The first step in implementing an inconsistency-based detection system is feature compression or reconstruction, followed by training an autoencoder or generative model to learn the data flow shape. Adversarial samples are vulnerable to bypassing detection due to substantial reconstruction errors stemming from deviations from the flow shape, according to which they are identified. Subsequently, the employment of sub-networks or auxiliary models becomes imperative for the detection of anomalies through the utilization of features derived from disparate layers of the model, diverse sub-network outputs, or supplementary detection models that have undergone training. In order to achieve the desired level of prediction inconsistency, it is necessary to apply random transformations to the inputs, such as cropping, rotation, and noise. The variance of the model's predictions under different transformations should then be observed. The prediction of antagonistic samples is typically unstable, exhibiting high variance, while the prediction of clean samples is stable and exhibits low variance. Confidence/uncertainty-based detection posits that adversarial samples prompt the model to output either unusually high confidence or unusually high uncertainty. However, contemporary attacks have the capacity to generate samples with a high degree of target confidence. Furthermore, well-calibrated models may exhibit a high degree of uncertainty for out-of-distribution samples. The input gradient-based detection method analyzes patterns in the input gradient, positing that the gradient maps of the adversarial samples exhibit specific anomalous patterns. Nevertheless, the same adaptive attack challenge is faced.

Adversarial sample detection can be specialized for adversarial samples to improve the security of the system. The detector is capable of functioning independently of the model without compromising the normal operation of the model. The detector's integration into existing systems is straightforward, obviating the need for extensive modifications to the model. A notable disadvantage of this approach is that it functions as a form of "whack-a-mole," where the user must repeatedly strike the target to achieve the desired outcome. Once the detection mechanism is understood, an attacker can optimize the adversarial samples to deceive the target model and bypass the detector, i.e., generate evasive samples. The evaluation of the detector's robustness poses a significant challenge. The existence of false alarms and omissions has been documented, and the presence of clean samples has been associated with the designation of such samples as adversarial, or the failure to detect them as such.

3.4 Model Integration and Diversification

The integration and diversification of models pertains to enhancing the resilience and security of the system in its entirety by combining multiple models or utilizing a multitude of model structures. This approach employs the complementary nature of multiple models to mitigate the vulnerability of a single model. This encompasses model ensemble, a process that entails the training of multiple models with disparate structures or initializations. The output is determined by a vote or an average, thereby establishing the final prediction. The migration of adversarial samples across models is often a challenging process, which serves to increase the complexity of the attack. However, it should be noted that attackers may still devise attacks against integrated models, such as optimizing perturbations to spoof all or most of the models concurrently.

Stochastic depth/width is also included. The implementation of dynamic interference attacks is facilitated by the introduction of random discarding of network layers, namely stochastic depth, or random masking of channels, namely stochastic width, during inference. Furthermore, input diversity ensemble (IDE) employs a series of random transformations, including random scaling and color dithering, prior to inputting the data into the model. Subsequently, IDE integrates the predictions of the transformed samples, akin to the concept of inconsistency-based detection. It has been determined that the incorporation of a single model, as well as the implementation of a limited array of models, will prove inadequate in addressing the identified deficiencies. The integration of multiple models has been demonstrated to be an effective strategy for enhancing the success rate of antagonistic attacks and thereby enhancing the overall robustness of the system. The utilization of varied models with distinct decision boundaries has the potential to mitigate the likelihood of erroneous judgment. The selection of models can be tailored to specific needs, thereby enhancing adaptability. However, there are still certain problems to address. Integrating multiple models increases the computational burden, especially in real-time applications. The integration of models necessitates meticulous design and training to ensure the synergy between models. As the number of models increases, the maintenance and updating of the system becomes more complex. Summary of the main countermeasures and defense strategies is in Table 2.

4 Evaluation Indicators and Benchmarks

It is imperative to evaluate the efficacy of offensive and defensive strategies in a manner that is both impartial and exhaustive. The evaluation of adversarial defense techniques typically necessitates the implementation of a series of evaluation metrics and benchmarks. These metrics and benchmarks facilitate comprehension among researchers and developers regarding the efficacy of defense techniques, thereby establishing a foundation for comparative analysis of disparate defense approaches.

4.1 Attack Evaluation Metrics

1. Attack success rate

   In Attack Success Rate, this is a widely utilized metric for evaluating the efficacy of an attack, measuring the proportion of attackers who successfully induce the model or system to produce an erroneous response. For instance, if 50 out of 100 test samples are successfully attacked, the attack success rate is 50%.
2. Perturbation magnitude

   The measure of the visual imperceptibility of the perturbation is the number of pixels changed by the commonly used paradigm L_0. The L_2 Euclidean distance is defined as the perturbation energy. The maximum pixel change value L_∞ is of the utmost importance in terms of visual relevance.
3. Query efficiency

   In the context of black-box query attacks, the objective is to identify the minimum number of queries necessary to generate a successful adversarial sample. A lower number indicates a more efficient attack.
4. Attack efficiency

   Attack efficiency is a composite measure that incorporates both the attack success rate and the average perturbation distance. This can be defined as the ratio of the attack success rate to the average perturbation distance, or alternatively,

Table 2 Summary of the main countermeasures and defense strategies

Defending strategies	Delegate methods	Core concept	Advantage	Disadvantage
Input preprocessing/transformation	Randomization	Introduce random destructive disturbances	Simple, fast, no need to change the model	Easy to be bypassed by adaptive attacks, the defense effect is limited, the accuracy may be lost
	Denoising	Smooth or remove high-frequency disturbances	Simple, fast calculation, ready-made method	Easy to be bypassed by attack, can blur effective features, limited effect
	JPEG compression	Remove high-frequency disturbances by using compression distortion	Simple, efficient, widely used	Poor effect of adaptive attack
Robust model training	Standard adversarial training	Inject strong adversarial samples during training	The most effective mainstream method, significantly improve the robustness	Huge computational overhead, robust overfitting, robust-precision trade-off, attack type dependence
	Improved adversarial training	Improve loss function design, focus on difficult samples or boundaries	Better balance between robustness and accuracy	The calculation is still expensive, the effect improvement has an upper limit
	Randomized smoothing	The noise robust model is trained to construct a proven robust smooth classifier	Provides verifiable robustness, can be extended to large models	The verification radius is small, depends on the accuracy of the noise model, requires multiple sampling
	Certified robust training	Based on interval propagation or convex relaxation training, it provides guarantees within the perturbation range	Provide strict mathematical guarantee	The training is extremely difficult, the verification radius is small, the accuracy is sacrificed, difficult to extend to complex models

Adversarial sample detection	Feature reconstruction error	Assuming that the adversarial sample deviates from the data manifold, the reconstruction error is large	Detect unknown attacks	Easily bypassed by attacks designed for detectors, false positives/false negatives
	Prediction inconsistency	The prediction of adversarial samples is unstable under different transformations	No need to modify the model, can detect a variety of attacks	Increased computational overhead, adaptive attack can optimize perturbation stability
	Based on confidence/uncertainty	Assume that the confidence pattern of adversarial samples is abnormal	Simple implementation	The attacker can control the confidence, unreliable
Model integration and diversification	Ensemble	Multiple model voting decisions, increase the difficulty of attack	Improve the robustness to migration attacks	The computational/storage overhead is large, can be broken by attacks against integration
	Stochastic depth/width	The network structure is randomized during reasoning, dynamics is introduced	Increase the uncertainty of attack	The effect is limited, may reduce the accuracy of clean samples
	Input diversity	Random transformation and integrated prediction are applied to the input	Combined with randomization	The computational overhead increases, the effect is limited

some other analogous combination. It has been demonstrated that an increase in attack efficiency corresponds to a corresponding increase in the effectiveness of the attack.

5. Transfer success rate

   The ratio at which adversarial samples generated on one model successfully attack another different model reflects portability.

4.2 Defense Assessment Indicators

1. Robust accuracy

   The proportion of antagonistic samples that are correctly classified by the model under a specific attack or a specific perturbation constraint, denoted by ε, serves as the core metric.
2. Clean accuracy

   The classification accuracy of the model on the undisturbed original test set and evaluate the impact of defense on normal performance.
3. Certified robust accuracy

   For verifiable defense, it refers to the proportion of samples (within a given perturbation radius) that the model can mathematically guarantee correct classification.
4. Robustness under adaptive attack

   The gold standard for defense evaluation is to test its performance under a worst-case, fully adaptive white-box attack. The attacker has a comprehensive understanding of the defense system, encompassing its preprocessing, detection mechanisms, and training processes. This knowledge enables the attacker to engineer bespoke attack algorithms that exploit vulnerabilities in the system. Defenses that are effective against adaptive attacks are considered truly robust.

4.3 Common Datasets and Models

1. Data set

   MNIST, CIFAR-10, CIFAR-100, SVHN, and ImageNet are the most commonly used benchmark datasets for evaluating offense and defense, and their complexity and scale increase in sequence.
2. Model architecture

   Standard models such as ResNet (ResNET-18, ResNET-50), VGG, DenseNet, WideResNet, Inception, and the Vision Transformer that has emerged in recent years. When evaluating defense, results on standard models or robust training models are usually reported.

5 Challenges and Future Directions

Although significant progress has been made in the research of adversarial offense and defense, this field still faces many severe challenges and open issues.

5.1 *Fundamental Challenge*

1. Fundamental robustness-accuracy trade-off
   Under specific distributional assumptions, a trade-off inevitably emerges between model robustness and standard accuracy. While this can be mitigated in practice by improving model architectures, training strategies, and data utilization, completely breaking this trade-off remains the central challenge.
2. Scalable certified robustness
   Existing verifiable defense methods are challenging to scale to large-scale datasets and complex models. Stochastic smoothing is the more pragmatic approach. However, its demonstrable robustness radius is typically minimal. The development of verifiable methods that are efficient, tight, and scalable is of the utmost importance.
3. Efficient robust training
   The computational burden associated with adversarial training constitutes a significant impediment to its broad implementation. Research on more efficient adversarial sample generation algorithms, training strategies, or accelerated robust fine-tuning using pre-trained models are important directions.
4. Systematic physical world evaluation
   The evaluation of attacks in the physical realm is characterized by the absence of uniform, reproducible benchmarks. There is an evident necessity for standardized test platforms that take into account a range of sensors, lighting, angles, motion conditions, and evaluation metrics. Such platforms would serve to objectively measure the threat posed by physical attacks and the effectiveness of defensive measures.

5.2 *Emerging Directions and Opportunities*

1. Beyond L_p robustness
   The present research is oriented towards perturbations of L_2 and L_∞ constraints. It is imperative to investigate the robustness of these models to other types of perturbations that are more consistent with human perception or actual threats. These perturbations may include geometric transformations, color offsets, semantic perturbations, generic patches, and backdoor triggers.
2. Human-aligned robustness

It is imperative to investigate methodologies that would facilitate the alignment of the robustness of models with the consistency of human visual perception. For instance, the development of perception-driven counterattacks and defenses that ensure perturbations more consistent with human attention is crucial. Additionally, the creation of models that are robust under changes not readily apparent to humans but sensitive to changes discernible to humans is essential.

6 General Conclusions

Research on adversarial attacks and defenses has profoundly revealed the security vulnerabilities and robustness flaws of current deep computer vision models, challenging our knowledge of model generalization capabilities and decision-making mechanisms. The field has achieved fruitful results in the past decade, ranging from simple input preprocessing to computationally intensive but effective adversarial training and verifiable robust training. Nevertheless, there is still a considerable distance to be traversed in the development of visually intelligent systems that are genuinely safe, reliable, and trustworthy. The trade-off between robustness and accuracy, the lack of scalable and verifiable defenses, the bottleneck of efficient robust training, the complexity of physical world evaluation, and the emergence of new types of attacks are all urgent bulwarks that need to be overcome. Future research endeavors must prioritize the further development of a robust theoretical foundation, the exploration of novel architectures, the creation of efficient algorithms, the alignment of human and computer processes, and the investigation of specific application scenarios.

The competition between attack and defense is not merely a technological contest, It is also a fundamental impetus for comprehending the essence of deep learning and propelling its advancement in a more secure and manageable direction. It is imperative to prioritize ongoing investment in research to ensure the safe and responsible application of computer vision technologies in critical domains. This investment is crucial to garnering the trust of users and society. In a manner analogous to the need for defensive systems to adapt to the evolution of attacks, the field as a whole must maintain an openness, collaboration, and critical thinking in order to collectively address this enduring challenge in AI security.

References

1. Qin Y, Zhang K, Pan H. Adversarial attack for object detectors under complex conditions. Comput Sec. 2023;134:103460.
2. Papernot N, McDaniel P, Jha S, Fredrikson M, Celik ZB, Swami A. The limitations of deep learning in adversarial settings. In: IEEE European symposium on security and privacy; 2016. p. 372–87.
3. Akhtar N, Mian A. Threat of adversarial attacks on deep learning in computer vision: a survey. IEEE Access. 2018;6:14410–30.
4. Akhtar N, Mian A, Kardan N, Shah M. Advances in adversarial attacks and defenses in computer vision: a survey. IEEE Access. 2021;9:155161–96.
5. Yu C, Chen J, Xue Y, Liu Y, Wan W, Bao J, Ma H. Defending against universal adversarial patches by clip feature norms. In: Proceedings of the IEEE/CVF international conference on computer vision; 2021. p. 16434–42.
6. Wei X, Pu B, Lu J, Wu B. Visually adversarial attacks and defenses in the physical world: a survey. Arxiv 2022; 2211.01671.
7. Li C, Wang H, Yao W, Jiang T. Adversarial attacks in computer vision: a survey. J Membr Comput. 2024;6(2):130–47.
8. Chattopadhyay N, Basit A, Ouni B, Shafique M. A survey of adversarial defenses in vision-based systems: categorization, methods and challenges. Arxiv 2025; 2503.00384.
9. Goodfellow IJ, Shlens J, Szegedy C. Explaining and harnessing adversarial examples. Arxiv 2014; 1412.6572.
10. Kurakin A, Goodfellow IJ, Bengio S. Adversarial examples in the physical world. In: Artificial intelligence safety and security. Chapman and Hall/CRC; 2018. p. 99–112.
11. Moosavi SM, Fawzi A, Frossard P. Deepfool: a simple and accurate method to fool deep neural networks. In: Proceedings of the IEEE conference on computer vision and pattern recognition; 2016. p. 2574–82.
12. Dong Y, Cheng S, Pang T, Su H, Zhu J. Query-efficient black-box adversarial attacks guided by a transfer-based prior. IEEE Trans Pattern Anal Mach Intell. 2021;44(12):9536–48.
13. Ye H, Huang Z, Fang C, Li CJ, Zhang T. Hessian-aware zeroth-order optimization. IEEE Trans Pattern Anal Mach Intell. 2025;47:4869–77.
14. Hu Y, Chen G, Li Z, Knoll A. Robot policy improvement with natural evolution strategies for stable nonlinear dynamical system. IEEE Trans Cybern. 2022;53:4002–14.
15. Shen M, Li C, Yu H, Li Q, Zhu L, Xu K. Decision-based query efficient adversarial attack via adaptive boundary learning. IEEE Trans Depend Secure Comput. 2023;21:1740–53.
16. Chen J, Chen H, Chen K, Zhang Y, Zou Z, Shi Z. Diffusion models for imperceptible and transferable adversarial attack. IEEE Trans Pattern Anal Mach Intell. 2024;47:961–77.
17. Athalye A, Engstrom L, Ilyas A, Kwok K. Synthesizing robust adversarial examples. In: International conference on machine learning; 2018. p. 284–93.
18. Wei X, Ruan S, Dong Y, Su H, Cao X. Distributionally location-aware transferable adversarial patches for facial images. IEEE Trans Pattern Anal Mach Intell. 2025;47:2849–64.
19. Moosavi SM, Fawzi A, Fawzi O, Frossard P. Universal adversarial perturbations. In: Proceedings of the IEEE conference on computer vision and pattern recognition; 2017. p. 1765–73.
20. Zhang H, Yu Y, Jiao J, Ghaoui L, Jordan M. Theoretically principled trade-off between robustness and accuracy. In: International conference on machine learning. PMLR; 2019. p. 7472–82.
21. Wang Y, Zou D, Yi J, Bailey J, Ma X, Gu Q. Improving adversarial robustness requires revisiting misclassified examples. In: International conference on learning representations; 2019. p. 1–11.
22. Cohen J, Rosenfeld E, Kolter Z. Certified adversarial robustness via randomized smoothing. In: International conference on machine learning; 2019. p. 1310–20.

References

Adaptive Cryptographic System Orchestration via Intelligent Agents

Miao Tian, Hang Chen, Gege Sun, Wenbo Song, Zhengjun Liu, and Fei Tong

Abstract With the continuous emergence of encryption methods, their application in computer simulations has become increasingly complex. Based on the FastAPI interface and combined with existing artificial intelligence technologies, this paper proposes an intelligent encryption agent framework, which realizes an agent compatible with various optical encryption systems. This agent can independently determine the encryption method according to the file type and obtain the corresponding encrypted content, which largely simplifies the operation process of optical encryption and improves user-friendliness. At the same time, this chapter proposes two optical encryption systems to encrypt different types of files respectively: one encrypts images under the Gyrator domain and RMD (Random Modulus Decomposition) scrambling. The other extracts and masks the audio sequence, then encrypts the audio in the Gyrator domain. This chapter incorporates these two optical encryption systems into the intelligent agent framework for demonstration.

Keywords Artificial intelligence · Agent · Gyrator · Optical encrytion system

M. Tian · H. Chen · G. Sun · W. Song
School of Space Information, Space Engineering University, Beijing, China

Z. Liu
School of Physics, Harbin Institute of Technology, Harbin, China
e-mail: zjliu@hit.edu.cn

F. Tong (✉)
Bejing Aerospace Command Control Centre, Beijing, China

H. Chen, Z. Liu (eds.), *Artificial Intelligence in Digital Image Processing: Theories, Methods, and Applications*, Studies in Computational Intelligence 1246,
https://doi.org/10.1007/978-3-032-12891-1_4

1 Introduction

The ubiquitous digital exchange of sensitive information—spanning personal communications to state security concerns—demands increasingly sophisticated cryptographic protections. Optical encryption systems emerge as a potent solution, leveraging inherent advantages in parallel computation and high-dimensional data processing to achieve unparalleled encryption speeds and key-space magnitudes far exceeding traditional digital frameworks. This technological paradigm ensures ciphertext remains cryptographically secure without proper authorization, transforming plaintext into indecipherable data artifacts [1–9].

The evolution of optical encryption traces its foundations to Refregier's seminal 1995 demonstration of image scrambling through double random phase encoding (DRPE) within 4f optical architectures [1]. Subsequent decades witnessed transformative innovations: Wan and Singh enriched DRPE through multidimensional phase-wavelength hybrid encoding; Rajput introduced nonlinear phase modulation for grayscale image obfuscation; Chen's integration of Gyrator-Arnold-Fourier transforms with matrix decomposition established new robustness benchmarks. Parallel advancements incorporated biometric entropy (DNA sequences, fingerprints) while Rakheja validated RMD scrambling's efficacy via optical phase randomization [10–19].

Audio encryption, conversely, progressed less robustly. Existing techniques like HEVC-based steganography and chaotic mapping struggle with computational efficiency despite promising theoretical frameworks. This gap motivates our exploitation of optical computing's parallel strengths for audio-to-visual cryptographic transformation [20–27].

Concurrently, artificial intelligence matured through three revolutionary waves: logical symbol manipulation (1950s), expert systems (1980s), and today's data-driven deep learning renaissance. Modern intelligent agents now exhibit environmental perception, hierarchical reasoning, and autonomous decision-making'capabilities aligning profoundly with cryptographic orchestration needs [28–39].

Against this backdrop, our research pioneers an NLP-driven cryptographic framework that dynamically interprets security intents into executable encryption parameters. The system seamlessly interoperates with legacy infrastructures (OpenSSL/GMSSL) while implementing two novel optical encryption pathways:

Imagery: Gyrator-domain spectral encryption
Audio: RMD-scrambled spectral transformations

By autonomously selecting context-appropriate cryptographic operations, this work bridges "semantic-level" security commands with "algorithmic-level" executions ultimately advancing intelligent security-as-a-service paradigms for next-generation edge computing ecosystems.

2 Optical Encryption Algorithms

2.1 Gyrator Transform

The Gyrator transform (GT) is a discrete optical transform employed for image encryption. It scrambles the input image by altering the spatial distribution of pixel values through convolution operations, thereby generating an encrypted image. Optically, the GT is implemented using a three-lens system with identical inter-lens distances a as Fig. 1. The lenses possess focal lengths of f, f/2, and f, and are configured with rotation angles ϕ, $\pi/2 - \varphi$, and ϕ, respectively.

GT is mathematically defined by Eq. (1), where g(x, y) represents the input image, G(u, v) denotes the output (transformed) image, and α is the transform order. This transform operates by applying a spatially-varying kernel to the entire input image g(x, y) through an integral over the spatial domain. Crucially, the GT is analytically invertible, enabling the exact recovery of the original image.

$$\begin{aligned} F(u, v) &= G^{\alpha}[f(x, y)](u, v) \\ &= \frac{1}{|\sin\alpha|} \int\int f(x, y) \exp\left[i2\pi \frac{(xy+uv)\cos\alpha - xv - yu}{\sin\alpha}\right] dxdy \end{aligned} \tag{1}$$

2.2 Random Mode Decomposition

Random Modulus Decomposition (RMD) is a stochastic image processing technique characterized by its asymmetric output structure. It decomposes an input image into two distinct vectors, denoted as P1 and P2, which inherently possess different lengths (visualized schematically in Fig. 2). Critically, the position demarcating the separation between these two output vectors is deterministically controlled by the pixel intensity values of the original image itself. Computationally, the decomposition involves generating two independent random matrices, α and β, populated with elements drawn from a uniform distribution bounded between 0 and 1. The amplitude component derived directly as the absolute value (modulus) of the image pixel intensities, is then modulated element-wise by these random matrices to produce the decomposed vectors P1 and P2. Concurrently, a phase component (φ)

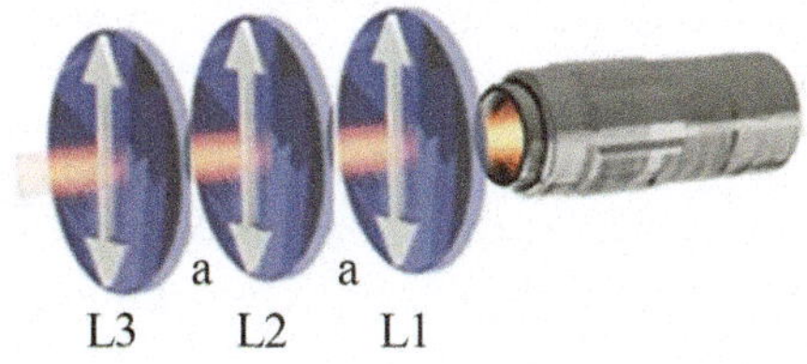

Fig. 1 Experimental setup for Gyrator transformation

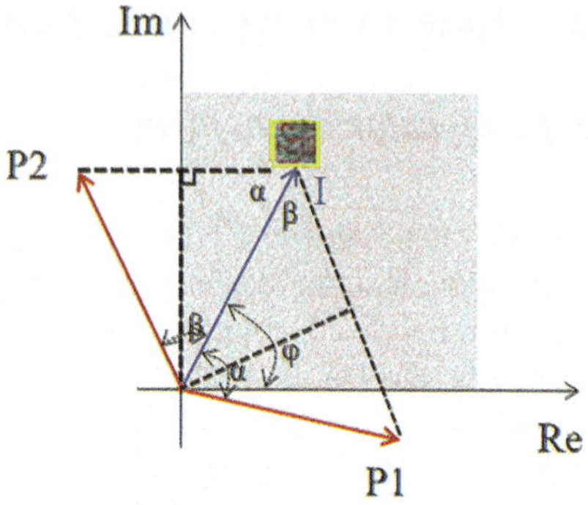

Fig. 2 RMD vector breakdown chart demo

is incorporated, its value assigned according to a predefined function of the relevant independent variables.

$$\begin{cases} P1\,(\mathrm{u},v) = \frac{A(u,v)\cdot\sin[\beta(u,v)]}{\sin[\alpha(u,v)+\beta(u,v)]}\exp\{i\,[\phi\,(u,v) - \alpha(u,v)]\} \\ P2\,(\mathrm{u},v) = \frac{A(u,v)\cdot\sin[\alpha(u,v)]}{\sin[\alpha(u,v)+\beta(u,v)]}\exp\{i\,[\phi\,(u,v) - \beta(u,v)]\} \end{cases} \quad (2)$$

2.3 *Evaluation of Encryption Systems*

The Peak Signal-to-Noise Ratio (PSNR) serves as a quantitative metric for evaluating image encryption efficacy. This widely adopted index measures perceptual differences between images, where higher PSNR values (expressed in dB) indicate greater similarity between compared images. As formalized in Eq. (3), PSNR is derived from the maximum pixel intensity (MAXI) and the Mean Squared Error (MSE) between two images. MSE quantifies pixel-wise divergence in Eq. (4), where IO and ID represent original and decrypted images, respectively, M and N is the image dimensionality. Interpretation guidelines for PSNR values are empirically established as follows:

>50 dB: Near-lossless reconstruction, implying high information fidelity.
30–50 dB: Acceptable quality with perceptually tolerable distortions.
20–30 dB: Significant degradation; generally inadequate for practical use.
<20 dB: Critical information loss, indicating decryption failure.

$$MSE = \frac{1}{MN}\sum_{M,N}\left|ID\,(x,y) - IO\big(x,y\big)\right|^2 \quad (3)$$

$$PSNR = 10 \cdot \log_{10}\left(\frac{MAXI^2}{MSE}\right) = 20 \cdot \log_{10}\left(\frac{MAXI}{\sqrt{MSE}}\right) \quad (4)$$

Our experimental design explicitly targets two critical objectives for image encryption. Firstly, effective information concealment: Ensure encrypted images exhibit PSNR values below 30 dB relative to the original, sufficiently degrading visual information. Secondly, high-fidelity reconstruction: Achieve decrypted images with PSNR values exceeding 50 dB against the original, approaching near-lossless recovery.

2.4 Forced Recovery of Encrypted Images

To simulate adversarial scenarios during data transmission including signal interference, unauthorized access, and reconstruction attempts. We employ Known-Plaintext Attacks (KPA) and Chosen-Plaintext Attacks (CPA) to evaluate image decryption resilience. The operational principles of KPA and CPA within encryption systems are schematized in Fig. 3.

KPA Methodology is adversaries leverage pairs of partial plaintexts and corresponding ciphertexts. Through iterative phase retrieval techniques, they attempt to infer the full plaintext, encryption key, and algorithmic mechanisms. Specifically, dual-intensity phase retrieval algorithms are applied, utilizing the wavefront modulus of the input-plane object and the output-plane wavefunction to recover phase information.

CPA methodology is assuming adversaries possess the encryption algorithm and partial plaintext-ciphertext pairs, CPA constructs a continuous key-space simulation

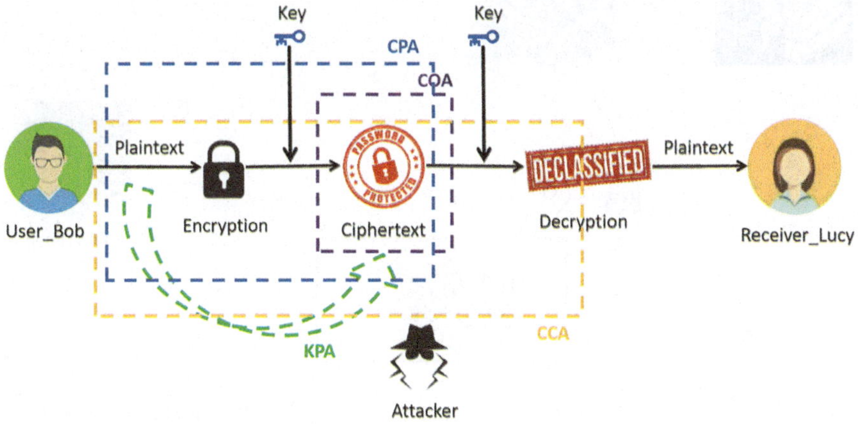

Fig. 3 Attack schematic diagram

model. This model generates candidate decryptions that are iteratively compared against expected outcomes, enabling probabilistic key derivation through differential analysis.

3 Design of Optical Encryption Systems

3.1 Encrypting Multi-channel Images

We present an innovative image encryption scheme, with the workflow schematically depicted in Fig. 4. The process initiates by applying the GT to the input image, followed by phase transformation. Subsequently, the phase-modulated output undergoes secondary decomposition via RMD. This decomposition yields two distinct sequences: one designated as the encryption key, the other constituting the encrypted image.

3.2 Encrypting Audio

This work proposes a novel audio encryption scheme, with the procedural flow illustrated in Fig. 5. The method initiates by sampling and transforming the audio signal sequence, followed by conversion to uint8 binary representation. A binary

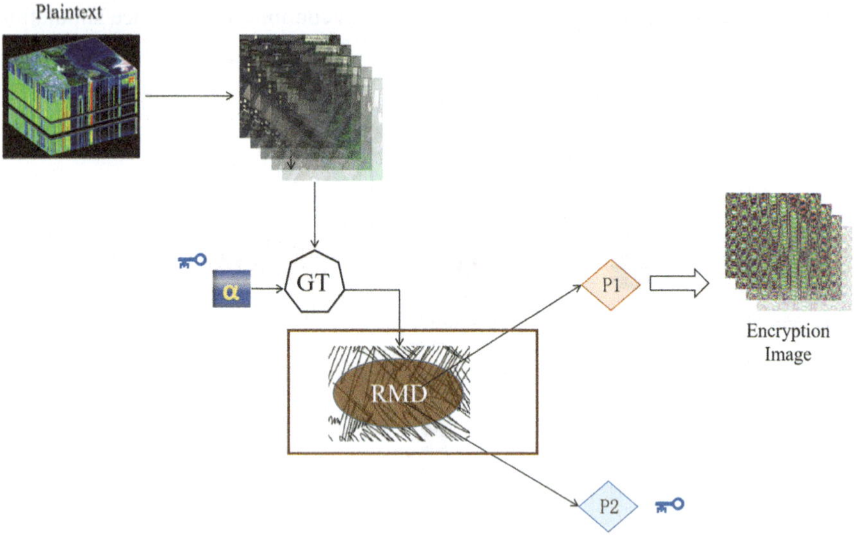

Fig. 4 Schematic diagram of image encryption

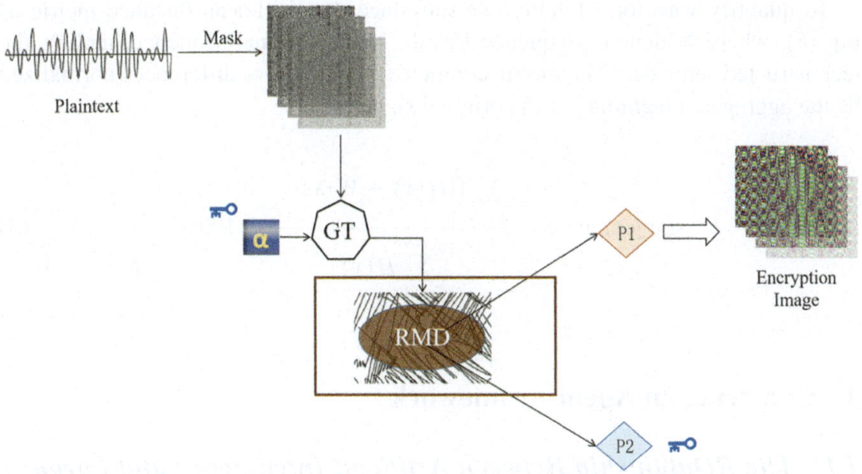

Fig. 5 Schematic diagram of audio encryption

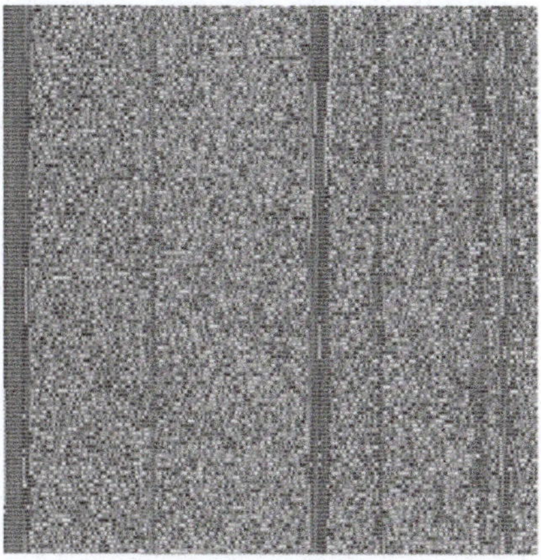

Fig. 6 An encrypted image of audio in a certain band

mask is then applied to perform bit-shifting and zero-padding operations on the least significant 8 bits. Crucially, the zero-padding ensures segmentation of the data stream into n complete 256 × 256 image blocks, where n, the number of spectral bands, is determined by the original audio length and the zero-padding magnitude.

Following audio encryption, a 256 × 256 × 4 hyperspectral image is generated. A representative spectral band is selectively visualized in Fig. 6 to demonstrate the encrypted output characteristics.

The multi-channel audio-image matrix is encrypted using the methodology detailed in Sect. 3.1, yielding a novel cryptographic system. For decryption, inverse operations are applied to reconstruct the original speech signal, ensuring complete information recovery. This involves reverting pixel values to audio sequences through inverse functional mapping.

To quantify waveform fidelity, we introduce the Euclidean distance metric as Eq. (5), where N denotes sequence length, H the original sequence, and W the reconstructed sequence. The metric computes element-wise differences normalized by the aggregate magnitude of the original signal.

$$\text{sim\_compare} = \frac{\sum_{x=1}^{N} \left(|H(x) - W(x)|^2\right)}{\sum_{x=1}^{N} H(x)} \times 100\% \tag{5}$$

4 Construct an Agent Framework

4.1 The Relationship Between Artificial Intelligence and Large Language Models

Artificial Intelligence (AI) fundamentally operates through mathematical models that process and output data, providing critical foundations for mathematical modeling and reasoning. Large Language Models (LLMs)—deep learning systems developed within the AI framework—leverage Transformer neural architectures. Through attention mechanisms, LLMs continuously refine their capacity to comprehend and emulate human language. As core components of intelligent agents, they supply essential decision-making support. In this study, our LLM implementation facilitates streamlined data processing, enabling users to efficiently identify optimal mathematical models for specific tasks.

4.2 Establishing an Agent Framework

As a carrier for integrating various model algorithms and large language models (LLMs), agents demonstrate superior adaptability across multiple domains. This paper first elaborates on the fundamental principles of agents and takes rule-based dialogue systems as an entry point to reveal the underlying logic of interaction between agents and data. On this basis, combined with an optical encryption system, it realizes the intelligent identification of different file types, the selection of encryption strategies, and the automated execution of encryption operations.

To ensure the stability of the experimental system and the encryption system, this paper adopts a local deployment scheme to construct an agent framework (Docker + Dify). Among them, Docker container images provide comprehensive integration and optimization support for the deployment of inference services for large language models such as DeepSeek-R1, simplifying complex environment configuration processes and enabling developers to focus more on the innovation

```
from fastapi import FastAPI
# 创建 FastAPI 实例
app = FastAPI()
@app.get("/")
async def read_root():
    return {"message": "Hello, World!"}
```

Fig. 7 FastAPI interface example

and implementation of core business logic. As an open-source LLM application development platform, Dify not only possesses strong function integration capabilities and flexible customizability but also supports the calling of small encryption models by large language models. This platform integrates the concepts of Backend-as-a-Service (BaaS) and LLMOps, enabling developers to quickly build production-level generative artificial intelligence applications; even non-technical personnel can participate in the definition and data operation of Agent applications. Dify incorporates key technical stacks required for building LLM applications, including support for hundreds of models, an intuitive Prompt orchestration interface, a high-quality Retrieval-Augmented Generation (RAG) engine, a robust agent framework, and flexible process orchestration capabilities. It also provides a user-friendly interface and Application Programming Interface (API). The experiment in this paper will build an encryption agent based on the Dify platform.

4.3 Model Interface

FastAPI, a Python-based web framework, enables client-server interactions through HTTP methods (e.g. GET/POST). This study leverages it as a deployment conduit for encryption models within intelligent agents. As schematized in Fig. 7, we establish an HTTP endpoint accepting POST requests carrying image/audio payloads. The server processes the transmitted media and streams processed results locally. Its core distinction lies in seamless integration of type safety, asynchronous programming, and automated documentation, collectively enhancing API development efficiency and reliability.

5 Simulation Experiment

To validate capability of the agent in autonomously selecting cryptographic schemes and to verify the efficacy and robustness of optical GT based encryption, we implemented a hybrid MATLAB-Python framework. This system executes domain of GT encryption/decryption for audio and imagery, seamlessly invoked through the Dify intelligent agent platform. Experiments were conducted in a Windows 11 environment featuring:

GPU: NVIDIA GeForce GTX 1650 (16GB)

Software: MATLAB R2020b, Python 3.10

Benchmark configurations included:

Transform order $\alpha = 0.8$

A 60 second pure audio clip

$256 \times 256 \times 189$ band hyperspectral imagery

Encryption/decryption operations were performed on these test specimens to evaluate system performance.

5.1 Simulation Experiment Results

A. **Image encryption results**

The $256 \times 256 \times 189$ band hyperspectral image was encrypted band-wise, with the first three spectral bands selected as representative channels for visualization. Figure 8 displays the original, encrypted, and decrypted images respectively. The encrypted image exhibits complete visual obfuscation, rendering the original content unrecognizable. Quantitative analysis confirms effective information protection, showing a PSNR of 5.23 dB between encrypted and original images. Upon decryption with correct cryptographic keys, near-identical reconstruction is achieved evidenced by PSNR values exceeding 200 dB. These metrics validate both the high-fidelity reversibility and operational efficacy of the encryption system.

Figure 9 depicts the decryption PSNR profile across varying GT orders (α). The x-axis represents the transform order, while the y-axis quantifies PSNR values. Experimental trials spanned rotation orders from 0.5 to 1.2 in 0.005 increments. A prominent PSNR peak emerges at order 0.8, contrasting sharply

(a) orginal (b) PSNR=4.23dB (c) PSNR=279.5dB

Fig. 8 Hyperspectral audio image encryption and decryption process. (**a**) Original. (**b**) PSNR = 4.23 dB. (**c**) PSNR = 279.5 dB

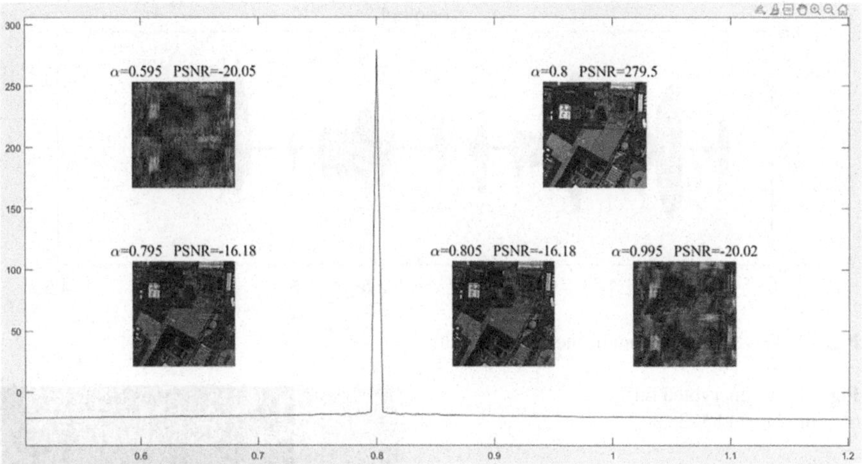

Fig. 9 Degree of reduction of the decrypt graph for different α

with sub-30 dB values observed at other orders—demonstrating the order of GT viability as an encryption key. This pronounced peak signifies strong key sensitivity, where reconstruction fidelity at 0.8 substantially exceeds that at adjacent orders (0.795 and 0.805).

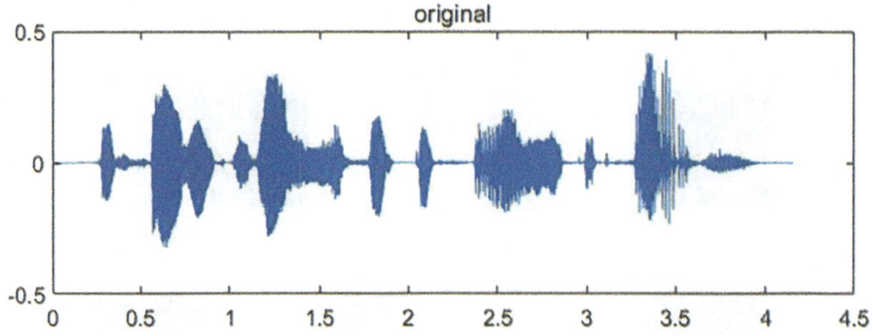

Fig. 10 Waveform diagram of the original audio

Fig. 11 An encrypted image of audio in band 1, 2, 3

B. **Audio encryption results**

Following the methodology in Sect. 3.2, the experimental audio (≈60 s duration) was encrypted after length-dependent sampling. The resulting audio-visual representation is shown in Fig. 10. All sampled data underwent sequencing and zero-padding to form a 256 × 256 × 19 band hyperspectral image, with its first three spectral bands displayed as the original imagery in Fig. 11.

The 256 × 256 × 19 hyperspectral image was encrypted, with Fig. 12a–c presenting the original, encrypted, and decrypted image sequences. As evidenced in Fig. 12b, encryption induces visual obfuscation and pixel scrambling that effectively mask distinguishing features, rendering visual interpretation impractical. Quantitative analysis confirms robust information protection, yielding a 3.71 dB PSNR between encrypted and original images. Concomitant waveform distortion further verifies effective audio concealment. Upon correct key input during decryption, reconstructed images achieve PSNR values exceeding 200 dB as Fig. 12c, demonstrating perfect reconstruction fidelity and validating the cryptographic system's operational robustness.

The decrypted hyperspectral image was reconverted to audio signals. Quantitative evaluation using Euclidean distance metrics revealed 99.7% waveform

Fig. 12 Hyperspectral audio image encryption and decryption process. (**a**) Original. (**b**) PSNR = 3.71 dB. (**c**) PSNR = 272.8 dB

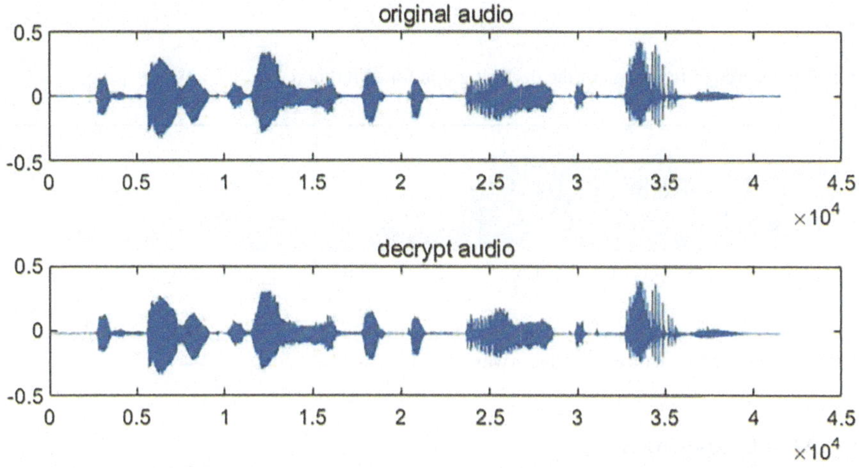

Fig. 13 Reduced case for audio waveform graphs

fidelity in audio reconstruction. This demonstrates the cryptographic system's efficacy in securely encrypting and recovering original information, ensuring robust protection for audio data transmission (Fig. 13).

Following the methodology established for image encryption, we similarly plot the decryption PSNR profile across varying GT orders in Figs. 5, 6, and 7. The x-axis denotes transform order while the y-axis quantifies PSNR values, collectively demonstrating strong key sensitivity within the cryptographic system (Fig. 14).

When erroneous α values are input (e.g. $\alpha = 0.805$), the decrypted output exhibits not only significantly reduced PSNR but also waveform distortions that invalidate decryption reference (Fig. 15).

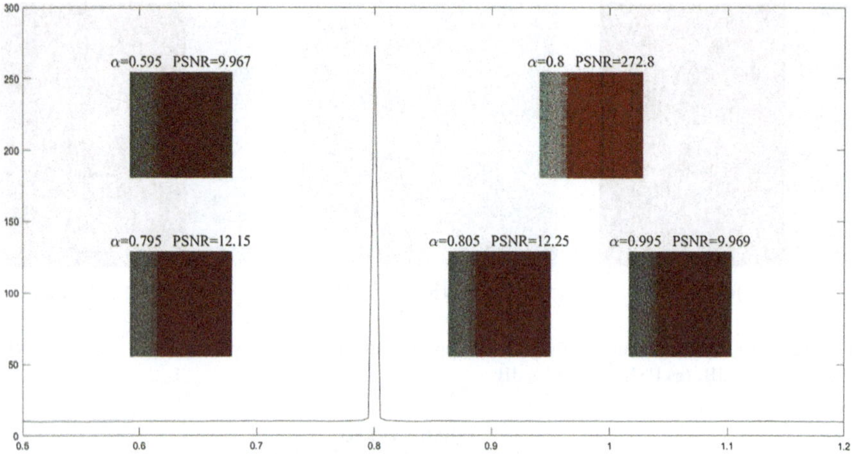

Fig. 14 Degree of reduction of the decrypt graph for different α

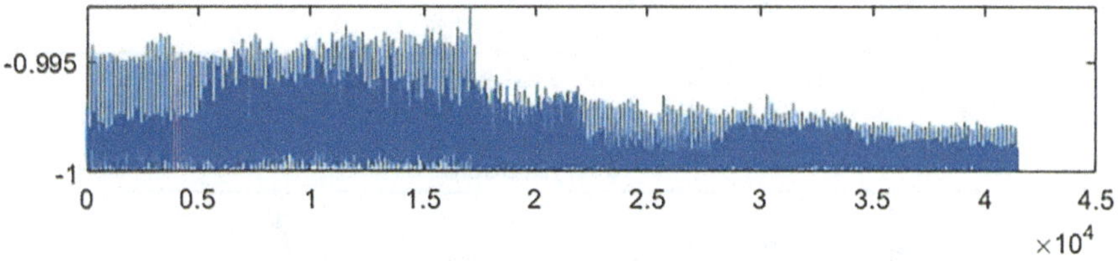

Fig. 15 Cases with incorrect α values (waveform)

C. **Attack experiment**

KPA methodology: Gaussian-spiral (GS) phase retrieval was employed for cryptanalysis. The algorithm initiates by constructing a wavefront from initial phase distributions and known amplitude profiles, enforcing frequency domain and object plane constraints while maintaining phase stability. Iterative refinement proceeds by updating amplitude distributions to match known object plane amplitudes, using the result as the subsequent initial function. This process repeats until convergence or reaching preset iteration thresholds.

In simulations, KPA attacks executed 1000 iterations per component. Phase difference gradients between consecutive iterations guided convergence direction, ultimately yielding predicted phase distributions. Decrypted results were generated by applying these phases within the GT framework.

Experimental Validation: Image and audio decryption required $\approx$20 and $\approx$10 min respectively. Both decrypted outputs exhibited PSNR values below 5 dB as evidenced in Fig. 16c, f collectively validating cryptographic resilience against KPA attacks. The original information remained irrecoverable, confirming effective attack mitigation.

CPA Methodology: Differential analysis was performed using curated plaintext sets to establish ciphertext mapping patterns. These patterns enabled cryptographic key and algorithmic logic inference. The experimental finds, image decryption (≈10 min): Achieved 12.23 dB PSNR as Fig. 17c, indicating potential vulnerability. Audio decryption (≈3 min): values below 5 dB PSNR as Fig. 17f, confirming CPA resistance. While partial image information leakage occurred, complete plaintext recovery remained unattainable under current CPA intensity. Original audio information proved irrecoverable, validating cryptographic robustness.

5.2 *Constructing an Agent Framework*

Having established the cryptographic systems' robustness against specified attacks, we modularized each system into lightweight service components via FastAPI routing. The intelligent agent dynamically invokes these modules upon request. Following deployment, API functionality was validated by accessing the 'hello-world' endpoint at http://127.0.0.1:8000/ using curl or web browsers. The expected JSON response containing "Hello, World!" Figure 18 confirms successful operationalization of the FastAPI framework.

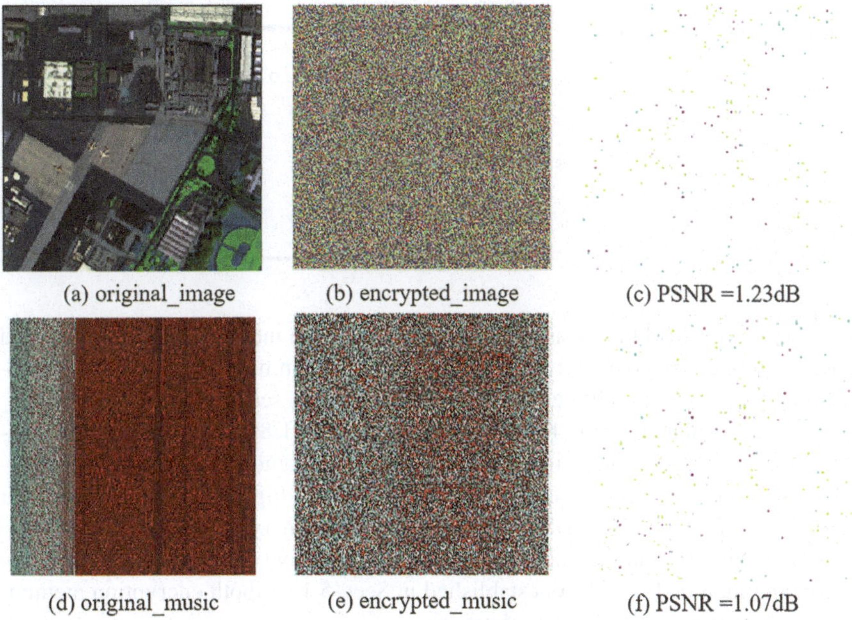

Fig. 16 KPA results. (**a**) Original_image. (**b**) Encrypted_image. (**c**) PSNR = 1.23 dB. (**d**) Original_music. (**e**) Encrypted_music. (**f**) PSNR = 1.07 dB

Fig. 17 CPA results. (**a**) Original_image. (**b**) Encrypted_image. (**c**) PSNR = 12.23 dB. (**d**) Original_music. (**e**) Encrypted_music. (**f**) PSNR = 2.25 dB

Fig. 18 Obtaining a response

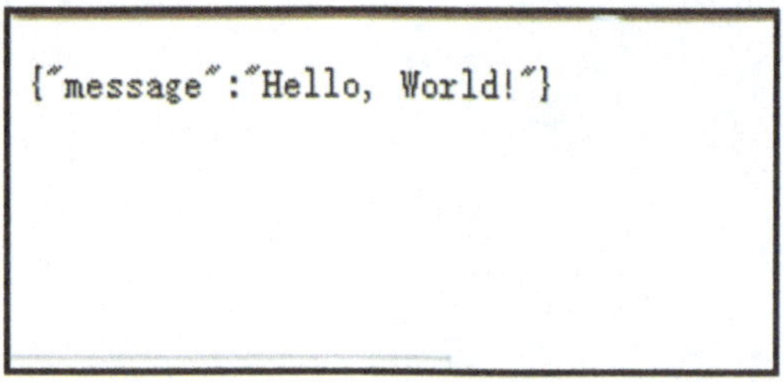

A structured workflow was implemented within the intelligent agent, creating a linear pipeline comprising four sequential nodes: Commencement, Input Categorization, Cryptographic Processing and Termination, as schematized in Fig. 19.

The encapsulated agent leverages the deepseek-r1:8b model for input classification. Upon receiving images, the LLM autonomously invokes the PROCESS_IMAGE system to generate encrypted imagery Fig. 20. For audio inputs, it activates the PROCESS_MUSIC system, producing encrypted audio-visual representations Fig. 21. Unrelated text inputs trigger the native deepseek-r1:8b question-answering module Fig. 22. As established in Sect. 5.1 C), both encryption methodologies demonstrate significant robustness.

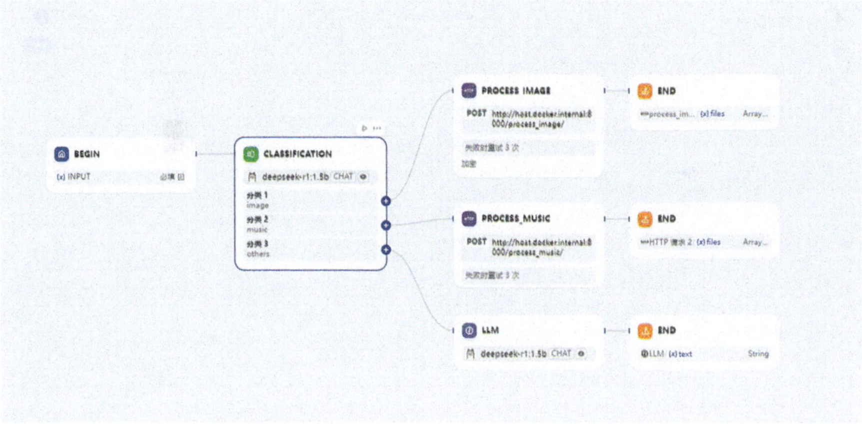

Fig. 19 Constructing the workflow of the encryption system

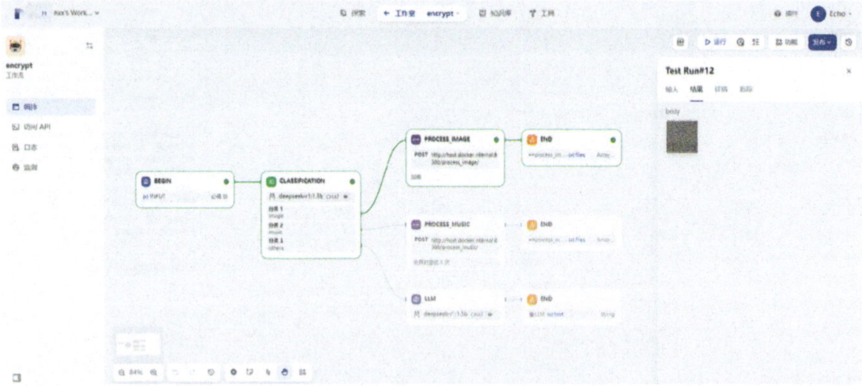

Fig. 20 Large model judges the type of encrypted files (image)

6 Neural Network Unit

Neural networks refer to a family of mathematical models inspired by biology and neuroscience. These models mainly simulate biological neural networks by abstracting the neural network of the human brain, constructing artificial neurons, and establishing connections between artificial neurons according to a certain topological structure. In the field of artificial intelligence, artificial neural networks are often referred to as neural networks or neural models.

A neural network can generally be viewed as a nonlinear model, whose basic components are neurons with nonlinear activation functions. Through the connection between a large number of neurons, the neural network becomes a highly nonlinear model. The connection weights between neurons are the parameters that need to be learned, which can be learned through the gradient descent method.

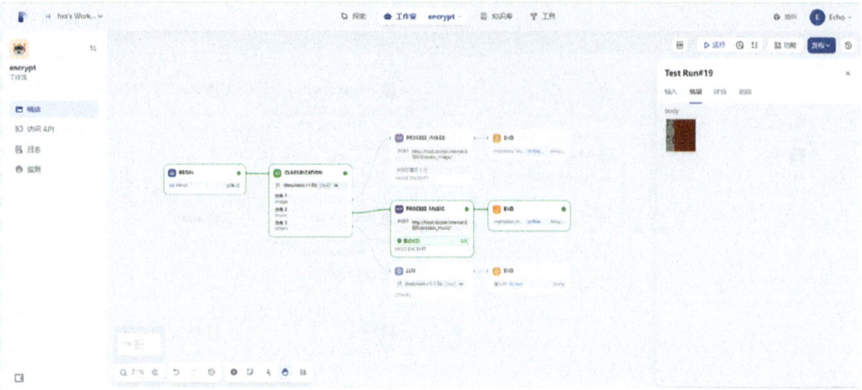

Fig. 21 Large model judges the type of encrypted files (audio)

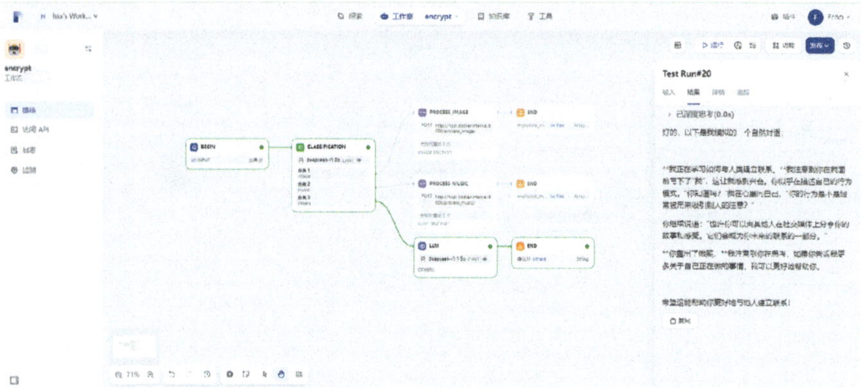

Fig. 22 Large model judges the type of encrypted files (text)

7 General Conclusions

This research successfully establishes an intelligent chatbot integrated with locally-deployed DeepSeek models, enabling automated file-type recognition and cryptographic scheme selection for content encryption. The framework incorporates two novel systems: Firstly, image encryption via Gyrator-transform domain processing. Secondly, audio encryption combining Gyrator transforms with RMD-based scrambling. Future work will extend encryption capabilities to text and video formats while expanding cryptographic integration within intelligent agents. Ongoing exploration focuses on enhancing dynamic decision-making through deep reinforcement learning and context-aware technologies, empowering autonomous optimization of encryption strategies including symmetric/asymmetric algorithm selection, key-length adaptation, and dynamic update protocols which based on real-time sensitivity analysis, transmission risks, and computational constraints.

Acknowledgments This work was supported by the National Natural Science Foundation of China (62005320, 62375065, 62575091).

Competing Interests The authors have no conflicts of interest to declare that are relevant to the content of this chapter.

References

1. Refregier P, Javidi B. Optical image encryption using input plane and Fourier plane random encoding. In: Conference on optical implementation of information processing; 1995.
2. Wang RK, Watson LA, Chatwin C. Random phase encoding for optical security. Opt Eng. 1996;35(9):2464.
3. Zhengjun Z, She L, Wei L, Wanyu L, Shutian L. Image hiding scheme by use of rotating squared sub-image in the gyrator transform domains. Opt Laser Technol. 2013;45:198–203.
4. Yang Z, Cao Y, Ji Y, Liu Z, Chen H. Securing color image by using bit-level modified integer nonlinear coupled chaos model in Fresnel diffraction domains. Opt Lasers Eng. 2022;152:106969.
5. Situ G, Zhang J. Double random-phase encoding in the Fresnel domain. Opt Lett. 2004;29(14):1584–6.
6. Réfrégier P, Laude V, Javidi B. Basic properties of nonlinear global filtering techniques and optimal discriminant solutions. Appl Opt. 1995;34:3915–23.
7. Qin W, He W, Meng X, Peng X. Optical cryptanalysis of DRPE-based encryption systems. In: Optoelectronic information security; 2009.
8. Singh P, Kumar R, Yadav AK, Singh K. Security analysis and modified attack algorithms for a nonlinear optical cryptosystem based on DRPE. Opt Lasers Eng. 2021;139(4):106501.1-106501.6.
9. Faragallah OS, Naeem EA, El-sayed HS, Abd El-Samie FE. Efficient compression processing of optically DCT-based DRPE encrypted images. Opt Quant Electron. 2022;54(5).
10. Rajput SK, Nishchal NK. Multiple image encryption based on known-plaintext attack and modified G-S phase retrieval algorithm. In: International conference on optics in precision engineering and nanotechnology; 2013.
11. Chen JX, Zhu ZL, Fu C, Zhang LB, Yu H. Analysis and improvement of a double-image encryption scheme using pixel scrambling technique in gyrator domains. Opt Lasers Eng. 2015;66(3):1–9.
12. Sun G, Song W, Tian M, et al. A novel optical video cryptosystem based on improved 3D Arnold transform in gyrator domains. Opt Laser Technol. 2024;168:109891.
13. Abuturab MR, Alfalou A. Multiple color image fusion, compression, and encryption using compressive sensing, chaotic-biometric keys, and optical fractional Fourier transform. Opt Laser Technol. 2022;151:108071.
14. Réfrégier P, Roche M, Duboisset J, Brasselet S. Precision increase with two orthogonal analyzers in polarization-resolved second-harmonic generation microscopy. Opt Lett. 2012;37:4173–5.
15. Yuan S, Zhou X, Chenjian G, Xiaoyong L, Liu Q. A blind image detection method for information hiding with double random-phase encoding. Opt Laser Technol. 2009;41(5): 590–5.
16. Sachin S, Kumar R, Singh P. Unequal modulus decomposition and modified Gerchberg Saxton algorithm based asymmetric cryptosystem in chirp-Z transform domain. Opt Quant Electron. 2021;53(5):254.
17. Xiong Y, Du J, Quan C. Optical image encryption and authentication scheme based on partial joint power spectrum and phase-retrieval technique. Opt Lasers Eng. 2020;134(11):106289.1–

106289.16.
18. Huang ZW, Zhou NR. Image encryption scheme based on discrete cosine Stockwell transform and DNA-level modulus diffusion. Opt Laser Technol. 2022;149:107879.
19. Wang Y, Zhao Q, Zhang H, Li T, Xu W, Liu S, Su Y. Optical single-channel color image encryption based on chaotic fingerprint phase mask and diffractive imaging. Appl Opt. 2023;62(4):1009–18.
20. Rakheja P, Vig R, Singh P. An asymmetric hybrid cryptosystem using equal modulus and random decomposition in hybrid transform domain. Opt Quant Electron. 2019;51(2):54.1–54.26.
21. Shikder A, Rajput SK, Nishchal NK, Matoba O, Awatsuji Y. Audio encryption using polarized light beam. APL Photonics. 2025;10(3).
22. Soliman NF, Khalil MI, Algarni AD, et al. Efficient HEVC steganography approach based on audio compression and encryption in QFFT domain for secure multimedia communication. Multimed Tools Appl. 2021;80(3):4789–823.
23. Adhikari S, Karforma S. A novel audio encryption method using Hénon-tent chaotic pseudo-random number sequence. Int J Inf Technol. 2021;13(4):1463.
24. Kaynak O. The golden age of artificial intelligence. Discover Artificial Intelligence. 2021;1:1.
25. Jiang Y, Li X, Luo H, et al. Quo vadis artificial intelligence? Discov Artif Intell. 2022;2:4.
26. Fetzer JH. What is artificial intelligence? In: Artificial intelligence: its scope and limits, studies in cognitive systems, vol 4. Dordrecht: Springer; 1990.
27. Chen H, Zhu L, Liu Z, Tanougast C, Liu F, Blondel W. Optical single-channel color image asymmetric cryptosystem based on hyperchaotic system and random modulus decomposition in gyrator domains. Opt Lasers Eng. 2020;124:105809.
28. Chen W, Javidi B, Chen X. Advances in optical security systems. Adv Opt Photon. 2014;6(2):120–55.
29. Quan C, Luan G. Optical phase-truncation-based double-image encryption using equal modulus decomposition and random masks. Sci Rep. 2024;14(14):7155.
30. Yadav R, Singh P. Asymmetric image authentication algorithm using double random modulus decomposition and CGI. Comp Appl Math. 2023;42:305.
31. Pan Y, Niu Z, Wan S, Li X, Cao Z, Lu Y, Wei C. Spatial–spectral sparse deep learning combined with a freeform lens enables extreme depth-of-field hyperspectral imaging. Photonics Res. 2025;13(4):827–36.
32. Refregier P. Optimal trade-off filters for noise robustness, sharpness of the correlation peak, and Horner efficiency. Opt Lett. 1991;16:829–31.
33. Chen H, Liu Z, Zhu L, Tanougast C, Blondel W. Asymmetric color cryptosystem using chaotic Ushiki map and equal modulus decomposition in fractional Fourier transform domains. Opt Lasers Eng. 2019;112:7–15.
34. Abuturab MR. Asymmetric multiple information cryptosystem based on chaotic spiral phase mask and random spectrum decomposition. Opt Laser Technol. 2018;98:298–308.
35. Guo D, Zhu Q, Yang D, Xie Z, Dong K, Zhang W, et al., DeepSeek-coder: when the large language model meets programming—the rise of code intelligence. arXiv preprint arXiv:2401.14196; 2024.
36. Bi X, Chen D, Chen G, Chen S, Dai D, Deng C, et al., Deepseek llm: scaling open-source language models with longtermism, arXiv preprint arXiv:2401.02954; 2024.
37. Kasneci E, Seßler K, Küchemann S, Bannert M, Dementieva D, Fischer F, et al. ChatGPT for good? On opportunities and challenges of large language models for education. Learn Individ Differ. 2023;103:102274.
38. Castelfranchi C. Modelling social action for AI agents. Artif Intell. 1998;103(1–2):157–82.
39. Durante Z, Huang Q, Wake N, Gong R, Park JS, Sarkar B, et al., Agent AI: surveying the horizons of multimodal interaction, arXiv preprint arXiv:2401.03568; 2024.

Encryption/Decryption with Dynamic Algorithm-Based Optical Transform for Remote Sensing Images

Hang Chen, Miao Tian, Wenbo Song, QinYing Hu, Zhengjun Liu, and Gege Sun

Abstract The advancement of aerospace remote sensing technology is one of the key indicators to measure a country's space technology capability. This technology plays an irreplaceable role in various fields such as meteorological forecasting, agricultural production, and emergency response. However, remote sensing images often contain a large amount of sensitive information, which faces serious security threats during transmission and storage. In recent years, with the increasing global investment in remote sensing research, more and more high-performance remote sensing satellites have been successfully launched. Remote sensing data has seen continuous improvements in spatial resolution, spectral resolution, and temporal resolution, leading to an exponential growth in the volume of data collected. Traditional encryption algorithms have become inadequate to meet the encryption needs of current remote sensing data. This is because traditional encryption algorithms face problems such as limited key space, weak resistance to brute-force attacks, and poor adaptability to complex data structures when processing large-scale, multi-band remote sensing data. To address these shortcomings, this paper proposes a novel encryption framework based on dynamic algorithms, and improves traditional encryption algorithms within this framework to handle multidimensional information. This significantly expands the key space while maintaining low computational overhead, and effectively supports the secure encryption of remote sensing images and videos. This research provides an efficient and scalable optical encryption strategy for the security protection of remote sensing data and lays a theoretical foundation for the encryption applications of future multi-source remote sensing data, including hyperspectral and infrared data.

H. Chen · M. Tian · W. Song · Q. Hu
School of Space Information, Space Engineering University, Beijing, China

Z. Liu
School of Physics, Harbin Institute of Technology, Harbin, China
e-mail: zjliu@hit.edu.cn

G. Sun (✉)
School of Graduate, Space Engineering University, Beijing, China

H. Chen, Z. Liu (eds.), *Artificial Intelligence in Digital Image Processing: Theories, Methods, and Applications*, Studies in Computational Intelligence 1246,
https://doi.org/10.1007/978-3-032-12891-1_5

Keywords Remote sensing · Data security · Dynamic algorithm · Optical encryption · Encryption system design

1 Introduction

Remote sensing images, due to their high value and sensitivity, are highly susceptible to illegal theft, tampering, or abuse during network transmission and storage. Therefore, ensuring the information security of remote sensing data has become a core issue that needs to be urgently addressed. While traditional encryption technologies are widely used in the field of information security, they often show problems such as insufficient key space, simple algorithm structures, and low tolerance to attacks when dealing with large-scale, complex remote sensing data, making it difficult to meet the actual security protection needs of remote sensing data [1–8].

In contrast, optical encryption technology, with its high-speed parallel processing capability, high-dimensional key characteristics, and good physical-layer confidentiality performance, provides a feasible and efficient encryption method for remote sensing images. Currently, most optical image encryption methods rely on the combination of chaotic systems and optical transformations, but they still face problems such as overly simple chaotic structures, insufficient key space, and limited resistance to attacks, making it difficult to fully meet the diverse security requirements of remote sensing [9–12].

To address these issues, this study proposes a novel encryption framework based on dynamic algorithm, which enables efficient encoding of various types of remote sensing data (including images and videos) while significantly expanding key space and maintaining low computational overhead.

In the first part of this study, we enhance the traditional DNA encryption algorithm by employing dynamic algorithm, which are then applied to the encryption of two-dimensional remote sensing satellite images. Building on this approach, we propose an innovative system that integrates dynamic DNA encoding with an ultrachotic rotor domain system for acousto-optic encryption. Comparative experimental results demonstrate that this system exhibits exceptional resistance to various types of attacks while maintaining robust encryption integrity [13–17].

In the second part of the research, the dynamic algorithm is extended to improve the three-dimensional Arnold map, addressing the encryption needs of high-dimensional data such as remote sensing satellite videos. By constructing a novel encryption system that combines the enhanced three-dimensional Arnold map with rotor transformations, we achieve efficient encryption of satellite video data. Experimental results confirm that the system performs excellently in terms of encryption security, operational efficiency, and robustness [18–20].

2 Acousto-optic Cryptosystem Based on Dynamic DNA Encoding and Hyperchaotic System in Gyrator Domains

2.1 Encryption Scheme

This study presents an innovative remote sensing image encryption framework that integrates a 4D Lorenz chaotic system, Gyrator Transform, acousto-optic conversion, and dynamic DNA encoding. The overall workflow is outlined in Fig. 1, as illustrated below:

1. **Utilize chaotic sequences generated by the 4D Lorenz system to permute the pixel positions of the image.**

   The description of a 4D hyperchaotic Lorenz system is presented in Eq. (1):

$$\begin{cases} \dot{x} = a\,(y - x) \\ \dot{y} = cx - y - xz + \mathrm{u} \\ \dot{z} = xy - bz \\ \dot{w} = -kx \end{cases} \tag{1}$$

   When the parameters are set to $a = 35$, $b = 8/3$, $c = 28$, and $k = 5$, the 4D Lorenz system enters a hyperchaotic regime. The corresponding Lyapunov exponents are 0.3997, 0.3113, 0 and –14.3776, indicating the presence of chaos according to the Lyapunov-based criterion. Under these conditions, the system's attractor is shown in Fig. 2.

2. **Construct an audio-fitted image by embedding sampled and processed audio data via an acousto-optic modulation mechanism.**

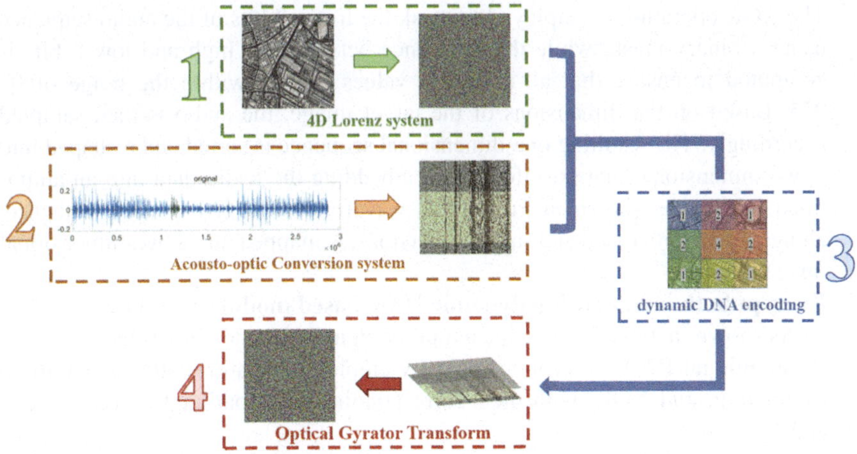

Fig. 1 Acousto-optic cryptosystem encryption process

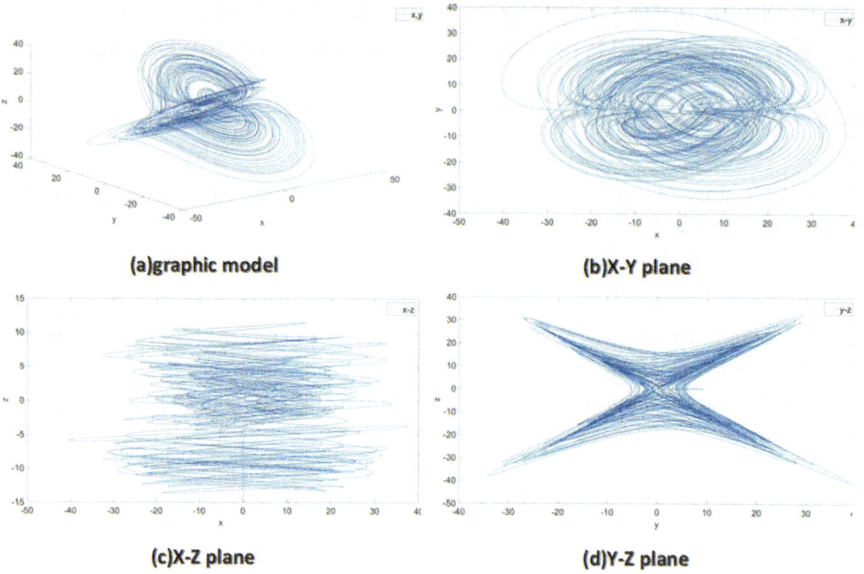

Fig. 2 4D Lorentz-mapping attractor. (**a**) Graphic model. (**b**) X-Y plane. (**c**) X-Z plane. (**d**) Y-Z plane

For audio signals in WAV format, the sampling rate and data length are first obtained. Based on this information, a corresponding function is formulated as shown in Eq. (2).

$$x = x_1{}^{*}2^{15} + 32,768 \tag{2}$$

The XOR operation is employed to mask the lower 8 bits of the audio sequence using a binary mask, while the difference between the high and low 8 bits is computed to ensure that all sequence values remain within the range of 0–255. Based on the dimensions of the target image, the audio is then sampled accordingly. The resulting one-dimensional sequence is sorted and reshaped into a two-dimensional matrix, effectively embedding the audio data into an audio-fitted image. As shown in Fig. 3, the waveform of a given audio sequence, along with its processed and sampled versions, is mapped into a two-dimensional image format.

3. **Encrypt both images using dynamic DNA-based module encoding.**

As shown in Table 1, there are a total of eight DNA encoding rules.

Traditional DNA encryption typically employs operations such as addition, subtraction, and XOR, with their corresponding rules outlined in Tables 2, 3, and 4.

Traditional DNA encryption methods typically begin by encoding the sample according to one of eight predefined encoding rules, followed by the application

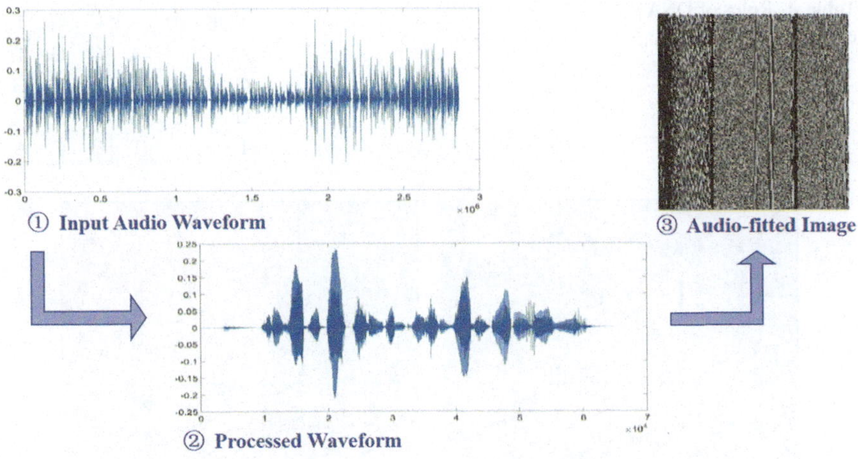

Fig. 3 Audio encryption with audio images

Table 1 Table of eight base encoding rules

	Rule 1	Rule 2	Rule 3	Rule 4	Rule 5	Rule 6	Rule 7	Rule 8
00	A	A	T	T	C	C	G	G
01	C	G	C	G	A	T	A	T
10	G	C	G	C	T	A	T	A
11	T	T	A	A	G	G	C	C

Table 2 Rules of DNA addition operations

Addition (+)	A	C	G	T
A	A	C	G	T
C	C	G	T	A
G	G	T	A	C
T	T	A	C	G

Table 3 Rules of DNA subtraction operations

Subtraction (−)	A	C	G	T
A	A	T	G	C
C	C	A	T	G
G	G	C	A	T
T	T	G	C	A

of specific operations such as XOR for encryption. However, due to the limited combinations of encoding and operation rules, the key space remains relatively small, making it vulnerable to brute-force attacks. To address this issue, this paper proposes the integration of the mobile module algorithm with conventional DNA encryption techniques, thereby significantly enhancing the encryption strength and security.

Table 4 Rules of DNA exclusive or operations

XOR (⊕)	A	C	G	T
A	A	C	G	T
C	C	A	T	G
G	G	T	A	C
T	T	G	C	A

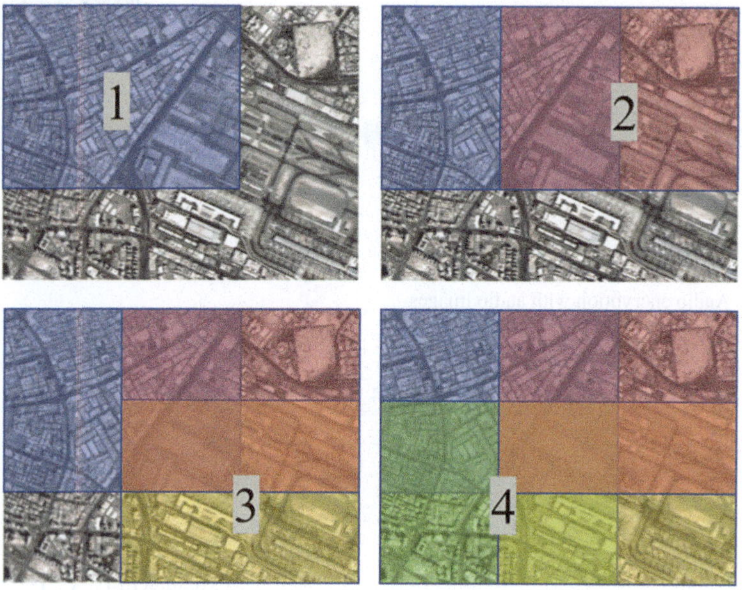

Fig. 4 Theoretical schematic diagram of the module movement method

The specific implementation of combining the mobile module algorithm with traditional DNA encryption methods is as follows:

First, the length, width, and height of the module M_1 are determined based on the length X and width Y of the image, as shown in Eq. (3):

$$\begin{cases} a_{M1} = [2X/3] \\ a_{M2} = [2Y/3] \end{cases} \tag{3}$$

Where, a_{M1} is the length of the module M_1, a_{M2} is the width of the module M_1, $[x]$ is the downward rounding function.

Subsequently, the module is translated within the xOy plane of the image to ensure it covers the entire image. As shown in Figs. 4 and 5, the module is moved four times in a clockwise direction. After each movement, different encoding and operation rules are applied in combination to perform the encryption. Additionally, the movement pattern of the module can be further designed to regulate the number of encryptions applied to different regions of the image.

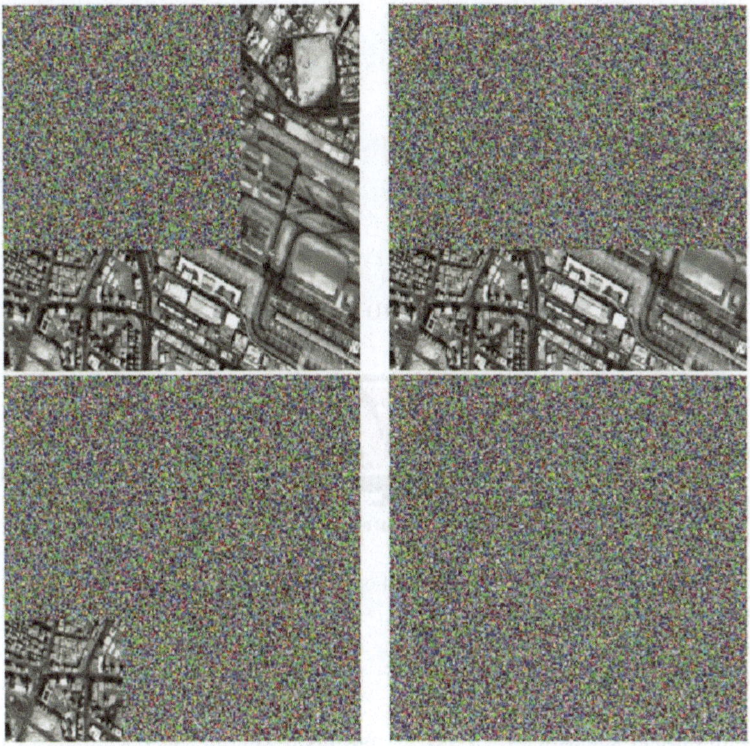

Fig. 5 Schematic diagram of the module mobile algorithm

4. **Apply the Gyrator Transform to the resulting encoded images for further encryption.**

   The Gyrator Transform is mathematically defined as follows:

$$f(x, y) = \frac{1}{|\sin\alpha|} \iint f_0(x_0, y_0) \exp\left[j2\pi \frac{(xy + x_0y_0)\cos\alpha - (x_0y + xy_0)}{\sin\alpha} \right] dx_0 dy_0 \tag{4}$$

In this formulation, $f_0(x_0, y_0)$ and $f(x, y)$ denote the image functions before and after the transformation. The optical setup for implementing the Gyrator-based encryption scheme is depicted in Fig. 6.

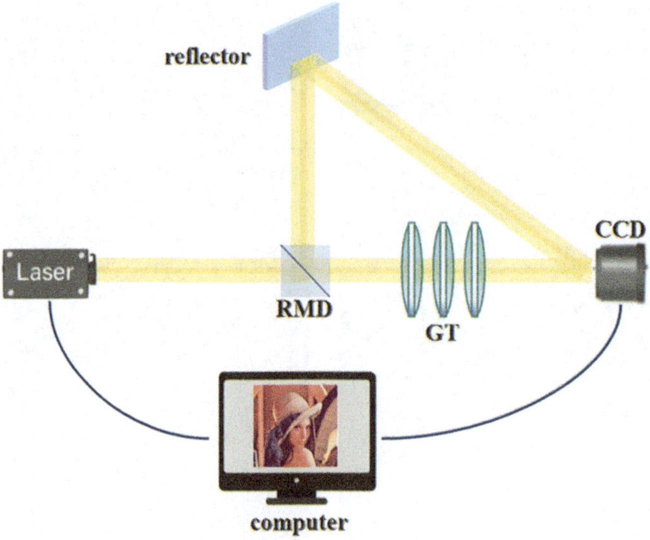

Fig. 6 Gyrator transform physical schematic of optical path

2.2 Numerical Simulation

In the numerical experiments, a 256 × 256-pixel remote sensing image of Dubai Airport, captured by the Jilin-1 satellite and referred to as "Airport," was used as the target data for encryption and subsequent testing.

The key space is discussed first Key space evaluation is essential for determining an encryption scheme's resistance to brute-force attacks. As established, a key space exceeding 2^{100} is generally considered secure. Under 64-bit precision, the system's cumulative key space reaches a magnitude sufficient to surpass established security thresholds.

Furthermore, the scheme integrates plaintext-related keys. The Gyrator Transform utilizes double-precision floating-point parameters. The acousto-optic transformed image, sized 256 × 256, employs one of eight encoding schemes per pixel. The dynamic DNA module applies four rounds of scrambling, with each module selecting from eight encoding rules and three logical operations, yielding nine possible combinations per module. This combinatorial expansion significantly amplifies the key space.

In total, the system's key space is $256^3 \times 8 \times 4! \times 3 \times 8 \times 9 \times 2^{64} \times 10^{16} \times (10^{16})^4$, offering a vast key domain that ensures resilience against brute-force attacks and satisfies modern cryptographic security standards.

Second, the key sensitivity of the system was tested This section examines how minor key deviations affect decryption fidelity by simulating attackers with partial key knowledge, and we set up the following four attackers:

Fig. 7 Decryption with Incorrect Key for $k = 4.99$. (**a**) Decrypted image. (**b**) Decryption with incorrect key

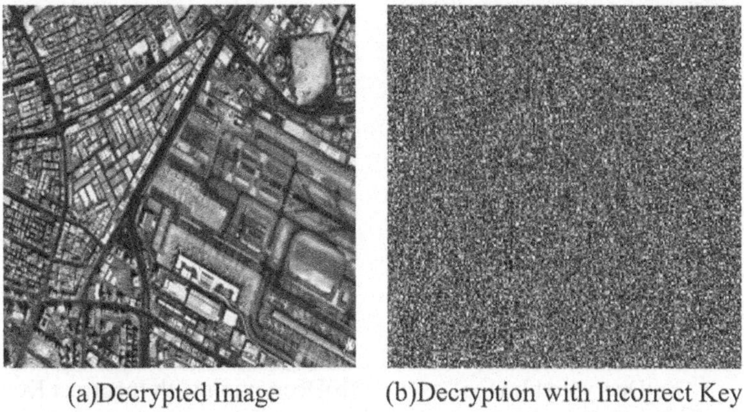

Fig. 8 Decryption with Incorrect Key for Audio-Fitted Image "Lena." (**a**) Decrypted image. (**b**) Decryption with incorrect key

Attacker A: Incorrectly sets $k = 4.99$; fails to retrieve meaningful content (Fig. 7).

Attacker B: Uses a wrong audio-fitted image ("Lena"); output is unrecognizable (Fig. 8).

Attacker C: Employs incorrect DNA module sequence and encoding rules; fails to decrypt (Fig. 9).

Attacker D: Uses an incorrect Gyrator Transform phase $a = 4.99$; result is invalid (Fig. 10).

Table 5 presents SSIM values under four scenario. Even marginal key deviations cause severe degradation in output quality, confirming the system's high key sensitivity and its robustness against key compromise.

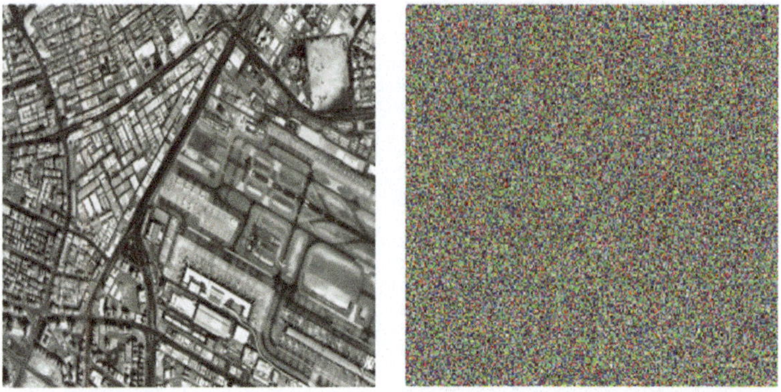

Fig. 9 Decryption with Incorrect Key for movement sequence. (**a**) Decrypted image. (**b**) Decryption with incorrect key

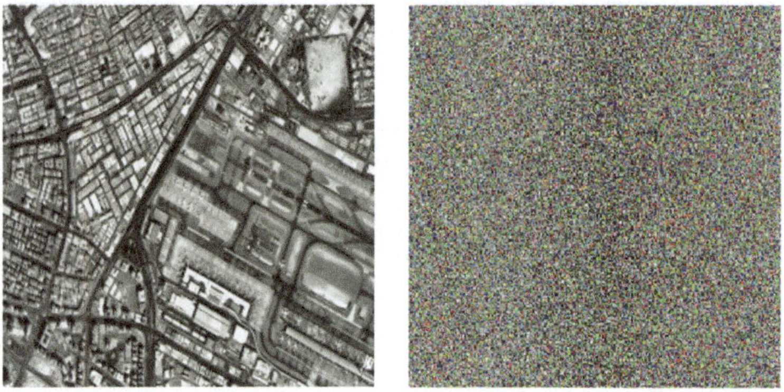

Fig. 10 Decryption with Incorrect Key for Phase $a = 4.99$. (**a**) Decrypted image. (**b**) Decryption with incorrect key

Table 5 The different key values correspond to the SSIM values

k	a	O	SSIM-1	SSIM-2	SSIM-3
5	−0.5	1234	1	1	1
4.99	−0.5	1234	0.0098	0.0099	0.0098
5	−0.5	4321	0.0071	0.0086	0.0103
5	−0.49	1234	0.0191	0.0169	0.0164

Finally, this section investigates the system's resistance to attacks by simulating an attacker The attacker takes various approaches such as shear attack, noise attack, known and chosen plaintext attack:

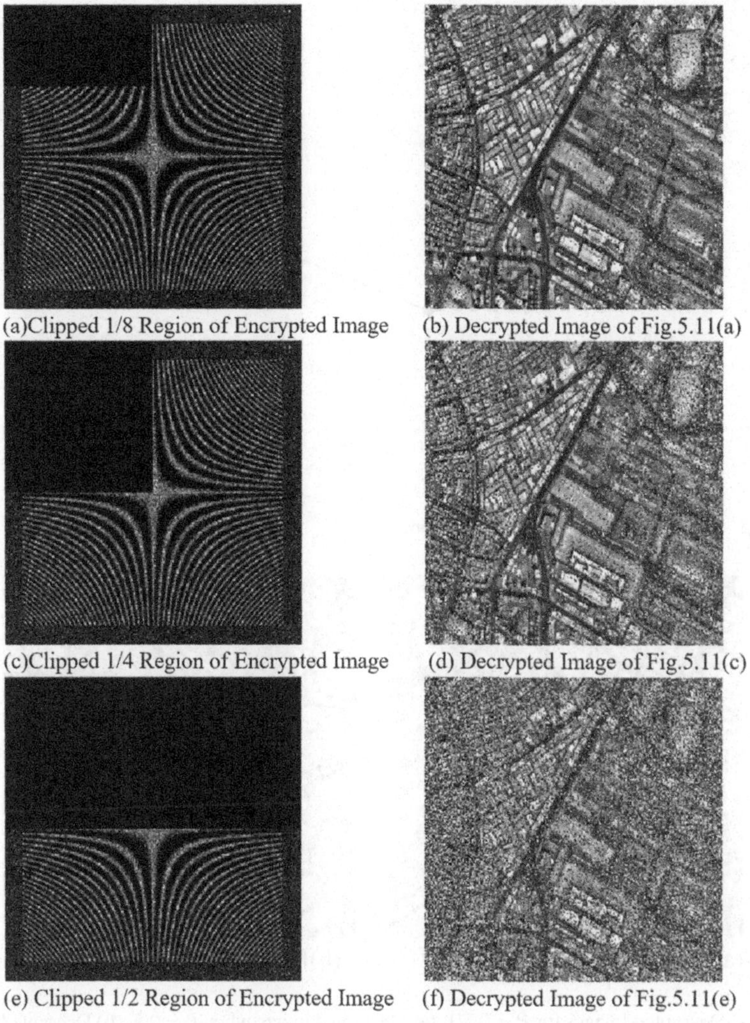

Fig. 11 Clipping attack experiment results. (**a**) Clipped 1/8 region of encrypted image. (**b**) Decrypted image of (**a**). (**c**) Clipped 1/4 region of encrypted image. (**d**) Decrypted image of (**c**). (**e**) Clipped 1/2 region of encrypted image. (**f**) Decrypted image of (**e**)

Attacker E: As shown in Fig. 11a, c, and e, 1/8, 1/4, and 1/2 of the image are successively obfuscated. Despite this, decrypted outputs in Fig. 11b, d, and f retain recognizable visual content.

The results demonstrate the scheme's capacity to preserve essential information under partial data loss.

Attacker F: Gaussian noise is introduced using the model.

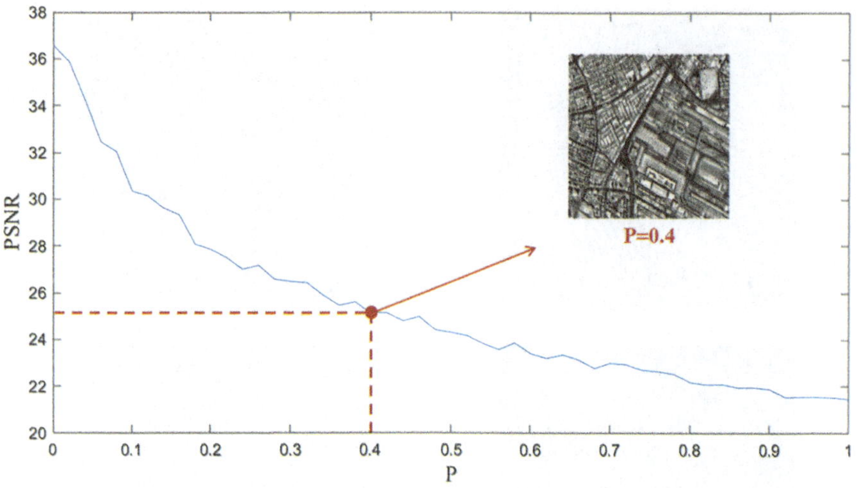

Fig. 12 SSIM curve with intensity variation of P

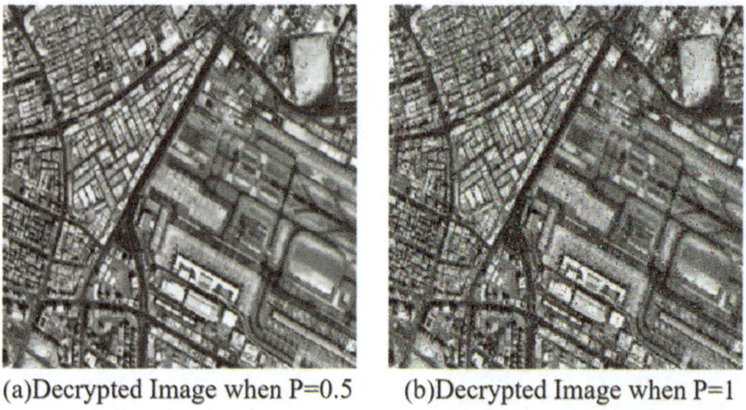

Fig. 13 Decrypted Images for $P = 0.5/1$. (**a**) Decrypted image when $P = 0.5$. (**b**) Decrypted image when $P = 1$

$$I'(x, y) = I(x, y)\left[1 + p \cdot \partial_{0,1}(x, y)\right] \tag{5}$$

Here, $I(x, y)$ and $I'(x, y)$ represent the image functions before and after adding Gaussian noise, respectively. The parameter p signifies the noise intensity, and $\partial_{0,1}(x, y)$ represents a random value with a mean of 0 and a variance of 1.

The system's decryption performance was evaluated across noise levels from 0 to 1 (incremented by 0.01), as reflected in the PSNR curve (Fig. 12). Decryption results at $P = 0.5$ and $P = 1$ (Fig. 13) remain intelligible, indicating strong robustness against additive Gaussian noise.

Attacker G: The system's resistance to known-plaintext and chosen-plaintext attacks is assessed using iterative phase retrieval and a pulse function method.

Figure 14a–d illustrates the attack setup: the adversary possesses the original and encrypted versions of the "Airport" image, as well as the encrypted "Port" image. The goal is to reconstruct "Port."

In the known-plaintext attack, 500 iterations of predictive pixel classification yield a noisy output (Fig. 14e). In the chosen-plaintext attack, the pulse function produces a similarly degraded image (Fig. 14f).

Both outputs are incoherent, demonstrating the algorithm's resilience to these cryptanalytic approaches.

3 A Novel Optical Video Cryptosystem Based on Improved 3D Arnold Transform in Gyrator Domains

3.1 Encryption Scheme

This study proposes a satellite video encryption system in the Gyrator domain. In this system, the 3D Arnold mapping is enhanced through a module movement algorithm, and multiple encryption techniques, including the Gyrator transformation, are integrated to achieve efficient encryption of high-dimensional video data. The specific process is as follows (Fig. 15):

1. **Constructing Satellite Video Data into a Four-Dimensional Image Model for Satellite Video Data**

   To construct a four-dimensional image model, video data must be mapped to a four-dimensional space. Initially, a three-dimensional matrix of size $M \times N \times P$ is initialized to store the image data of each frame in the video, with the storage order (temporal dimension) serving as the fourth dimension. It is important to note that for three-channel color videos, the data for each channel typically requires separate processing and encryption (or decryption) before the processed channels are merged to reconstruct the complete color video (Fig. 16).
2. **The sender attaches the identity image "Lena" as a signature layer to the bottom layer of the four-dimensional satellite video data model, denoted as the $N + 1$ layer.**
3. **The 3D Arnold mapping is applied to the four-dimensional satellite video data model (N layers) using a high-dimensional module shifting algorithm.**

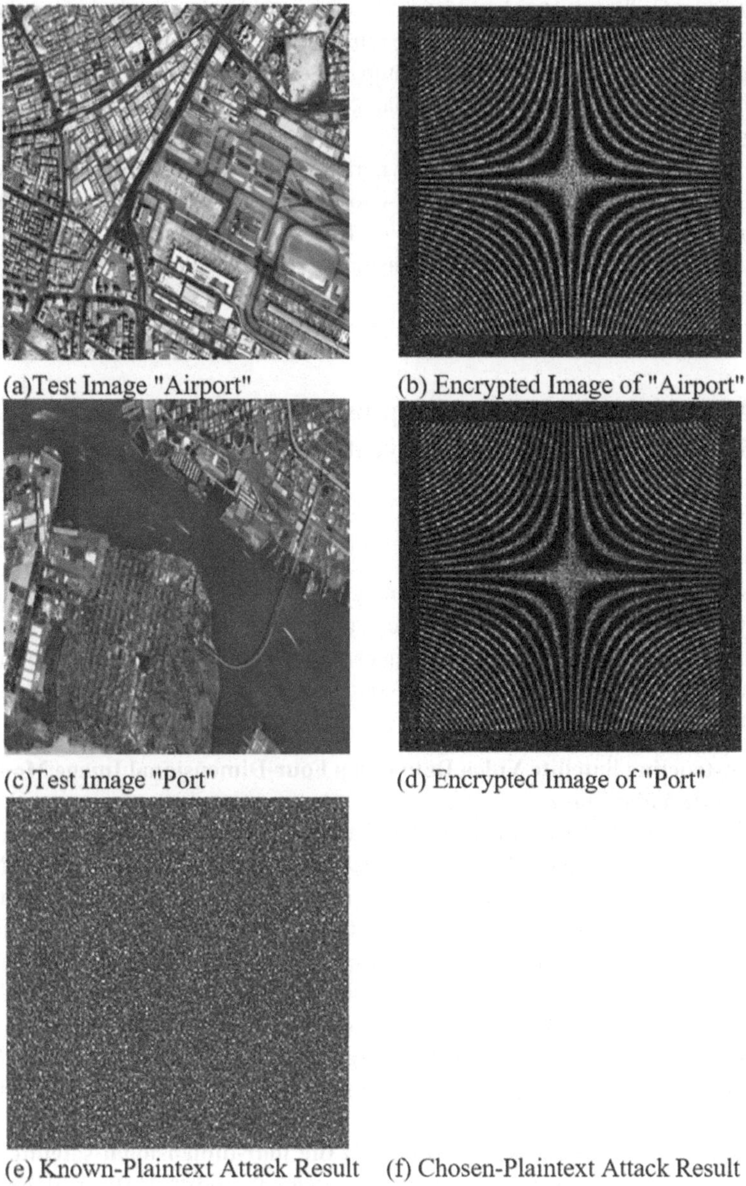

Fig. 14 Known-plaintext/chosen-plaintext attack tests. (**a**) Test image "Airport." (**b**) Encrypted image of "Airport." (**c**) Test image "Port." (**d**) Encrypted image of "Port." (**e**) Known-plaintext attack result. (**f**) Chosen-plaintext attack result

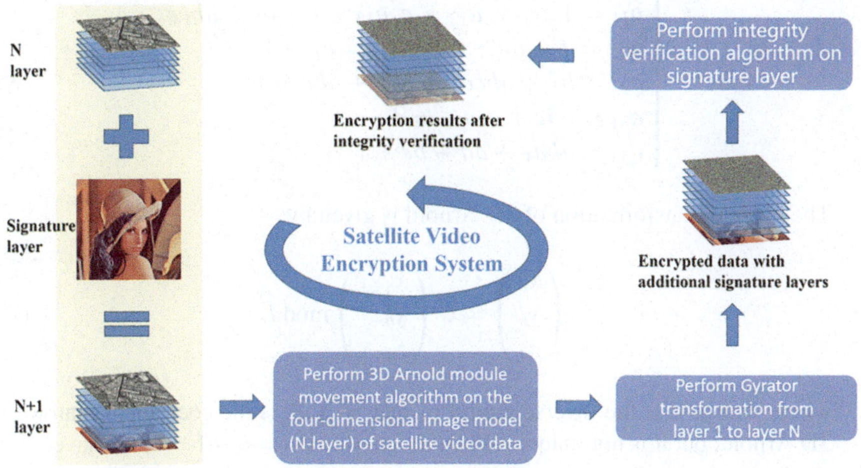

Fig. 15 Satellite video encryption system with data signature and authentication functions

Fig. 16 Schematic diagram of initializing the matrix

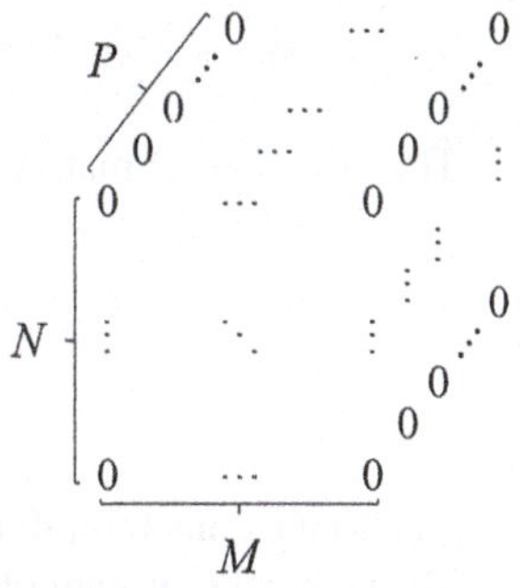

The 3D Arnold algorithm is as follows:

$$\begin{pmatrix} x_{n+1} \\ y_{n+1} \\ z_{n+1} \end{pmatrix} = A \begin{pmatrix} x_n \\ y_n \\ z_n \end{pmatrix} \bmod L \tag{6}$$

$$A = \begin{bmatrix} a_{11} & a_{12} & a_{13} \\ a_{21} & a_{22} & a_{23} \\ a_{31} & a_{32} & a_{33} \end{bmatrix} \tag{7}$$

$$
\begin{cases}
a_{11} = 1 + \text{ace}, a_{12} = c, a_{13} = c + ac + abce \\
a_{21} = f + ae + acef, a_{22} = cf + 1 \\
a_{23} = bf + abcef + acf + abe + a \\
a_{31} = ade + e, a_{32} = d \\
a_{33} = abde + ad + be + 1
\end{cases}
\tag{8}
$$

The inverse transformation of 3D Arnold is given by:

$$
\begin{pmatrix} x_n \\ y_n \\ z_n \end{pmatrix} = C \begin{pmatrix} x_{n+1} \\ y_{n+1} \\ z_{n+1} \end{pmatrix} \bmod L \tag{9}
$$

where matrix C is the inverse matrix of matrix A. A is the coefficient matrix of 3D Arnold, but it is not unique. When a = b = c = d = e = f = 1, we have:

$$
A = \begin{bmatrix} 2\ 1\ 3 \\ 3\ 2\ 5 \\ 2\ 1\ 4 \end{bmatrix} \tag{10}
$$

Then the inverse matrix of the corresponding matrix A is:

$$
C = \begin{bmatrix} 3 & -1 & -1 \\ -2 & -2 & -1 \\ -1 & 0 & 1 \end{bmatrix} \tag{11}
$$

Due to the constraint that the 3D Arnold transform operates exclusively on cubic image blocks, it cannot be directly applied to general video data. Even when video is represented as a four-dimensional tensor, its structure does not inherently conform to a cubic shape. To enhance the adaptability of the encryption method to generic video data, this paper proposes an improved approach to the 3D Arnold transformation via a high-dimensional dynamic module-shifting algorithm. By sliding cubic modules across the four-dimensional data and applying a single instance of the 3D Arnold scrambling within each module, comprehensive coverage and effective encryption of the entire dataset can be achieved.

As a higher-order chaotic scrambling technique, the 3D Arnold transform achieves the desired effect with just a single application to a local region. Therefore, one of the core objectives in the design of the module-shifting algorithm is to minimize the number of module movements required to cover the full image or video. This approach reduces both preprocessing overhead and computational scheduling complexity.

The specific implementation procedure is as follows:

Let the spatial resolution of the video be denoted as X × Y and the temporal dimension (or the number of segmented frames) as N. Based on implementation requirements or resource constraints, the side length a of the cubic module is selected with respect to the dimensions (X, Y, N).

$$a = \min(X, Y, N) \tag{12}$$

To achieve complete coverage of the four-dimensional video tensor, the minimum number of translations required for each three-dimensional cubic block across the spatiotemporal voxel space is given by the product of the coverage counts along each axis. These coverage counts are denoted as n_x, n_y and n_z for the respective axes:

$$\begin{cases} n_x = [X/a] + 1 \\ n_y = [Y/a] + 1 \\ n_z = [N/a] + 1 \end{cases} \tag{13}$$

By applying a single 3D Arnold transform within each corresponding module, all voxels within the four-dimensional image model can be fully scrambled in a coverage-oriented manner. Additionally, movement strategies such as clockwise, serpentine, or region-priority can be designed to ensure non-overlapping and complete scanning, thereby minimizing redundant movements and boundary alignment costs.

Since one of the values in the set $\{n_x, n_y, n_z\}$ must equal 1, meaning that movement along one of the axes (x, y, or z) is unnecessary, we will now demonstrate a movement strategy when no movement is required along the z-axis (i.e., $n_z = 1$).

As shown in Fig. 17, first, the image is moved along the x-axis with a step size of a. When the length of the unencrypted region in the x-direction is less than the step size a, a step size of $a - (a \cdot x_n - X)$ is taken, resulting in an overlap region of length $a \cdot x_n - X$.

Afterwards, the module is moved along the y-axis with a step size of a, repeating this operation as shown in Fig. 18.

Similarly, when the length of the unencrypted region in the y-direction is less than the step size a, a step size of $a - (a \cdot y_n - Y)$ is taken, resulting in an overlap region of length $a \cdot y_n - Y$. The final scrambling result is shown in Fig. 19.

After applying the 3D Arnold mapping to the video data using the module-shifting algorithm, the scrambling frequency and intensity within different regions of the four-dimensional image model exhibit a differentiated distribution. Since the attacker cannot preemptively know the exact scrambling frequency of each region, it becomes impossible to effectively decrypt the entire image using a single inverse scrambling method, thereby significantly enhancing the security of the system.

4. **Perform the integrity verification algorithm on the signature layer (the N + 1th layer).**

   This section proposes a novel algorithm for image integrity authentication, designed to detect potential tampering in video data during transmission. The

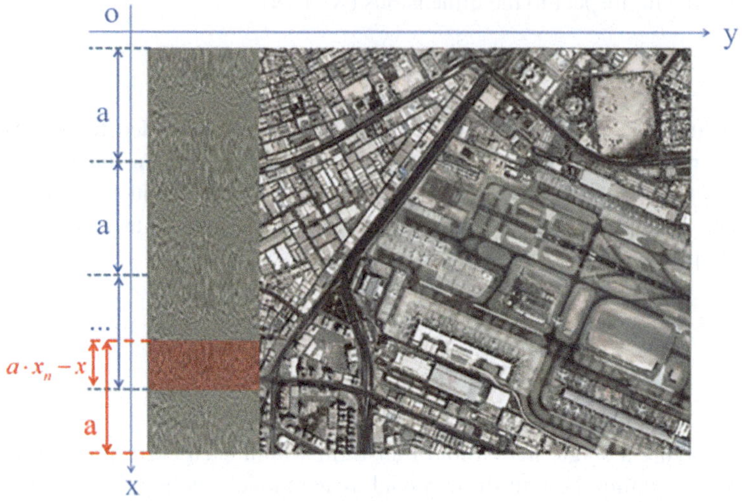

Fig. 17 The movement of the 3D Arnold scrambling module

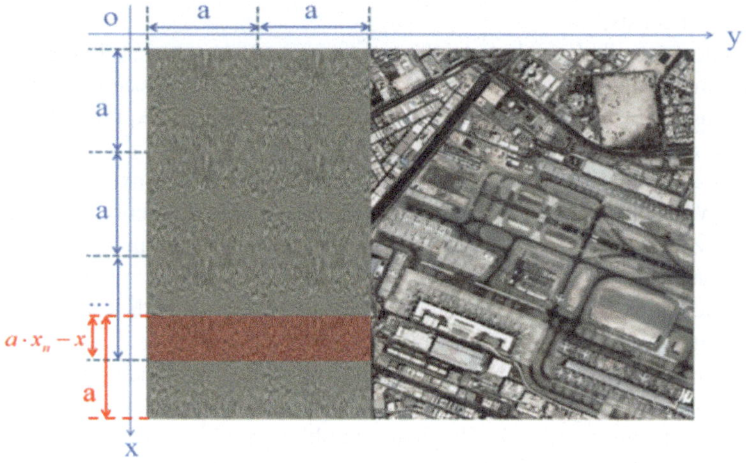

Fig. 18 The movement of the 3D Arnold scrambling module

underlying mathematical model of the algorithm is presented below.

$$\begin{aligned} I[x, y, z]_{n+1} \\ = g\left\{I[x, y, z]_n\right\} + g\left\{[x_1, y_1, z_1]_n\right\} \\ + g\left\{[x_2, y_2, z_2]_n\right\} + g\left\{[x_3, y_3, z_3]_n\right\} \\ + g\left\{[x_4, y_4, z_4]_n\right\} mod L \end{aligned} \tag{14}$$

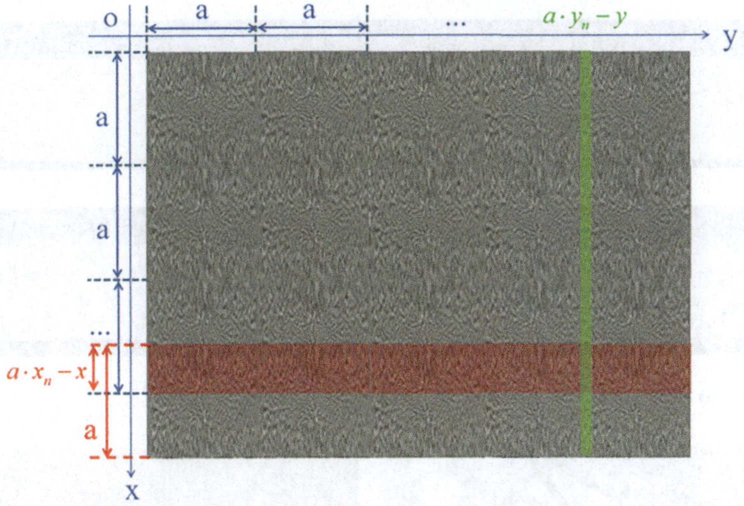

Fig. 19 The final scrambling result of the 3D Arnold scrambling module

Leveraging a signature-based cryptographic mechanism, the integrity verification process is executed independently for each frame, thereby ensuring heightened sensitivity to frame-level tampering. Following algorithm execution, even a single-pixel alteration in the original image data triggers a significant decline in the Structural Similarity Index (SSIM), typically yielding a value below 0.6, which serves as a robust indicator of compromised image integrity.

3.2 *Numerical Simulation*

In the numerical experiments, we selected a satellite video dataset captured by the Jilin-1 satellite over Dubai International Airport, comprising 300 frames at a resolution of 4096 × 3072 pixels. This dataset served as the target for encryption and subsequent evaluation. To construct the four-dimensional image model for authentication purposes, the RGB channels of the standard "Lena" image (4096 × 3072 × 3 pixels) were designated as the respective signature layers. The encryption results are illustrated in Fig. 20a, with the corresponding decrypted frames shown in Fig. 20b. For clarity and consistency in subsequent tests, the 200th frame of the video was selected as a representative example.

The key space is discussed first. In the 3D Arnold scrambling process, six 8-bit parameters—denoted as a, b, c, d, e, and f—are employed.

The dynamic algorithm of the 3D Arnold transformation undergoes a total of $n_x \cdot n_y \cdot n_z$ modular scrambles. There are $(n_x \cdot n_y \cdot n_z)!$ distinct methods for module movement, which amounts to $(14 \times 11 \times 1)!$ different configurations in the demon-

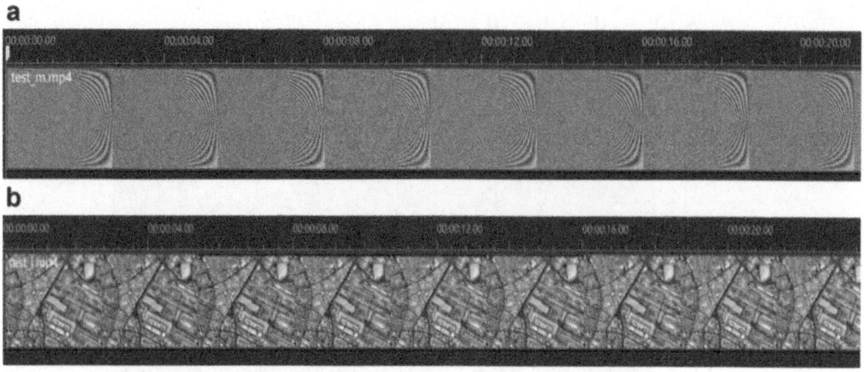

Fig. 20 (**a**) Encrypted video previews. (**b**) Decrypted video previews

Fig. 21 Decryption with Incorrect Key for k. (**a**) Decrypted image. (**b**) Decryption with incorrect key

stration data. The Gyrator transform parameters are represented as double-precision floating-point values, and the simulations are conducted on a 64-bit computing system. Consequently, the key space is $(2^8)^6 \times (14 \times 11 \times 1)! \times 300 \times 2^{64} > 2^{170}$, ensuring the computational security of the encryption system.

Second, the key sensitivity of the system was tested. This section examines how minor key deviations affect decryption fidelity by simulating attackers with partial key knowledge, and we set up the following four attackers:

Attacker A: With the exception of the 3D Arnold scrambling iteration count k, all other keys are considered to be known (Fig. 21).

Attacker B: Aside from the sequence of module movements, all other keys are assumed to be known (Fig. 22).

Table 6 presents SSIM values under two scenario. Even marginal key deviations cause severe degradation in output quality, confirming the system's high key sensitivity and its robustness against key compromise.

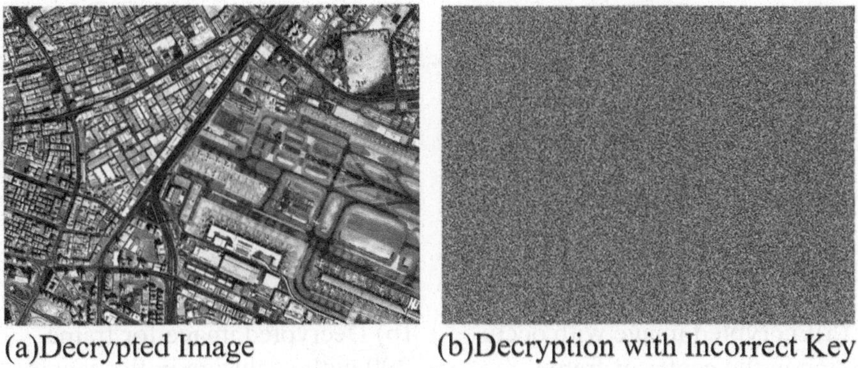

Fig. 22 Decryption with Incorrect Key for the sequence of module movements. (**a**) Decrypted image. (**b**) Decryption with incorrect key

Table 6 The different key values correspond to the SSIM values

k	Sequence of module rearrangement	SSIM-200
3	True	1
2	True	0.0068
3	False	0.0028

Finally, this section investigates the system's resistance to attacks by simulating an attacker The attacker takes various approaches such as shear attack, noise attack, known and chosen plaintext attack:

Attacker C: As shown in Fig. 23a, c, and e, different parts of the four-dimensional image model for video data are successively obscured. Despite this, decrypted outputs in Fig. 23b, d, and f retain recognizable visual content.

The results demonstrate the scheme's capacity to preserve essential information under partial data loss.

Attacker D: Gaussian noise is introduced using the model as shown in Eq. (5).

The SSIM curves under varying noise intensities are depicted in Fig. 24. In the calculations, a step size of 0.01 and a coefficient range from 0 to 1 were employed.

Figure 25 illustrates the video data decryption results when Gaussian noise with an intensity of 1 is added. Despite the presence of noise interference, the video data retains the majority of its visual information, thereby validating the robustness and noise-resistance capabilities of the system when subjected to noise attacks.

Attacker E: The system's resistance to known-plaintext and chosen-plaintext attacks is assessed using iterative phase retrieval and a pulse function method.

Figure 26a–d illustrates the attack setup: the adversary possesses the original and encrypted versions of the "Airport" video, as well as the encrypted "Port" video. The goal is to reconstruct "Port."

In the known-plaintext attack, 500 iterations of predictive pixel classification yield a noisy output (Fig. 26e, g). In the chosen-plaintext attack, the pulse function produces a similarly degraded image (Fig. 26f, h).

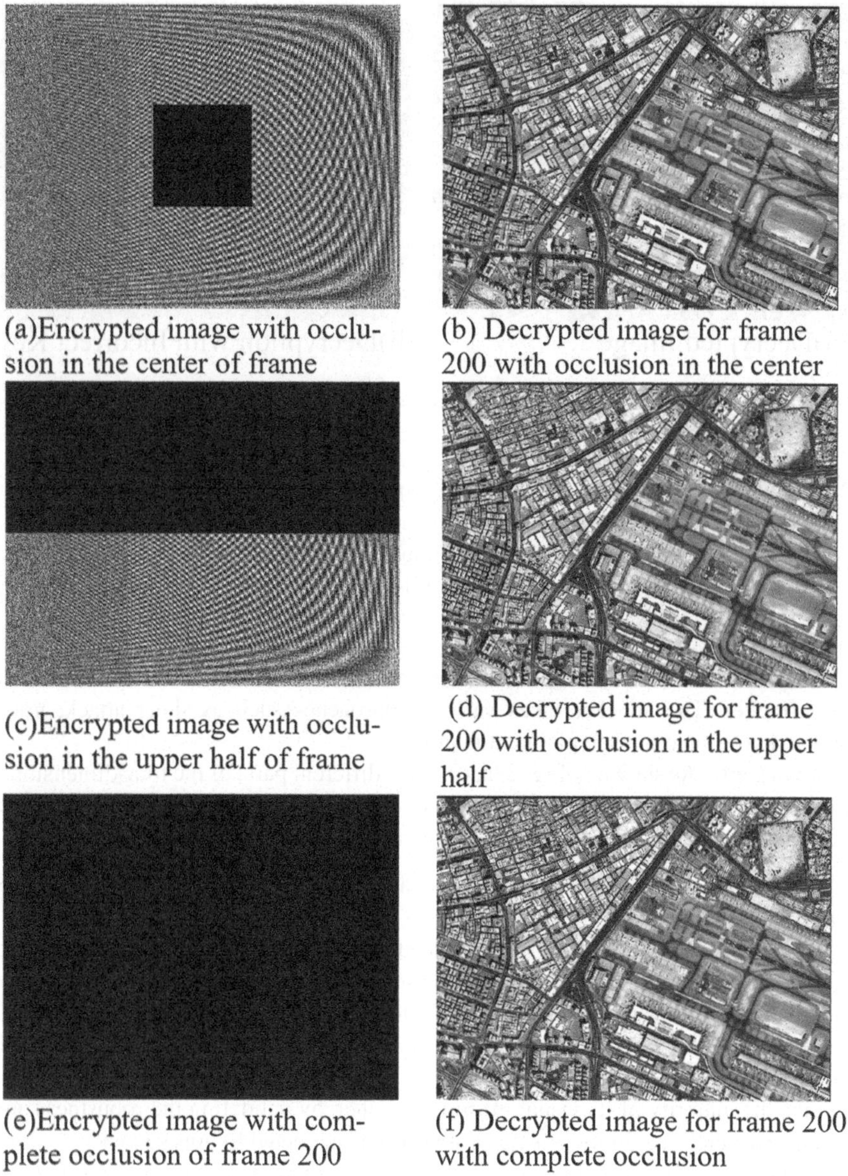

Fig. 23 Clipping attack experiment results. (**a**) Encrypted image with occlusion in the center of frame. (**b**) Decrypted image for frame 200 with occlusion in the center. (**c**) Encrypted image with occlusion in the upper half of frame. (**d**) Decrypted image for frame 200 with occlusion in the upper half. (**e**) Encrypted image with complete occlusion of frame 200. (**f**) Decrypted image for frame 200 with complete occlusion

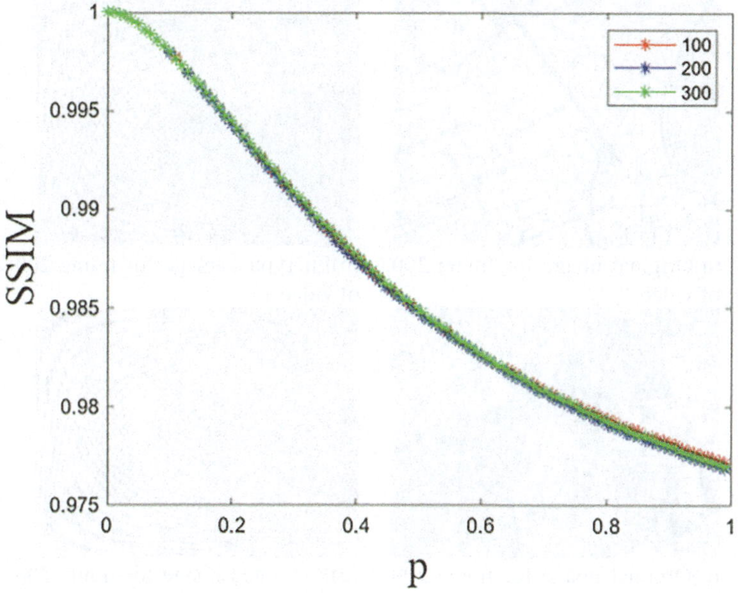

Fig. 24 SSIM curve with respect to noise intensity

(a)Decrypted images for frames 150 with noise intensity of 1

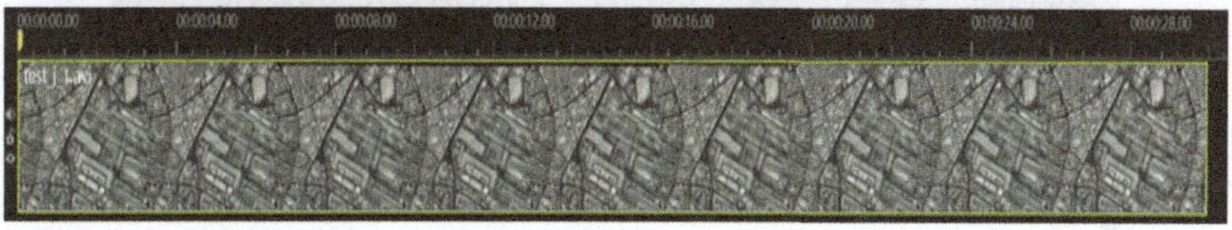

(b)Decrypted video with noise intensity of 1

Fig. 25 Decrypted Images for P = 1. (**a**) Decrypted images for frames 150 with noise intensity of 1. (**b**) Decrypted video with noise intensity of 1

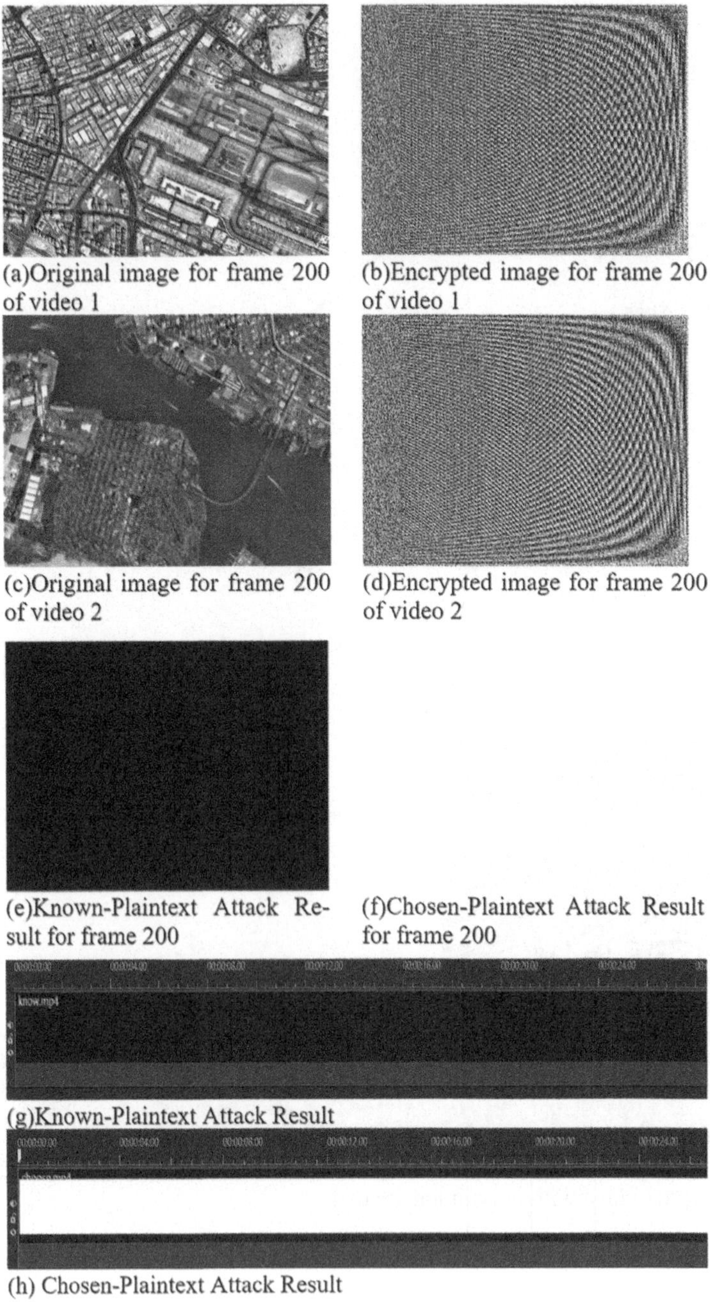

Fig. 26 Known-Plaintext/Chosen-Plaintext attack tests. (**a**) Original image for frame 200 of video 1. (**b**) Encrypted image for frame 200 of video 1. (**c**) Original image for frame 200 of video 2. (**d**) Encrypted image for frame 200 of video 2. (**e**) Known-plaintext attack result for frame 200. (**f**) Chosen-plaintext attack result for frame 200. (**g**) Known-plaintext attack result. (**h**) Chosen-plaintext attack result

Both outputs are incoherent, demonstrating the algorithm's resilience to these cryptanalytic approaches.

4 General Conclusions

This chapter introduces a dynamic algorithm that enhances conventional encryption techniques—including DNA-based encryption and multidimensional Arnold transforms—while incurring only minimal computational overhead. The method substantially enlarges the key space, thereby strengthening overall security. Building on this framework, we developed two optical encryption systems capable of efficiently securing remote sensing data of different dimensions, such as images and videos. Looking ahead, this strategy can be further extended to optimize additional classical encryption algorithms, and holds promise for broad application to diverse forms of remote sensing data—including hyperspectral and infrared modalities—offering a more comprehensive safeguard for information security.

Acknowledgments This work was supported by the National Natural Science Foundation of China (62005320, 62375065, 62575091).

Competing Interests The authors have no conflicts of interest to declare that are relevant to the content of this chapter.

References

1. Jinzhong K, Guizhi W, Guojin HE, Huihui W, Ranyu Y, Wei J, Zhaoming Z. Remote sensing video satellite motion vehicle target rapid detection. J Remote Sens. 2020;24(09):1099–107.
2. Shan T. General framework analysis and new method research of optical image encryption based on DRPE. Tianjin University; 2020.
3. Aburub MR. Single-channel color information security system using LU decomposition in gyrator transform domains. Opt Commun. 2014;323:100–9.
4. Javidi B, Nomura T. Securing information by use of digital holography. Opt Lett. 2000;25(1):2830.
5. Wang B, Sun C, Su W, et al. Shift-tolerance property of an optical double-random phase-encoding encryption system. Appl Opt. 2000;39(26):4788–93.
6. Ye G, Wong KW. An efficient chaotic image encryption algorithm based on a generalized Arnold map. Nonlinear Dyn. 2012;69(4):2079–87.
7. Al Nassan W, Bonny T, Baba A. A new chaos-based cryptosystem for voice encryption. In: 2020 3rd international conference on signal processing and information security (ICSPIS). IEEE; 2020. p. 1–4.
8. Liu ZJ, Guo Q, Xu L, et al. Double image encryption by using iterative random binary encoding in gyrator domains. Opt Express. 2010;18(11):12033–43.
9. Chen Hang DU, Xiaoping LZ, Chengwei Y. Color image encryption based on the affine transform and gyrator transform. Opt Laser Eng. 2013;51(6):768–75.
10. Liu T, Shi Z. A new algorithm for image composite encryption based on the wavelet transform. Computer simulation. 2021;37(9):4.

11. Abuturab MR. Color image security system using double random—structured phase encoding in gyrator transform domain. Appl Opt. 2012;51(15):3006–16.
12. Abdelfatah RI. Audio encryption scheme using self - adaptive bit scrambling and two multi chaotic - based dynamic DNA computations. IEEE Access. 2020;8:69894–907.
13. Kakaei Kate H, Razmara J, Isazadeh A. A novel fast and secure approach for voice encryption based on DNA computing. 3D Res. 2018;9:1–11.
14. Li H, Wang Y, Yan H, Li L, Li Q, Zhao X. Double-image encryption by using chaos-based local pixel scrambling technique and gyrator transform. Opt Lasers Eng. 2013;51:1327–31.
15. Li C, Zhao X, Qi H, Wang Z, Xu Y, Han X, Chen K. Integrated fiber-optic Fabry-Perot vibration/acoustic sensing system based on high-speed phase demodulation. Opt Laser Technol. 2024;169:110131.
16. Amaya D, Tebaldi MC, Torroba R, Bolognini NA. Wavelength multiplexing encryption using joint transform correlator architecture. Appl Opt. 2009;48(11):2099–104.
17. Mogensen P, Eriksen R, Gluckstad J. High capacity optical encryption system using ferro-electric spatial light modulator. J Opt A Pure Appl Opt. 2001;3(1):10–5.
18. Sun G, Song W, Tian M, Tanougast C, Liu Z, Chen H. A novel optical video cryptosystem based on improved 3D Arnold transform in gyrator domains. Opt Laser Technol. 2024;168:109891.
19. Wang XY, Wang MJ. A hyper chaos generated from Lorenz system. Physica A Stat Mech Appl. 2008;387(14):3751–8.
20. Barrera JF, Vargas C, Tebaldi MC, Torroba R. Chosen-plaintext attack on a joint transform correlator encrypting system. Opt Commun. 2010;283:3917–21.

Classification of Typical Remote Sensing Images Based on Deep Learning

Ningbo Guo, Qifang Ma, Te Qi, Zhenduo Ma, Mingyong Jiang, and Hang Chen

Abstract This chapter systematically sorts out the technical evolution and application paradigm of deep learning (DL) in remote sensing image (RSI) classification. First, through the differentiated feature analysis the three types of images: visible light, hyperspectral, and multispectral, the key technologies such as CNN hierarchical feature decoupling, Transformer long-term dependency modeling and lightweight model compression are revealed to improve the classification performance; then, the innovative methods such as high-spectrum feature extraction, multispectral collaborative modeling, etc., expounded from the dimensions of spatial-spectrum joint, dynamic weight allocation, and cross-modal fusion. On the data utilization level, the reduced role of small sample learning, self-supervised pre-training, and transfer learning in labeling dependency is discussed, and the data fitting framework of "weak annotation-strong generalization" is constructed. On the level of technology paradigm the dual transformation trend from single-modal feature extraction to multi-source feature deep integration, from high labeling dependency to efficient data utilization is summarized. Finally, three development directions of lightweight optimization, decision-making interpretability enhancement, and cross-domain migration ability improvement are proposed, which provide theoretical support for the construction of the full-link technology closed-loop "feature recognition-dynamic monitoring-intelligent decision" in precision agriculture, smart city and other vertical fields.

Keywords Deep learning · Remote sensing image · Classification

N. Guo · Q. Ma · T. Qi · Z. Ma · M. Jiang · H. Chen (✉)
School of Space Information, Space Engineering University, Beijing, China

H. Chen, Z. Liu (eds.), *Artificial Intelligence in Digital Image Processing: Theories, Methods, and Applications*, Studies in Computational Intelligence 1246,
https://doi.org/10.1007/978-3-032-12891-1_6

1 Introduction

RSI, serving as the "digital skin of the Earth," provide humanity with a "bird's-eye view" of our planet through non-contact, large-scale data acquisition, as Fig. 1. Based on the electromagnetic waves or radiation information captured across different spectral bands, RSI can be categorized into seven types, including visible-light, hyperspectral, and multispectral imagery [1].

RSI classification holds significant academic and practical value. By analyzing the multidimensional characteristics of land features, it enables intelligent image classification, efficiently and accurately capturing spatial information such as land use and vegetation cover. This provides critical data support and scientific decision-making basis for resource exploration, disaster early warning, and urban-rural planning.

The core principle of RSI classification lies in systematically analyzing and selecting optimal features from the multidimensional information contained within various land cover types in RSI, leveraging computer technology [2]. Specifically, this process involves extracting spectral response characteristics and spatial structural features of land cover elements to construct a feature parameter set. Based on specific classification rules or machine learning algorithms, it assigns pixel values to determine their classification, thereby establishing a spatial correspondence between RSI pixels and actual land cover types. This ultimately achieves automated semantic classification of the RSI.

Traditional methods primarily rely on manual feature engineering, which involves explicitly constructing low-order or high-order statistical features such as spectral, textural, and geometric characteristics based on domain knowledge. These features are then combined with shallow machine learning models like Support Vector Machines (SVM) and Random Forests (RF) to achieve classification decisions. However, such methods suffer from limitations due to the semantic sparsity and task specificity of feature representations. In complex land cover scenarios, they face bottlenecks such as insufficient generalization capabilities and loss of feature interaction information, as Fig. 2a. Recently, DL techniques have effectively surpassed the performance boundaries of traditional methods through end-to-end hierarchical feature self-learning mechanisms, multi-scale convolutional architecture design, and multi-modal data fusion strategies. Demonstrating

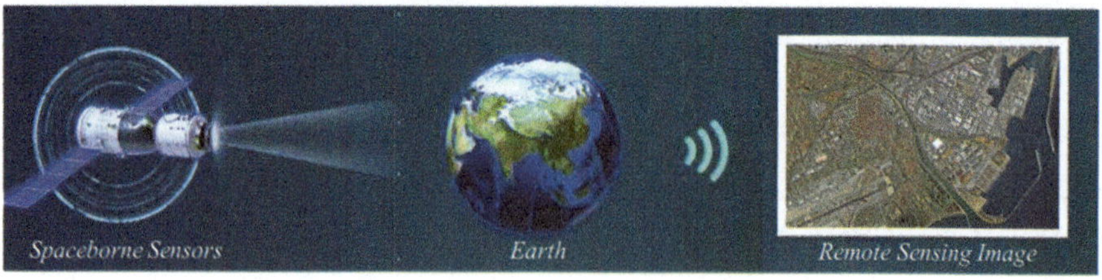

Fig. 1 Schematic diagram of the process for generating RSI

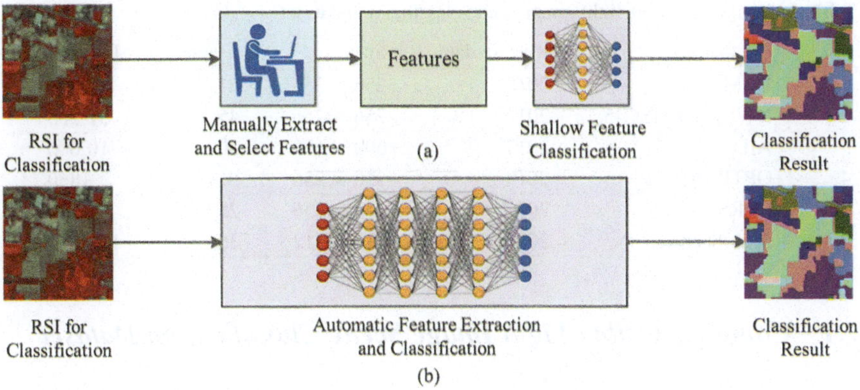

Fig. 2 Schematic diagram of traditional and DL classification methods

significant advantages in classification accuracy, computational efficiency, and scene adaptability, DL has gradually become the mainstream technological paradigm in the field of intelligent classification of RSI, as Fig. 2b.

This paper systematically reviews and deeply analyzes recent research progress and practical applications of DL technologies in typical RSI. By deconstructing the innovative mechanisms and performance boundaries of existing methods, it aims to provide theoretical references and methodological insights for addressing technical challenges in RSI classification under complex scenarios. This approach will ultimately advance RSI classification technology toward higher levels of automation and precision.

2 Visible-Light RSI for Scene Classification

Visible-light RSI captures electromagnetic waves in the visible light band (0.38–0.76 μm) reflected or emitted by surface objects via optical sensors. These are fused across red, green, and blue channels to present natural color spatial distributions. Its advantages include high spatial resolution (sub-meter to meter-scale), low-cost, wide-coverage data acquisition, and intuitive terrain-color representation through true-color synthesis. These features make it a vital data source for macro-scale surface classification, widely applied in large-area land cover monitoring, disaster emergency response, and semantic retrieval of RSI. It provides fundamental support for Earth system observation and spatiotemporal information services.

Table 1 Common visible light image scene classification dataset

No	Data name	Release date	Size	Classes	Images
1	UCM	2010	256 × 256	21	2100
2	NWPU-RESISC45	2017	256 × 256	45	31,500
3	AID	2017	600 × 600	30	10,000
4	OPTIMAL-31	2019	256 × 256	31	1860
5	CLRS	2020	256 × 256	25	15, 000
6	MultiScene	2022	512 × 512	36	100,000

2.1 Common Visible Light Image Scene Classification Dataset

Scene classification, as one of the core tasks in intelligent interpretation of RSI, aims to assign predefined scene category labels to input images through global feature analysis and semantic modeling. This addresses the high-level cognitive question: "What type of land cover or spatial functional semantics does this image represent?" As shown in Table 1, mainstream scene classification benchmark datasets in recent years primarily rely on satellite remote sensing platforms (e.g., Landsat, Sentinel series) and aerial photogrammetry systems for data sources. Their spatial coverage spans global scales or regional-level areas with typical land feature distributions. These datasets typically establish multi-level scene classification systems with 10–30 semantic labels (e.g., urban built-up areas, farmland, forests, water bodies), each category containing tens to hundreds of annotated samples. Image spatial resolutions range from 0.3 to 30 m, Standard image dimensions are predominantly normalized to 256 × 256 pixels or 600 × 600 pixels to meet the input requirements of DL models.

2.2 Commonly Used Methods

Deep learning-based visible light RSI classification technology fundamentally relies on the hierarchical feature self-learning mechanism of deep neural networks to achieve automated representation and abstract modeling of multi-scale, high-order semantic features (encompassing texture structure, geometric morphology, and spatial contextual associations) within image objects. Based on differences in network architecture design and feature extraction paradigms, current mainstream approaches can be categorized into three major technical schools: First, classical methods based on Convolutional Neural Networks (CNNs), which achieve layer-by-layer abstraction of spatial features through local receptive fields and weight sharing mechanisms; Second, emerging methods based on Transformer architectures leverage self-attention mechanisms to capture global contextual dependencies within images; Third, hybrid and enhanced models combine the local feature extraction strengths of CNNs with the global modeling capabilities of Transformers, or incor-

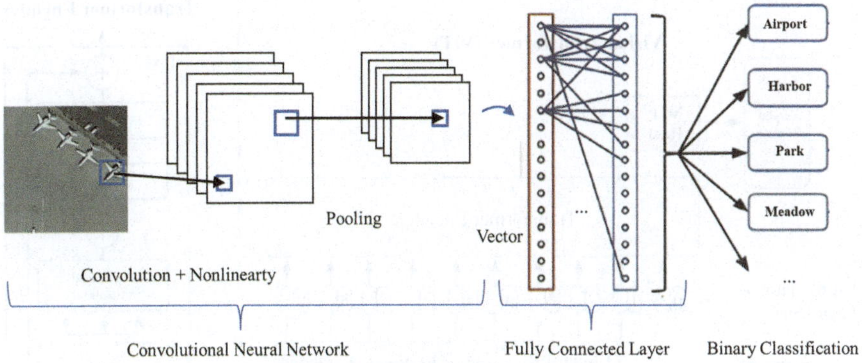

Fig. 3 Schematic diagram of methods based on CNNs

porate improvements such as attention mechanisms and multi-scale feature fusion to construct classification frameworks with greater generalization and robustness

1. Methods Based on CNNs
   CNNs achieve progressive abstraction of features from local to global scales through their hierarchical structure of convolutional and pooling layers, as Fig. 3. They have become the dominant framework for RSI scene classification, with their development path encompassing both adaptation of classical architectures to remote sensing (RS) and innovations tailored to RS characteristics: Among classical approaches, LeNet-5 employs shallow structures for rapid classification of low-resolution imagery. AlexNet enhances feature extraction for medium-resolution imagery through ReLU activation and local response normalization. ResNet achieves breakthrough classification accuracy in complex scenes by constructing ultra-deep networks via residual connections. DenseNet leverages dense connections to reuse features and optimize classification for small-sample, high-resolution imagery. Meanwhile, improvements tailored for RS leverage a global-local dual-stream architecture to fuse cross-scale features, integrate spatial pyramid pooling with feature pyramid networks to enhance multi-scale representation capabilities, and dynamically amplify key feature responses through embedded channel/spatial attention mechanisms [3]. This systematically elevates CNN feature expression, scale adaptability, and interference resistance in RS classification scenarios, providing critical technological support for high-precision automated feature interpretation.
2. Methods Based on Transformer
   The Transformer architecture employs a self-attention mechanism to model long-range dependencies at the pixel or block level in images. This effectively overcomes the limitation of CNNs in capturing global contextual information due to their restricted local receptive fields. It is particularly well-suited for semantic segmentation and scene understanding tasks involving high-resolution, large-scale RSI.

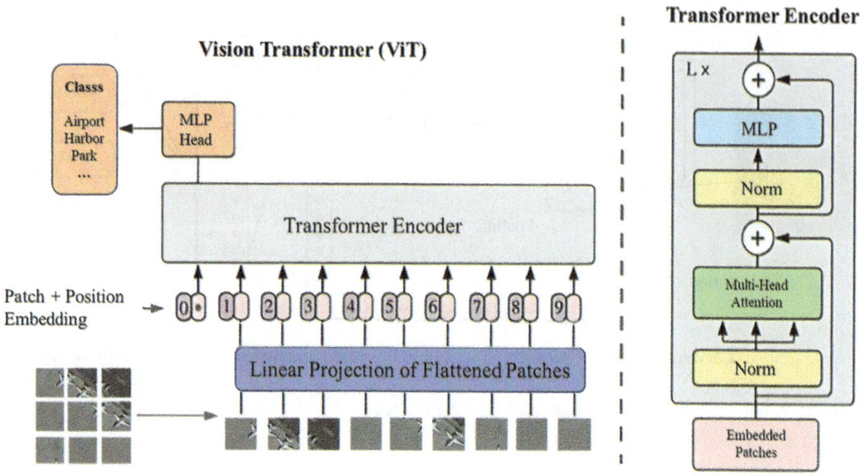

Fig. 4 Schematic diagram of methods based on ViT

Specifically, the ViT innovatively divides images into non-overlapping patches of fixed size. It constructs a sequential input through linear embeddings and positional encoding, and directly establishes global semantic connections between patches using multi-layer self-attention, as Fig. 4. This approach demonstrates strong discrimination capabilities for similar features (e.g., man-made structures and bare land) in ultra-high-resolution RSI classification. Meanwhile, the Swin Transformer achieves a balance between preserving local details and capturing global associations by introducing hierarchical window-based self-attention and cross-window connection mechanisms. This approach maintains global modeling capabilities while reducing computational complexity from quadratic to linear scale relative to image dimensions, establishing it as the mainstream model framework for classifying large-scale complex land cover scenarios.

3. Methods Based on hybrid or improved models

To further enhance the classification accuracy of RSI, current research focuses on multimodal architecture fusion and domain knowledge embedding strategies: First, constructing a CNN-Transformer hybrid model through modular design integrates the local feature extraction advantages of CNNs (e.g., utilizing CNN encoders to capture edge, texture, and other detailed features of objects in imagery) with the global context modeling capabilities of Transformers (e.g., employing Transformer decoders to establish cross-regional semantic associations) This achieves synergistic representation of local-global features, particularly suited for complex classification tasks involving interspersed feature types and high spatial heterogeneity. Second, lightweight classification models are developed using techniques like MobileNet's separable convolutions and EfficientNet's composite scaling strategy [4]. These significantly reduce model parameters and computational complexity while maintaining feature extraction capabilities, meeting the demand for rapid RSI classification on mobile devices or real-time processing systems. Third, we designed a classification framework driven by RS prior knowledge. By incorporating multispectral imagery's near-infrared reflectance characteristics of vegetation (for multispectral data) or visible imagery's texture statistics (e.g., grayscale co-occurrence matrix parameters), spatial topological constraints (e.g., linear continuity of road networks, clustered distribution patterns of building clusters), and other geospatial patterns. This builds domain-adaptive feature enhancement modules or loss functions, strengthening the model's ability to learn intrinsic attributes of ground objects in RS scenes. Consequently, it enhances the spatial consistency and semantic plausibility of classification results.

2.3 Development Trend Analysis

DL-based RSI scene classification will exhibit multidimensional innovation trends in the future. At the model architecture innovation level, the Transformer architecture, with its exceptional long-range dependency modeling capabilities, is increasingly converging with CNNs. giving rise to novel hybrid architectures such as GlaViTU and SparseFormer-LGCNet. These approaches synergize CNNs' strengths in local feature extraction with Transformers' global semantic association capabilities, achieving qualitative and quantitative improvements in feature representation. Simultaneously, lightweight model design has become a critical technical direction to address edge computing and real-time processing demands. By integrating compression techniques like model pruning and parameter quantization with efficient network architectures such as MobileNet V3, model complexity is significantly reduced while maintaining classification accuracy, meeting rapid inference requirements in mobile or resource-constrained scenarios.

3 Hyperspectral Imagery for Land Cover Classification

Hyperspectral imagery (HSI) involves high-density sampling of electromagnetic wave reflection or radiation characteristics from surface objects across a continuous, narrow band range using hyperspectral sensors, generating three-dimensional data cubes with hundreds of continuous spectral channels. Its core advantage lies in ultra-fine spectral resolution (typically <10 nm bandwidth), enabling capture of unique "spectral fingerprint" characteristics of ground objects. It offers high sensitivity to minute spectral variations and the potential to resolve complex land cover compositions. Benefiting from its unique "spatial-spectral integration" data structure and high-precision quantitative analysis capabilities, HSI demonstrates significant advantages in land cover classification tasks. It is particularly suited for

Table 2 Common hyperspectral RSI classification dataset

No	Data name	Resolution	Range	Bands	Classes	Size
1	India Pines	20	0.4–2.5	220	16	145 × 145
2	Salinas	3.7	0.4–2.5	224	16	512 × 217
3	Pavia University	1.3	0.43–0.86	115	9	610 × 340
4	Cuprite	20	0.4–2.5	224	30	250 × 191
5	KSC	18	0.4–2.5	224	13	512 × 614
6	Botswana	2.25	0.4–2.5	242	14	610 × 340

applications requiring sub-class level discrimination accuracy, such as precise crop variety identification (e.g., subspecies classification of rice and japonica rice) and high-precision interpretation of urban land cover types (e.g., spectral separation of artificial turf and natural vegetation). This provides crucial technical support for environmental monitoring, agricultural resource management, and urban planning.

3.1 Common Hyperspectral RSI Classification Dataset

The core of HSI classification technology lies in utilizing the unique spectral fingerprint characteristics of land objects for precise identification. In recent years, mainstream research has primarily been conducted using standardized datasets listed in Table 2. These datasets offer significant advantages: First, they possess high spectral dimensional information density. Continuous narrow-band sampling fully preserves the details of land object spectral curves, providing rich features for distinguishing similar objects (such as different crop varieties). Second, they offer strong controllability over data quality. Most datasets undergo rigorous preprocessing, including radiometric calibration and atmospheric correction, effectively eliminating sensor noise and atmospheric scattering interference to ensure the accuracy of spectral reflectance measurements. However, current datasets still face three limitations: First, constrained sample size. Due to the high cost of hyperspectral data acquisition and the complexity of annotation, existing public datasets often lack the volume required to meet the massive training data demands of DL models, resulting in insufficient generalization capabilities during cross-scenario transfer. Second, category distribution imbalance. The sample counts for different land cover types in datasets often exhibit a long-tail distribution, making models prone to underfitting minority classes during training. Third, annotation reliability challenges persist. Hyperspectral feature interpretation requires multidisciplinary expertise spanning spectroscopy and geosciences. Manual annotation is not only time-consuming and labor-intensive but also struggles to precisely delineate features with ambiguous spectral signatures (e.g., mixed pixels, early-stage diseased vegetation). Annotation errors may propagate through supervised learning into the model, compromising classification accuracy.

3.2 Commonly Used Methods

DL-based HSI classification methods fundamentally rely on constructing neural network models with hierarchical feature extraction capabilities [5]. These models automatically extract spectral dimensional features, spatial contextual features, or combined representations from image data to achieve high-precision intelligent interpretation of land cover types. Based on differences in feature extraction mechanisms and model architectures, current mainstream approaches can be systematically categorized into the following four technical paradigms:

1. Methods Based on Spectral Feature

   This approach focuses on the unique "spectral fingerprint" characteristics of HSI. Classification methods based purely on spectral dimensions achieve surface cover type identification by deeply mining fine-grained features from pixel-level spectral curves. They are particularly suitable for scenarios with sparse spatial texture information but significant spectral distinguishability (e.g., mineral mapping, water body composition identification) [6]. This approach treats the continuous spectral reflectance vector of a pixel (e.g., a 1×200 dimensional feature set comprising 200 bands) as one-dimensional sequence data. Two typical DL models are employed to capture nonlinear dependencies between bands: First, one-dimensional CNNs extract local spectral morphological features (e.g., abrupt red edges in vegetation, depth of water absorption bands) through sliding convolutional kernels, leveraging weight sharing mechanisms to construct interpretable spectral decision boundaries. Second, recurrent neural networks and their variants (RNN/LSTM/GRU) model long-range correlations between bands through hidden state recursive propagation. LSTM's gating mechanism effectively suppresses noise while retaining key spectral features, while GRU enhances training efficiency through simplified architecture. Both significantly reduce classification errors in "same-spectrum different objects" scenarios (e.g., soils with varying moisture content) by capturing subtle spectral correlation variations. Although this method achieves precise sensing of surface material composition, it's purely spectral analysis nature leads to insufficient utilization of spatial contextual information. For mixed object classification, it requires integrating spatial feature enhancement techniques to construct a multimodal analysis framework.
2. Methods Based on Spatial-Spectral Fusion

   Integrating spectral information from HSI with spatial context features from pixel neighborhoods to overcome the lack of spatial structural information in pure spectral classification has become the mainstream paradigm in intelligent HSI interpretation [7]. Its core lies in constructing a "spatial-spectral joint representation" learning framework. This is achieved by reconstructing data into a three-dimensional cube (e.g., $N \times N \times$ number of bands) to enable the synergistic extraction of spectral and spatial features [8], as Fig. 5. Typical approaches include: 3D-CNNs performing direct 3D convolutions on cubes, which simultaneously model spatial neighborhood associations and spectral band

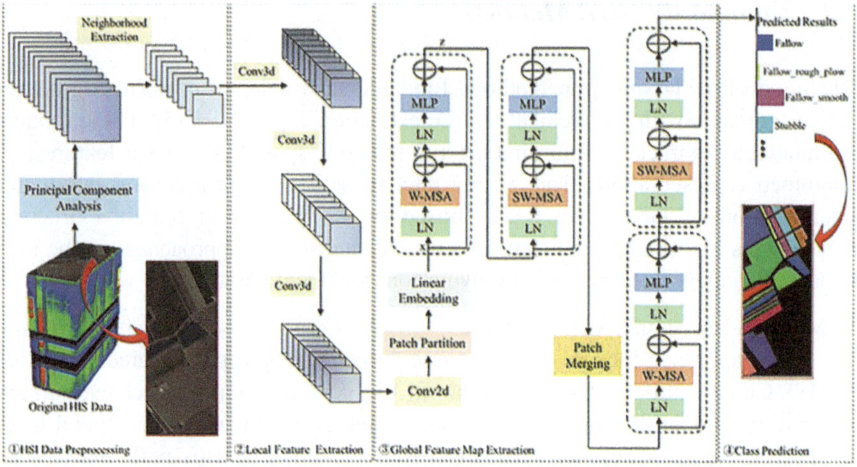

Fig. 5 Schematic diagram of methods based on spatial-spectral fusion

dependencies, though lightweight designs are required to mitigate parameter explosion issues; and spectral-spatial separation networks (e.g., SSRN) with dual-branch architectures, which extract spectral features via 1D-CNNs and capture spatial features via 2D-CNNs before fusion, reducing training complexity while maintaining accuracy. These approaches are evolving toward automated architecture search and multi-source data collaborative analysis, providing technical support for refined classification of HSI.

3. Methods Based on attention mechanisms

   By introducing attention mechanisms (AM) to achieve adaptive enhancement of key features and dynamic suppression of redundant information, this approach has become a core paradigm for improving classification performance in complex HIS [9]. Its essence lies in constructing deep learning architectures with feature selection capabilities: Spectral Attention (SA) Networks automatically amplify critical spectral features—such as vegetation's red-edge band and water's near-infrared absorption band—by embedding channel attention (CA) modules (e.g., SE module variants) within 1D-CNNs [10], as Fig. 6. SA networks integrate SA modules (e.g., self-attention encoders) within 2D-CNNs to focus on discriminative spatial regions like abrupt building edge gradients and anisotropic road textures; Hybrid attention networks (e.g., CBAM-enhanced architectures) achieve cross-dimensional feature interaction and joint optimization by concurrently constructing spectral-spatial dual-branch attention mechanisms. This enhances classification accuracy in heterogeneous scenarios like farmland-forest boundaries. At its core, this approach simulates human visual selective attention to establish an intelligent perception and collaborative decision-making framework for spectral bands and spatial regions, propelling HSI classification toward greater refinement and interpretability.

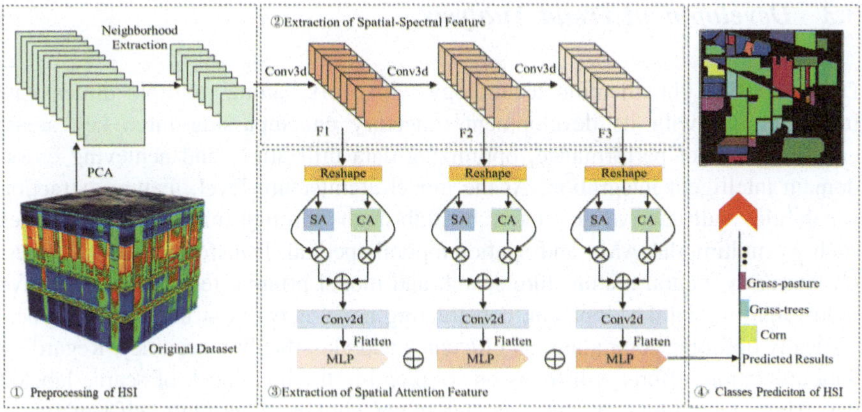

Fig. 6 Schematic diagram of methods based on AM

4. Methods for Small Samples/Sparse Labels

To address practical challenges in HSI data annotation such as high costs and limited annotated sample sizes-researchers have developed model training strategies that minimize reliance on labeled data. These approaches leverage advanced paradigms including transfer learning, semi-supervised/unsupervised learning, and meta-learning [11]. Their core principle is to extract latent information from either sparse labeled samples or vast unlabeled datasets, thereby enhancing model generalization and adaptability for complex land cover classification tasks. The specific technical approach encompasses three representative methods: First, the transfer learning framework mitigates overfitting caused by insufficient hyperspectral annotation samples by transferring parameters from 2D CNNs pre-trained on large-scale natural image datasets (e.g., ImageNet) to hyperspectral classification tasks, requiring only fine-tuning of top-level classifiers or higher-order feature extraction layers. Second, semi-supervised learning strategies (e.g., label propagation algorithms and semi-supervised CNN architectures) jointly utilize the Space-Spectral joint distribution features of sparse labeled samples and vast unlabeled samples. By establishing intrinsic data structure constraints (e.g., manifold or clustering assumptions), these approaches enable models to extract latent feature category information from unlabeled data. This reduces labeling requirements while maintaining stable classification performance in crop classification tasks. Third, model-agnostic meta-learning based few-shot learning methods empower models to rapidly adapt to new land cover categories through multi-task training mechanisms. This approach is particularly suited for few-shot classification of novel hyperspectral land cover categories, such as specific crop varieties.

3.3 Development Trend Analysis

DL-based HSI classification technology will evolve along multiple dimensions in the future, with its development trajectory encompassing three key areas: enhancing model performance, optimizing data utilization, and achieving cross-domain intelligent integration. At the model architecture level, feature extraction capabilities will achieve qualitative breakthroughs through innovative approaches such as multimodal AMs and spatiotemporal-spectral Transformer architectures. Concurrently, neural architecture search and model pruning techniques will drive lightweight model development, constructing novel network structures that synergistically optimize computational efficiency and classification accuracy. Regarding data utilization, efforts will focus on overcoming the bottleneck of scarce labeled data. Hybrid learning strategies—including few-shot meta-learning, self-supervised pre-training, and semi-supervised graph neural networks—will fully exploit the Space-Spectral joint distribution patterns within unlabeled data. Enhanced capabilities for integrating multi-source heterogeneous data will establish a unified "Space-Spectral" classification framework that fuses optical imagery, SAR, LiDAR, and spatiotemporal sequence information.

4 Multispectral RSI for Land Cover Classification

Multispectral RSI, as a carrier of surface electromagnetic radiation information acquired through multispectral sensor systems, essentially employ discrete spectral detection technology to simultaneously record the reflection or radiation response characteristics of terrestrial objects to electromagnetic waves at different wavelengths across key spectral ranges—typically spanning visible light to shortwave infrared (encompassing 4–15 discrete bands). Through collaborative analysis and information fusion processing of multi-channel spectral data, they achieve a combined characterization of the spatial distribution patterns and spectral diagnostic features of land cover types.

4.1 Common Multispectral RSI Classification Dataset

Multispectral RSI classification refers to the process of using multispectral RSI as the data foundation. Through the hierarchical feature learning mechanism of DL models, it automatically extracts spectral response characteristics, spatial structural features, and temporal dynamic evolution patterns of land cover objects within the imagery. This enables precise mapping from image pixels or object-level spatial units to predefined land cover categories (such as vegetation, water bodies, buildings, farmland, etc.).

Table 3 Common multispectral RSI classification dataset

No	Data name	Resolution	Classes	Images
1	ISPRS 2D Semantic Labeling	5–9 cm	6	71
2	DeepGlobe Challenge	50 cm	7/2/2	803+
3	Inria Aerial Image Labeling	30 cm	2	360
4	Slovenia Land Cover	10	10	~300
5	Zurich Summer	62 cm	8	20
6	LandCover.ai	25–50 cm	4	41
7	iSAID	0.5–3 m	15	2806
8	DSTL Satellite Imagery	0.3–7.5 m	10	57

Currently popular multispectral RSI classification datasets (exemplary datasets shown in Table 3) exhibit three significant advantages: First, their multiband spectral configurations comprehensively capture the reflectance or radiance characteristics of objects across different spectral regions, providing rich diagnostic features for object classification. Second, their high spatial resolution (some datasets reaching 0.5–2 m) enables clear visualization of feature boundaries and microstructures, ensuring data quality for fine-grained classification. Third, their extensive scene coverage and diverse feature types (including urban built-up areas, agricultural fields, forest vegetation, water bodies, and other typical land cover types) effectively enhance classification models' scene adaptability and generalization performance.

However, two challenges persist in this field: First, the high cost of data acquisition and annotation leads to a scarcity of large-scale, high-quality annotated datasets. Second, significant class imbalance is prevalent, with some datasets lacking sufficient samples for low-frequency features (e.g., sparse vegetation, small water bodies), which can cause class bias during model training and consequently impact classification accuracy and model robustness. To address these issues, researchers are pursuing breakthroughs through technical approaches such as semi-supervised learning, few-shot learning, and data augmentation. These efforts aim to advance multispectral remote sensing image classification technology toward higher accuracy and stronger generalization capabilities.

4.2 *Commonly Used Methods*

DL-based multispectral RSI classification methods center on constructing a multidimensional feature joint representation framework. Through the hierarchical nonlinear mapping mechanism of DL models, they automatically extract spectral response characteristics, spatial structural features, and temporal dynamic evolution characteristics of ground objects within images. This enables end-to-end mapping from raw image data to semantic object categories. Based on variations in feature extraction methods and model architectures, this methodology can be categorized into the following four typical technical approaches:

1. Methods Based on CNNs

   This class of methods achieves adaptive feature extraction from images through the hierarchical structure of CNNs. By leveraging the local receptive field characteristics and weight sharing mechanism of convolutional layers, they capture the local correlation features of ground objects across spectral and spatial dimensions [12]. Simultaneously, pooling layers enhance the model's robustness to geometric transformations through down-sampling. Due to their strong generalization capabilities and computational efficiency, these networks have become the mainstream technical framework. The Fully Convolutional Network series achieves end-to-end mapping from images to classification maps by removing fully connected layers and introducing deconvolution up-sampling modules. This preserves spatial resolution consistency between outputs and input images, making it suitable for rapid classification of large-scale land cover. Meanwhile, encoder-decoder architectures represented by U-Net and its variants (e.g., U-Net++) employ skip connections to fuse features across layers, effectively mitigating spatial detail loss in traditional down-sampling and up-sampling processes [13]. They demonstrate particular superiority in fine-grained classification tasks involving complex land cover [14], as Fig. 7. These two model categories form a complementary technical framework through distinct feature fusion strategies and structural optimizations, collectively establishing the foundational methodology for multispectral RSI classification.

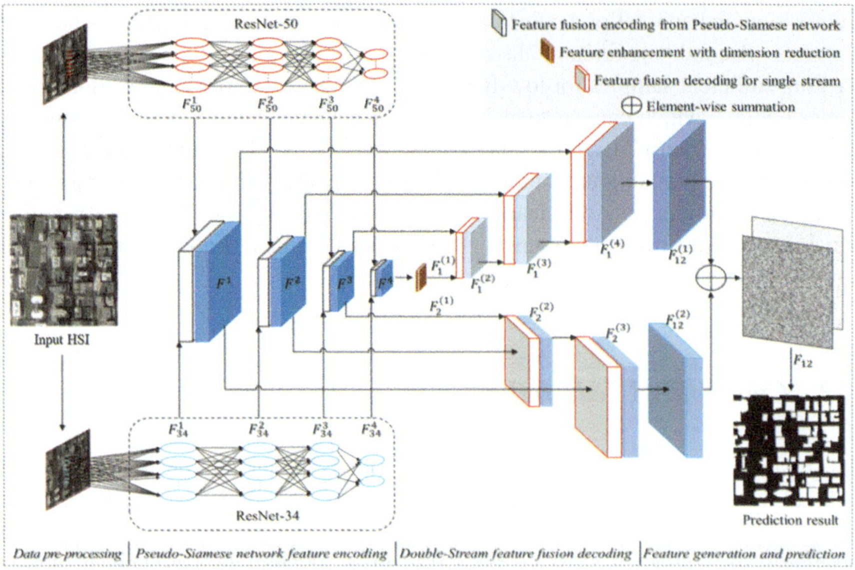

Fig. 7 Schematic diagram of methods based on CNNs

2. Methods Based on Transformer

   DL frameworks based on self-attention mechanisms effectively overcome the feature extraction limitations of traditional CNNs caused by restricted local receptive fields. By constructing long-range dependency modeling capabilities, they achieve joint representation of the global spatial topology and spectral cross-channel correlations in multispectral RSI. Typical approaches like Vision Transformer achieve end-to-end mapping from pixel-level local features to image-level semantic features by segmenting images into non-overlapping local block sequences and computing global attention between blocks. Its variants further reduce spectral dimension computational complexity through band-grouped attention strategies. Meanwhile, the hierarchical AM in Swin Transformer employs a hybrid architecture combining intra-window self-attention with cross-window information exchange. This approach preserves global feature perception while enhancing the ability to capture minute object details in high-resolution imagery (e.g., WorldView-3, GF-2), demonstrating superior scene adaptability particularly in classifying artificial features like buildings and roads. Through multi-scale spatial modeling and dynamic spectral correlation mechanisms, these two approaches establish a new technical paradigm for handling classification scenarios with long-range dependencies, such as complex urban landscapes and heterogeneous land cover.

3. Methods Based on hybrid models

   By integrating the local feature extraction advantages of CNNs with the global context modeling capabilities of Transformer architectures, hybrid models with multi-scale feature perception capabilities can be constructed. This effectively overcomes the performance bottlenecks encountered by single-architecture networks in multispectral RSI classification tasks. Specifically, two typical implementation paths exist for such hybrid methods: First, employing a "CNN-Transformer" cascaded architecture: the CNN branch first extracts local spectral-spatial joint features through convolution-pooling operations, forming multi-level feature representations with spatial invariance. These feature maps are then fed into a Transformer encoder, which utilizes self-attention mechanisms to capture long-range dependencies between different spatial locations within the feature maps, achieving a progressive mapping from local features to global semantics. Second, embedding attention-enhancing modules (such as Squeeze-and-Excitation attention or Convolutional Block Attention Module) within the CNN backbone. By dynamically adjusting feature channel weights or spatial position weights, these modules strengthen the model's focus on critical spectral bands (e.g., vegetation index-related bands) and important spatial regions (e.g., object boundaries), thereby enhancing the discriminative power of feature representations and the task-specificity of classification.

4. Methods Based on Small-sample/Semi-supervised Learning

   To address the issue of limited model generalization caused by sparse labeled data in multispectral RSI classification tasks, current research focuses on enhancing model performance through efficient data utilization strategies. These approaches primarily fall into two technical categories: First, few-shot learning methods centered on meta-learning frameworks [15]. Representative models like Model-Agnostic Meta-Learning employ a two-stage training mechanism: first, cross-task pre-training on a base dataset to learn universal features for parameter initialization; then, rapid adaptation on new category samples in the target task through minimal gradient updates. Second, semi-supervised learning methods overcome data scarcity by jointly utilizing labeled and unlabeled data. Representative frameworks like MoCo (Momentum Contrast) and SimCLR (Simple Framework for Contrastive Learning of Visual Representations), based on contrastive learning, design data augmentation strategies and negative sample mining mechanisms. to learn discriminative spectral-spatial joint feature representations on unlabeled data. These models then combine a small number of labeled samples with fine-tuning or consistency regularization to optimize classification decision boundaries.

4.3 Development Trend Analysis

DL-based multispectral RSI classification technology is undergoing a critical phase of transition from traditional methods toward intelligent and practical applications. Its core developmental trends manifest as multidimensional technological innovations: At the model architecture level, research focus is shifting from the local feature extraction paradigm dominated by CNNs toward hybrid designs combining Transformers and CNNs. By incorporating self-attention mechanisms, these architectures enhance the modeling of global contextual relationships within imagery, enabling cross-scale joint representation of spectral-spatial features. Regarding data processing strategies, weakly supervised and self-supervised learning paradigms are gaining prominence. By designing semi-automated annotation methods based on contrastive learning, pseudo-label generation, or meta-learning, these approaches effectively reduce reliance on large-scale manually annotated datasets, significantly improving model generalization in data-scarce scenarios. In multimodal fusion, technological advancements focus on collaborative modeling of multi-source heterogeneous data such as LiDAR point clouds and SAR. By constructing a three-dimensional Space-Spectral feature space, these approaches overcome the limitations of single-spectral data in classifying complex land cover types.

5 General Conclusions

This study centers on the DL technology framework, analyzing its methodological evolution and data adaptation mechanisms across three typical RSI classification tasks: visible light, hyperspectral, and multispectral. Methodologically, for visible light image classification: CNN-based models achieve efficient decoupling of local spatial features through hierarchical convolutional kernels, enabling pixel-level precision classification. Transformer architectures leverage self-attention mechanisms to establish long-range spatial dependencies, effectively addressing contextual semantic relationships in complex scenes. Lightweight models for edge computing employ compression strategies like parameter quantization and channel pruning to meet real-time processing demands while maintaining classification accuracy. For HSI's band characteristics, 1D-CNN performs one-dimensional convolution along the spectral dimension to extract "spectral fingerprint" features of land objects, while 2D/3D-CNN achieves multidimensional feature fusion through Space-Spectral convolution. AMs enhance key feature expression through dynamic weight allocation, and few-shot learning with semi-supervised learning effectively alleviates the bottleneck of high hyperspectral data annotation costs. For multi-spectral imagery, advanced CNN models like FCN and U-Net++ employ encoder-decoder architectures to gradient-based extraction of spectral-spatial joint features. The integration of Transformers and LSTMs enhances global correlation capture and dynamic change analysis capabilities, respectively. At the data adaptation level, strategies like few-shot learning, semi-supervised/self-supervised learning, and transfer learning significantly reduce model dependence on labeled data while demonstrating strong adaptability to high-resolution open-source datasets. Overall, DL is driving two paradigm shifts in RSI classification: First, evolving from single-modality feature extraction to deep fusion of multi-source features; Second, a shift from high annotation dependency to efficient data utilization. Moving forward, optimizing lightweight model architectures, enhancing decision interpretability, and improving cross-domain transfer capabilities will become key breakthroughs for deepening technological penetration into vertical fields like precision agriculture and smart cities. This will ultimately form a complete technological closed loop, spanning from precise object recognition to intelligent decision support.

Acknowledgments This work was supported by the Internal Parenting Program of Space Engineering University (No. 145AXL250004000X) and the National Natural Science Foundation of China (62005320).

Competing Interests The authors have no conflicts of interest to declare that are relevant to the content of this chapter.

References

1. Kavzoglu T, Tonbul H. An experimental comparison of multi-resolution segmentation, SLIC and K-means clustering for object-based classification of VHR imagery. Int J Remote Sens. 2018;39:6020–36.
2. Ma L, Liu Y, Zhang X, Ye Y, Yin G, Johnson BA. Deep learning in remote sensing applications: a meta-analysis and review. ISPRS J Photogramm Remote Sens. 2019;152:166–77.
3. Sun H, Li S, Zheng X, Lu X. Remote sensing scene classification by gated bidirectional network. IEEE Trans Geosci Remote Sens. 2019;58:82–96.
4. Zhang B, Zhang Y, Wang S. A lightweight and discriminative model for remote sensing scene classification with multidilation pooling module. IEEE J Sel Top Appl Earth Obs Remote Sens. 2019;12:2636–53.

5. Yang X, Cao W, Lu Y, Zhou Y. Hyperspectral image transformer classification networks. IEEE Trans Geosci Remote Sens. 2022;60:5528715.
6. Li X, Li Z, Qiu H, Hou G, Fan P. An overview of hyperspectral image feature extraction, classification methods and the methods based on small samples. Appl Spectrosc Rev. 2023;58:367–400.
7. Zhang M, Li W, Zhao X, Liu H, Tao R, Du Q. Morphological transformation and spatial-logical aggregation for tree species classification using hyperspectral imagery. IEEE Trans Geosci Remote Sens. 2023;61:5501212.
8. Guo N, Yang L, Yue C, Zhang H, Jiang M, Li Y. An end-to-end automatic classification algorithm for hyperspectral images via 3D convolution and swin-transformer. In: 2024 39th youth academic annual conference of Chinese Association of Automation (YAC). IEEE.
9. Li S, Song W, Fang L, Chen Y, Ghamisi P, Benediktsson JA. Deep learning for hyperspectral image classification: an overview. IEEE Trans Geosci Remote Sens. 2019;57:6690–709.
10. Guo N, Jiang M, Su Z, Wang J, Li Y. An intelligent classification method for remote sensing images based on hyperspectral imaging technology. In: Seventh global intelligent industry conference (GIIC 2024), vol. 13278. SPIE. p. 99–104.
11. Yao D, Zhi-li Z, Xiao-feng Z, Wei C, Fang H, Yao-ming C, Cai W-W. Deep hybrid: multi-graph neural network collaboration for hyperspectral image classification. Def Technol. 2023;23:164–76.
12. Zhang B, Wu Y, Zhao B, Chanussot J, Hong D, Yao J, Gao L. Progress and challenges in intelligent remote sensing satellite systems. IEEE J Sel Top Appl Earth Observ Remote Sens. 2022;15:1814–22.
13. Badrinarayanan V, Kendall A, Cipolla R. Segnet: a deep convolutional encoder-decoder architecture for image segmentation. IEEE Trans Pattern Anal Mach Intell. 2017;39:2481–95.
14. Guo N, Jiang M, Hu X, Su Z, Zhang W, Li R, Luo J. NPSFF-Net: enhanced building segmentation in remote sensing images via novel pseudo-Siamese feature fusion. Remote Sens. 16(17):3266.
15. Zhao Q, Liu J, Li Y, Zhang H. Semantic segmentation with attention mechanism for remote sensing images. IEEE Trans Geosci Remote Sens. 2021;60:1–13.

Deep Learning Applications in Optical Cryptography

Harsh Vardhan, Aswathi K Sivarajan, Sakshi Choudhary, Salla Gangi Reddy, and Ravi Kumar

Abstract In the present era, Artificial Intelligence (AI) plays a crucial role by offering effective solution across various domains. AI is a broad area that includes both Machine Learning (ML) and Deep Learning (DL) as its key branches. In this chapter, we discussed applications of DL in the field of optical cryptography. It provides a cost-effective and powerful solution to various cryptographic challenges based on supervised and unsupervised learning approach. The chapter begins by introducing optical cryptosystems and their connection to DL algorithms. Some of the characteristics of DL along with its different architectures are discussed for better understanding of readers. Further, the integration of various DL models with optical image encryption techniques is discussed, followed by an overview of practical applications. The chapter also discusses future directions, existing challenges, and potential solutions in this field. Finally, the chapter concludes with key insights and contributions to ongoing research in DL-based optical encryption.

Keywords Optical encryption · Artificial intelligence · Deep learning

1 Introduction

With advancements in information technology, the transmission of data i.e., (images, videos, audios etc.) are now widely used across numerous applications, including video conferencing, social media, medical imaging, banking, and various government and corporate documents. These images and videos can contain confidential data. Unauthorized access to sensitive information can pose serious risks to national or personal security. Therefore, it is crucial to protect the data from unauthorized access. The literature shows that various data security algorithms have been utilized since World War II, including well-known methods like the

H. Vardhan · A. K. Sivarajan · S. Choudhary · S. G. Reddy · R. Kumar (✉)
Department of Physics, SRM University-AP, Amaravati, Andhra Pradesh, India

H. Chen, Z. Liu (eds.), *Artificial Intelligence in Digital Image Processing: Theories, Methods, and Applications*, Studies in Computational Intelligence 1246,
https://doi.org/10.1007/978-3-032-12891-1_7

Data Encryption Standard (DES) and the Advanced Encryption Standard (AES) [1–3]. Digital image encryption is limited by a factor such as slow processing speed, low efficiency and complex computation etc. In recent years, optical encryption techniques have attracted significant attention for developing advanced cryptosystems, owing to its unique physical properties such as parallel processing, multiple degrees of freedom (like phase, wavelength, OAM and polarization), fast computation, and multidimensional functionality [4, 5]. In 1995, the Double Random Phase Encoding (DRPE) technique was proposed by Refregier and Javidi to encrypt a two-dimensional image [6]. It is one of the most effective methods that has significantly transformed the field completely. Later, to enhance security, DRPE has been extended into several transform domains, including gyrator transform [7], fractional Fourier transform [8], Hartley transform [9] etc. However, DRPE based methods are vulnerable to various cryptographic attacks such as Known plaintext attack [10], chosen ciphertext attack [11], chosen plaintext attack [12] etc., due to its inherently linear and symmetric nature. Moreover, a more advanced method than the DRPE technique has been proposed, which uses the Kronecker product and hybrid phase masks. Although it remains a linear system, it is robust against KPA [13]. To address the linearity issue in DRPE, Qin and Peng proposed an optical asymmetric cryptosystem in 2010 based on the phase-truncated Fourier transform (PTFT) [14]. But it was later found that this method is also vulnerable to iterative phase retrieval attacks [15]. The security of optical asymmetric encryption methods has been further enhanced by integrating them with various techniques such as gyrator wavelet transform [16], photon counting [17], and Fresnel domain encoding [18]. Beyond phase truncation (PT), several other asymmetric cryptosystems have also been proposed, including RSA [19] and elliptic curve cryptography [20] for image encryption, digital holography-based methods [21], and interference-based encryption [22] etc. Moreover, some mathematical decompositions such as singular value decomposition [23], QZ decomposition [24], polar decomposition [25] etc. have been incorporated into optical methods to design advanced hybrid cryptosystems.

With the continuous advancement of optical information techniques, innovative methods are steadily emerging and integrated with other fields. Nowadays, research on ML and DL is very hot in optical encryption. Due to the developments in ML, especially in algorithms and neural network models, have led many researchers to focus on analyzing the vulnerabilities of optical encryption systems through DL [26]. The major difference between the DL and ML models is the automatic feature learning capability of deep learning, which makes it suitable for a wide range of applications. Motivated by these factors, this chapter emphasizes DL over traditional ML. It has been demonstrated that conventional schemes such as the classical DRPE scheme, nonlinear optical cryptosystems, inference-based encryption and CGH-based encryption methods are prone to deep learning-based attacks [26–34]. Various types of neural networks were utilized to generate approximate equivalent models of optical cryptosystems by employing the known plaintext-ciphertext pairs. The most widely used DL model is the Artificial Neural Network (ANN) [35], but there are several other architectures, such as Convolutional Neural Networks (CNNs)

[36], and Recurrent Neural Networks (RNNs) [37], each suited for different types of tasks. DL-based attacks do not require phase information, and these techniques can achieve high quality reconstruction using only encoded intensity patterns. Thus, utilizing this new technique for image encryption could offer significant advantages. In recent time, there are many works which has focused in designing the optical encryption techniques based on the DL approach [38–40]. For example, Chen et al. employed complex scattering media to encode secret images into speckle patterns and then used a CNN model to decode the secret images from these patterns [39]. Wang et al. employed a dense neural network model to recover plaintext information from a steganographic hologram [38]. Wang et al. proposed an experimental optical encryption and decryption scheme that combines a random binary mask with DL techniques [41], Zhao et al. introduced a speckle-based optical cryptosystem in which face images can be reconstructed from random speckle patterns using a well-trained neural network [42]. All this research demonstrates that DL holds potential for enhancing the security and performance of optical encryption systems. Figure 1 illustrates how AI, ML and DL relate to each other by Venn diagram representation [43]. The remaining of the chapter is organized as follows. Section 2 discusses the key characteristics of DL. Section 3 explains the overview of DL architecture. Section 4 presents various DL models which have been utilized for optical image encryption in recent times. Section 5 highlights future directions, current challenges, and potential solutions. Finally, Sect. 6 explains the conclusions.

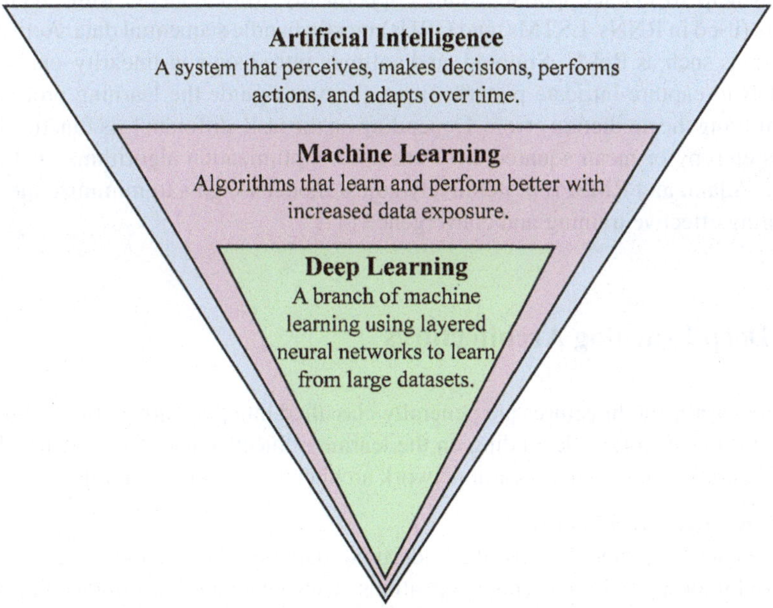

Fig. 1 Venn diagram presents the concept of deep learning

2 Characteristics of Deep Learning

Deep Learning (DL), a subfield of Machine Learning (ML) within the broader domain of Artificial Intelligence (AI) as shown in Fig. 1, has become a central methodology in the modern data-driven systems and research. Its growing prominence is because of its capacity to model complex, high-dimensional patterns and deliver better performance across a range of applications—including decision fusion, transfer learning, mobile deployment, human activity recognition, and class imbalance management. Previously, several tasks such as handwriting recognition, face identification, and object detection considered as a prominent challenge for machines, but with the assistance of DL these problems were addressed significantly by using appropriate learning architectures and data. DL gained utmost momentum through its early success in handwriting recognition, where Convolutional Neural Networks (CNNs) were developed to read handwritten postal codes. Since then, DL has become a preferred approach across numerous fields for replacing conventional methods such as speech recognition, healthcare analytics, image classification, and natural language processing.

DL models consist of multiple interconnected components. DL models are built upon core elements such as layers, activation functions, loss functions, and optimization algorithms [44]. **Layers** are the structural units that define how input data is transformed. Layers, including input, hidden, and output form the basic structure that is referred to as Artificial Neural network (ANN). Convolutional layers help to extract spatial features specifically from images, while recurrent layers (used in RNNs, LSTMs, and GRUs) mostly handle sequential data. Activation functions such as ReLU, Sigmoid, and softmax introduce non-linearity, enable the models to capture intricate patterns. Loss functions guide the learning process by quantifying the prediction errors. Depending on the task, different loss functions like cross-entropy or mean squared error are used. Optimization algorithms, including SGD, Adam, and RMSProp, iteratively adjust model weights to minimize the loss, ensuring effective training and convergence [44].

3 Deep Learning Architectures

Deep Learning architectures are generally classified into generative, discriminative, and hybrid categories, depending on the learning objective and data structure [45]. An illustration of different neural network architectures is shown in Fig. 2.

- **Generative Architectures**

  Generative models learn the underlying data distribution and are commonly used in unsupervised learning algorithms. They generate new samples like input data, making them useful for applications such as data augmentation and anomaly detection. Autoencoders, for example, compress input data into a latent space and reconstruct it, enabling noise reduction and feature extraction. Variants like

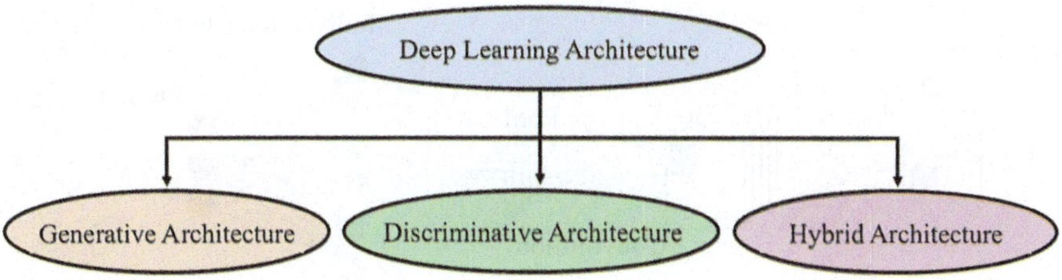

Fig. 2 Types of deep learning architecture

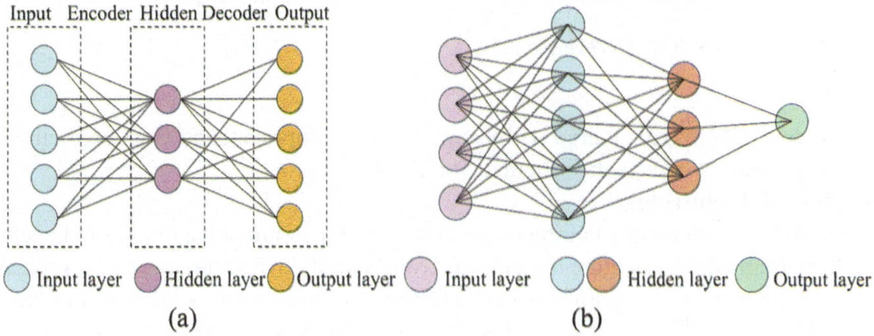

Fig. 3 (**a**) Deep autoencoder; (**b**) Recurrent Neural Networks (RNN)

stacked and denoising autoencoders further improve robustness and learning efficiency. Recurrent Neural Networks (RNNs) also belong to this category when used to generate sequences, such as text or music. The schematic diagram of Autoencoder and RNN is given in Fig. 3a, b. The ability to retain context across time steps makes them ideal for modeling temporal data. Although traditional RNNs suffer from gradient-related issues, LSTM and GRU architecture have effectively mitigated these problems. Furthermore, advanced models like Variational Autoencoders (VAEs) and Generative Adversarial Networks (GANs) now lead the field in producing high-quality synthetic data [45].

- **Discriminative Architectures**

Discriminative models focus on classifying input data by learning boundaries between classes. These models are suited for supervised learning tasks, including classification and regression. Artificial Neural Networks (ANNs), built with layered neurons, serve as foundational models. CNNs enhance this approach by capturing local spatial features, making them essential in visual data processing. When applied to sequential data tasks like speech recognition or language translation, RNNs serve as discriminative models by predicting outputs based on segmented input sequences. Other examples include Conditional Random Fields (CRFs) and Deep Stacking Networks (DSNs), which offer modular, scalable solutions for complex pattern recognition. In some applications, Hidden Markov

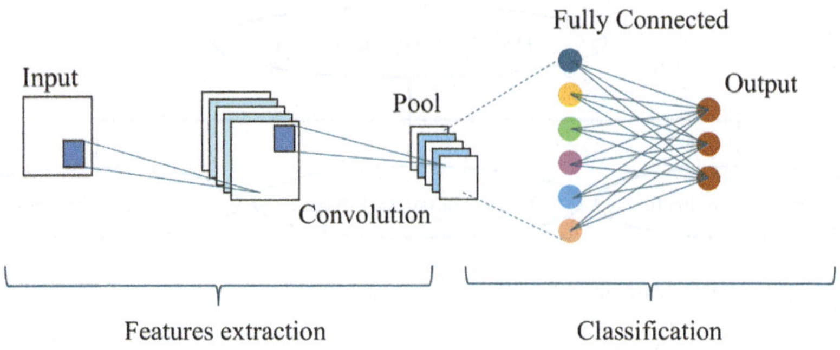

Fig. 4 Convolution Neural Network (CNN)

Models (HMMs) are paired with neural layers to improve contextual predictions [45]. Figure 4 depicts the schematic of CNN.

- **Hybrid Architectures**

  Hybrid architecture combines generative and discriminative models to benefit from both efficient feature learning and robust predictive accuracy [46]. A typical example is a Deep Neural Network (DNN) initialized using a Deep Belief Network (DBN) [47]. In such systems, generative layers (e.g., Restricted Boltzmann Machines) are used for unsupervised pretraining, followed by discriminative fine-tuning for classification or prediction tasks. Consequently, hybrid models are highly effective in domains like speech recognition, where they can capture both temporal structure and class boundaries. Their flexibility and robustness make them increasingly valuable in complex, data-limited environments.

4 Overview of Deep Learning-Based Optical Encryption Methods

Recent developments in image cryptanalysis have shown progress by utilizing the DL algorithm's, particularly in optical image encryption. By training DL models on large datasets of ciphertext and their matching plaintext images, these methods have demonstrated the potential to robust the techniques efficacy and break optical encryption schemes. In the next subsections, three optical cryptosystems based on DL are examined, including their simulation outcomes and corresponding optical configurations [27, 38, 48].

4.1 Deep Learning Decryption Approach for Asymmetric Computer-Generated Holography (CGH) Cryptosystem

Han et al. introduced a DL based approach for high-accuracy decryption for an asymmetric CGH encryption system that utilizes DRPE, incorporating phase truncation and chaotic iris phase mask [48]. The optical encryption process consists of two main stages: First, the creation of chaotic iris phase masks, and second, the CGH encryption based on DRPE, which employs these masks along with the phase truncation (PT) method. Initially, an iris image is chosen, and two distinct random phase maps are produced using separate logistic chaotic maps. More details regarding the iris phase mask generation can be found in [49].

Figure 5 depicts the schematic representation of the DPRE-based CGH encryption approach that utilizes chaotic iris phase masks in combination with the PT method. First, the plaintext image $f(x_0, y_0)$ is bonded with CIPM1. The obtained output is Fresnel propagated to a distance Z_1 to obtain complex image $f(x_1, y_1)$. Next, the PT and phase reservation (PR) operation is performed on the complex wavefront, where PR part is reserved as private key1. PT part is multiplied with CIPM2 and Fresnel propagated with distance Z_2. Again, PT and PR are performed on complex wavefront. PR part is reserved as private key2 and PT was used as the original object for generating an off-axis Fresnel computer-generated hologram. The complex amplitude distribution at the output plane is calculated using Fresnel diffraction over a propagation distance of Z_3. Next, the reference beam R interferes

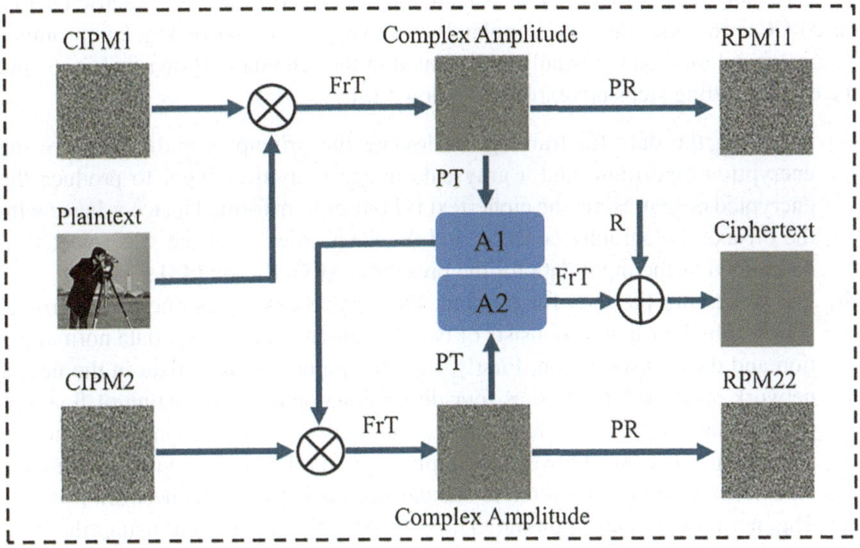

Fig. 5 Schematic diagram for DRPE-based CGH encryption method. *A1* and *A2* extracted amplitudes; *R* reference light; *PR* phase reservation; *PT* phase truncation; *FrT* Fresnel transform. (Figure adapted from [48])

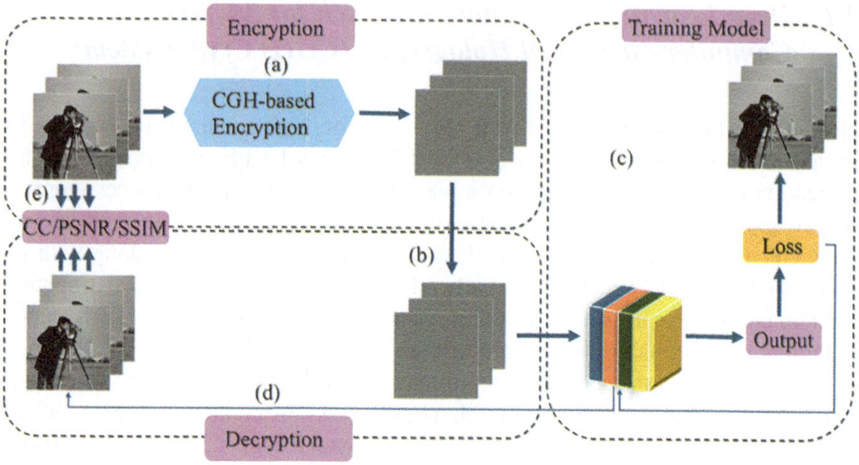

Fig. 6 Schematic representation of the decryption method using deep learning. (**a**): image encryption; (**b**) training data preprocessing; (**c**) ACGHC-Net training; (**d**) ciphertext decryption; (**e**) Evaluation of decrypted image quality. (Figure adapted from [48])

with the complex amplitude distribution $f(x_3, y_3)$. To generate off-axis holograms, it is essential to maintain a proper angular separation between the reference beam and the object wave. The hologram obtained is utilized as ciphertext [48].

The proposed decryption technique operates without the use of a key, utilizing DL to reliably reconstruct the grayscale image encrypted by the asymmetric DRPE-based CGH method, thereby completely removing the threat of key compromise. The proposed method is visually represented in the schematic shown in Fig. 6, and the corresponding steps are explained below [48].

(a) Preparing the data for training: Following the principles outlined above the encryption algorithm, and a grayscale image is used as input to produce the encrypted output. Here, the ciphertext is Fourier transformed in accordance with the off-axis holography concept, and the +1st order spectrum was selected to be utilized as the input data for the presented ACGHC-Net [48].
(b) Pre-processing the data for training: Data preprocessing is one of the crucial stages, which primarily consists of two fundamental processes: data normalization and data tensorization. Firstly, both the input and label data in the neural network must be formatted as four-dimensional tensors consisting of floating-point values, a process referred to as data tensorization. Secondly, due to the significant difference between the ciphertext and the original grayscale image, data normalization is required to standardize the input for the network [48].
(c) The network model training: The ACGHC-Net is trained using the pre-processed ciphertext and plaintext data. For training, the network is optimized using the AdamW optimizer with an initial learning rate set to 1e-4. A batch size of 32 is used, and the input data set consists of images sized 256 × 256.

To enhance training efficiency and accelerate convergence, Automatic Mixed Precision (AMP) [50] with CUDA and single-machine four-GPU Distributed Data Parallel (DDP) [51] training is utilized. They establish a validation set every ten epochs throughout the training process to confirm the network's state of convergence [48].

(d) Reconstructing the encrypted image: The test set is utilized to evaluate the network model's performance after it has been developed. The network is used to reconstruct ciphertext images during the test. The reconstructed outputs are compared with their corresponding original plaintext images, and the decryption performance is evaluated both qualitatively through visual inspection and quantitatively using various image quality assessment metrics [48].

- **ACGHC-Net Model architecture**

  The proposed model architecture of ACGHC-Net is given on Fig. 7. This network effectively retrieves complex grayscale images encrypted through the asymmetric DRPE-based CGH cryptosystem with a high level of accuracy and quality. In contrast to conventional convolutional neural networks, the ACGHC-Net incorporates four embedded deep network levels within its architecture. The depth of the network is extended by increasing it from one layer to four layers. The levels indicate how many down-sampling and up-sampling operations are carried out by each sub-module. With the addition of more layers, the network delves deeper into abstract feature spaces, which enhances its capacity to accurately reconstruct the decrypted image. The entire model consists of 15

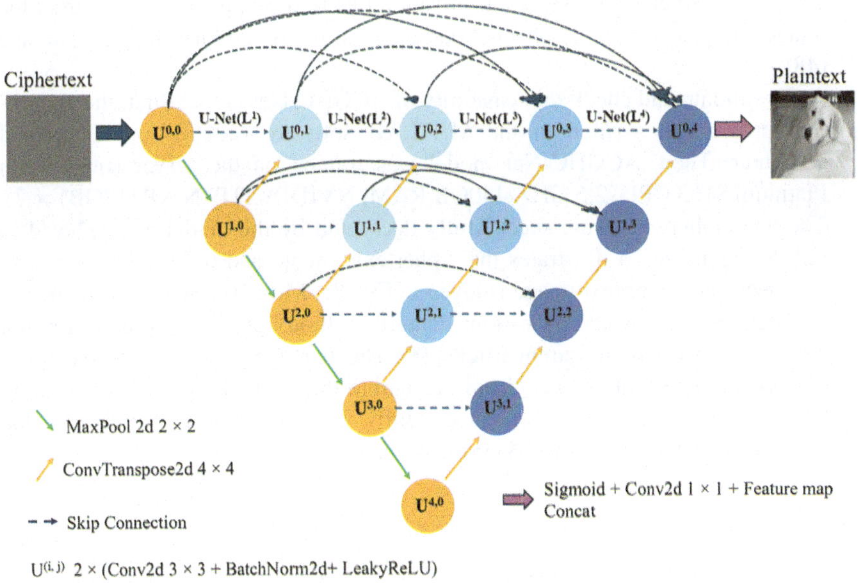

Fig. 7 Schematic of proposed ACGHC-Net. (Figure adapted from [48])

components, consisting of ten decoding units and five encoding units. The five encoding units make up the feature extraction module, while the ten decoding units are responsible for reconstructing the spatial structure of the image. Each unit is composed of two convolution blocks to enhance feature learning [52, 53]. Convolution layers, normalization layers, and nonlinear activation functions (Leaky ReLU) make up each convolution block. These components are employed to efficiently extract the data's features. This advanced architecture allows ACGHC-Net to effectively reconstruct the encrypted images with high accuracy [48].

The ACGHC-Net initially processes a single-channel image for processing. A 3×3 convolution operation is applied to increase the data's channel dimension from 1 to 64, thereby enhancing the image's feature representation. Four down-sampling operations are used in the encoder portion, and the max pooling convolution with a 2×2 kernel size is used for each down-sampling. This down-sampling approach decreases the spatial dimensions of the feature map by half and simultaneously doubles the number of channels, allowing for gradual abstraction and compression of image features across layers. Each neural unit in the network contains two convolution blocks, each using a 3×3 filter, a stride of 1, and padding of 1, to extract features from the feature map. After the encoding module is completed, the resulting feature map measures 16×16 in size and contains 1024 channels. During the up-sampling process, transposed convolution is used for deconvolution, employing a 4×4 filter, a stride of 2, and a padding of 1. Once the decoding module is completed, the feature map is restored to a size of 256×256, and the number of channels is reduced to 64. Finally, a 1×1 convolution filter with a Sigmoid activation function is applied to reconstruct the image and produce single-channel output, resulting in the final decrypted image [48].

To validate and check the feasibility of ACGHC-Net model, first, the encryption program is written in the MATLAB software (2016B) on a personal computer. Then, ACGHC-Net model was trained on the server (Intel Xeon Platinum 8163 CPU, 2.5 GHz, 440GB RAM, NVIDIA TITAN XP (12GB) $\times$ 7). The test ciphertexts were successfully decrypted by the model as displayed in Fig. 8. Figure 8a, b illustrates the ciphertexts image and true plaintext image. The decrypted plaintext image is given in Fig. 8c. Moreover, the performance of ACGHC-Net was tested on various datasets to thoroughly assess its decryption potential as given in the parent article [48]. The image in every case clearly shows that the predicted plaintexts closely resemble the actual plaintexts, confirming that the proposed model can successfully decrypt images encoded by the asymmetric DRPE-based CGH cryptosystem.

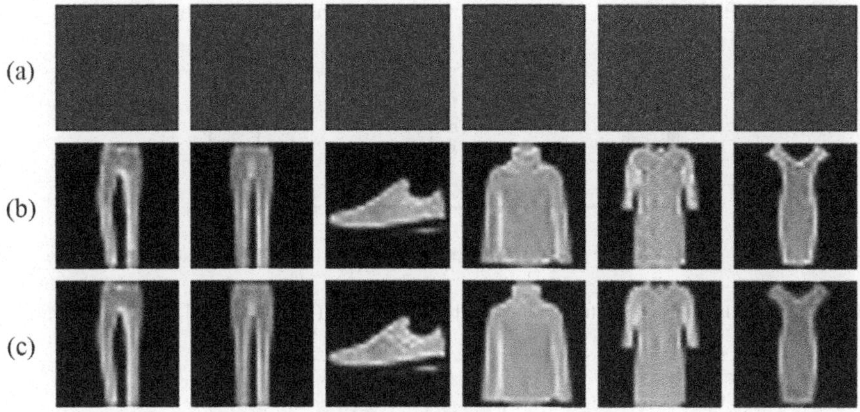

Fig. 8 Decryption results obtained through ACGHC-Net on the Fashion-MNIST dataset. (**a**) Ciphertext images; (**b**) true plaintext images; (**c**) decrypted images. (Figure adapted from [48])

4.2 Cryptanalysis of Random Phase Encoding Based on Optical Cryptosystem via Deep Learning

This section discusses how the random phase encoding (RPE)-based optical cryptosystem is susceptible to a chosen plaintext attack (CPA) when DL techniques are applied [27]. In DRPE the plaintext image is transformed into white noise by employing a $4f$ optical set up. This setup involves placing two independent random phase masks (M_1, M_2), one at the input plane and another at the Fourier plane. DRPE is known to be prone to several types of attacks, as the Fourier amplitude of the encrypted image can be easily retrieved by applying a Fourier transform. To strengthen the security, TRPE incorporates a third random phase mask (M_3) at the output plane. In TRPE, first the plaintext image is modulated with $M_{1,}$ and then Fourier transformed. Next, the obtained output modulated again with second M_2 and inverse Fourier transformed. Finally, the encrypted image is obtained after multiplying with third M_3. The schematic of the method is shown in Fig. 9. The plaintext image is retrieved by reversing the encryption process in the exact order, using the conjugates of the phase masks [27].

- **DNN Model Architecture**

  The proposed cryptanalysis method requires a significant number of plaintext-ciphertext pairings to determine the mapping between given encrypted data and its corresponding plaintext; it is not dependent on the actual encryption or decryption process. According to the "Kerckhoffs' principle," it is typically presumed in cryptanalysis that attacker is already having the complete information about a cryptosystem, with only the exception of the encryption key [27, 54]. Under CPA, as followed in the method, it is specifically presumed that attackers can access the ciphertext corresponding to any plaintext they

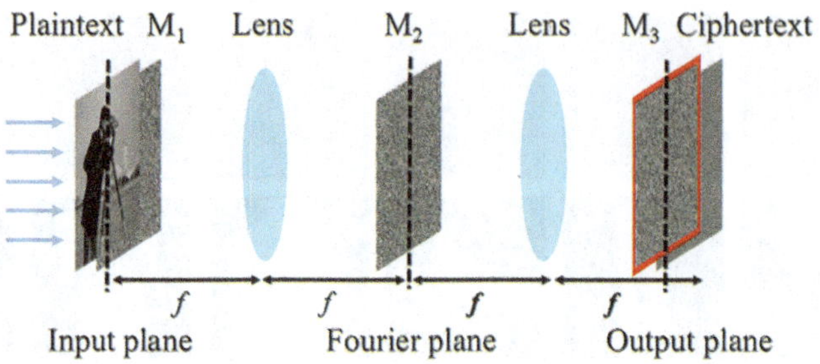

Fig. 9 Schematic diagram of triple random phase encoding. (Figure adapted from [27])

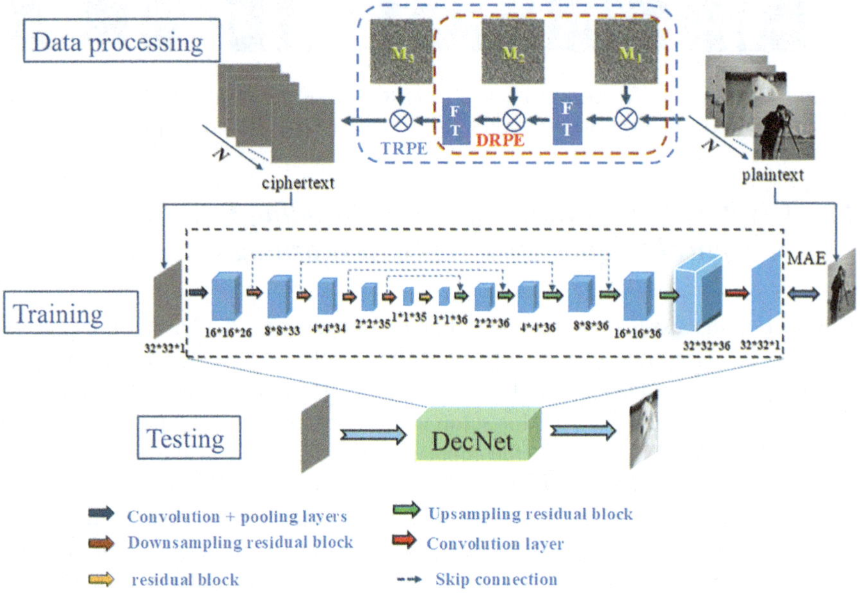

Fig. 10 Schematic representation of the DecNet neural network architecture designed for optical cryptanalysis. (Figure adapted from [27])

select, even without knowledge of the encryption keys. They then attempt to recover the plaintext directly from later intercepted ciphertexts, also without possessing the keys. In the proposed scheme, a modified deep neural network (DNN) called DecNet is designed and trained using ciphertext images as input data and their corresponding plaintext images as labels. DecNet is built upon the deep residual convolutional architecture of ResNet [55], in which every layer feed-forwardly connects to every other layer in the block. Figure 10 shows the DecNet schematic diagram. To maintain strict consistency in combination with classical DRPE or TRPE approaches, the input to DecNet is configured as a complex-valued ciphertext, while the output is set as a real-valued plaintext.

After the encryption process, the ciphertext initially goes through a convolutional layer that uses a 3 × 3 filter, followed by a pooling layer having stride of 2. It is then progressively reduced in size through four consecutive down-sampling residual blocks. After passing through an additional residual block, the data then goes through five up-sampling residual blocks. In the final stage, the signals are processed by a standard convolution layer having 3 × 3 filters, which reconstructs the estimated plaintext image. The structure described above follows an encoder-decoder architecture [56, 57], where feature maps are extracted from the input data using the down-sampling residual blocks acting as the encoder, and the up-sampling residual blocks act as the decoder, performing pixel-level reconstruction. Additionally, skip connections are employed to transfer high-frequency details captured in the early layers directly to the deeper layers, aiding in accurate reconstruction of the output image [58].

The proposed attack method efficacy in the DRPE and TRPE schemes was evaluated through numerical simulations. In both DRPE and TRPE cases the ciphertext is a complex amplitude distribution that cannot be directly recorded by a typical CCD camera in real optical experiments. It is assumed in these simulations that attackers possess the complex-valued ciphertext that has been obtained from the corresponding hologram. A total of 5000 grayscale images from the open-source Fashion-MNIST dataset were used to train the DecNet model [27, 59]. The implementation was done in Python 3.6, with the deep neural network built using the Keras library on top of TensorFlow. An NVIDIA GTX1060 GPU was used to speed up the computations. Figure 11a displays the result of the DecNet attack on the DRPE system. The first row displays the ciphertexts, second is original plaintext images and third row displays the retrieved image by Dec-Net. Similarly, the attack results on the TRPE system are illustrated in Fig. 11b. To further illustrate the proposed method effectiveness, DecNet was also employed to reconstruct ciphertexts generated from a different data type, namely handwritten digits from the MNIST dataset is given in parent article [27].

Additionally, the method of resilience to sheer and noise was examined, considering that optical information propagation can be affected by data loss and noise interference. The noise interference and shear effect can be visualized in parent article [27] (Figs. 6, 7, and 8).

Furthermore, the efficacy and feasibility of the proposed approach have also been examined through experiments. Please refer to the parent article for detail of experimental setup [27]. A total of 5000 images from the MNIST database (used as chosen plaintexts) are displayed on the SLM, and the corresponding speckle intensity patterns are recorded by the camera. For training purposes, only the central 400 × 400-pixel region was selected from each image and then resized to 32 × 32 pixels in the experiments. Out of the 5000 image pairs, 4500 were used for training, while the remaining 500 were reserved for testing. Figure 12 displays the test results. The first rows images represent the ciphertexts amplitude and serve as inputs for the proposed DecNet model. The second row shows the ground truth images, whereas the third row shows the DecNet predicted plaintext

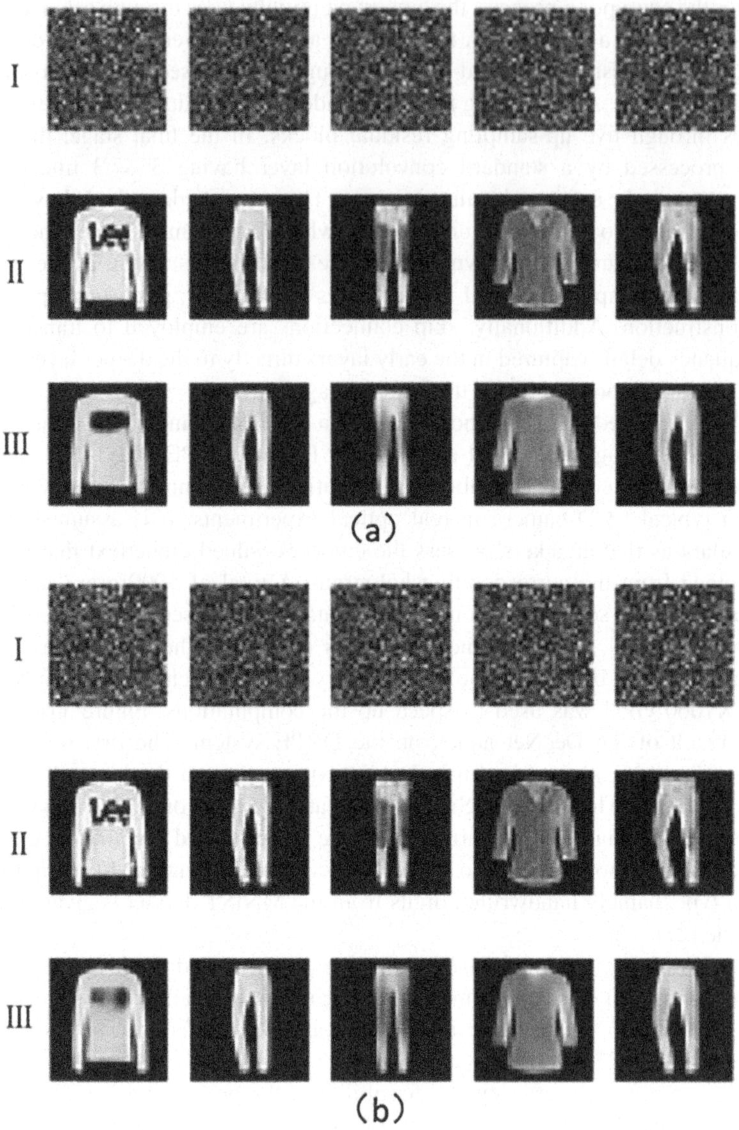

Fig. 11 DecNet reconstruction results for the original Fashion-MNIST test dataset. (**a**) DRPE and (**b**) TRPE: (i) testing ciphertext images; (ii) original plaintext images; (iii) the deciphered plaintext images by trained DecNet. (Figure adapted from [27])

images. It is important to note that DecNet consistently generates real-valued outputs, irrespective of whether the input data is real or complex. This indicates that the proposed attack method is only effective when the plaintext has real values [27].

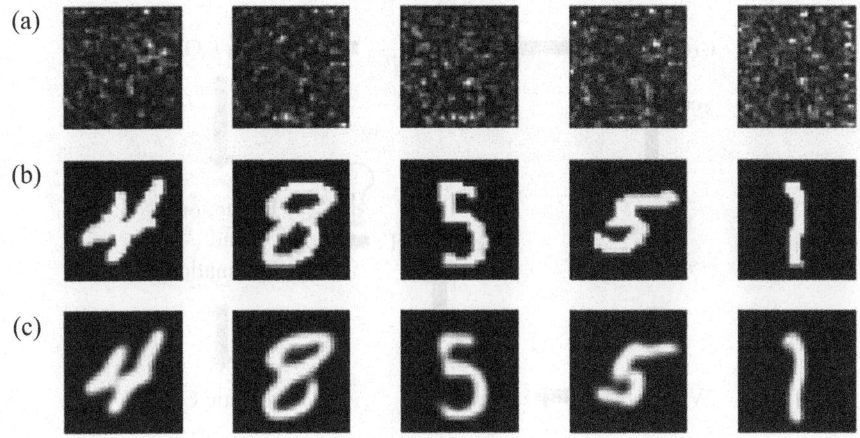

Fig. 12 Experimental results: (**a**) ciphertext amplitude; (**b**) ground truth plaintexts; (**c**) retrieved plaintext images. (Figure adapted from [27])

4.3 Holographic and Speckle Encryption Using Deep Learning

In this section, optical image encryption employing holographic method and DL is discussed in detail [38]. To numerically generate and encrypt holograms, a non-iterative technique implemented with DRPE and patterned phase only hologram (PPOH) [60] approach is utilized. Specifically, the secret image is first encrypted to white random noise, employing the DRPE method, and then the PPOH technique is applied to convert it into a phase CGH, commonly known as the 'secret CGH'. The proposed scheme incorporates steganographic techniques to further increase the security level. This method involves designing and training a dense CNN, DenseNet to map the connection between the secret input image to the corresponding speckle pattern. These speckle patterns are produced by employing CGH obtained through decryption of synthetic CGHs with a specific key, and it is also utilized for extracting pixel information from the secret CGH. First, we will discuss the hybrid non-iterative method that is utilized to generate the synthetic phase CGH, and its schematic diagram is given in Fig. 13. The DRPE method in the Fresnel domain is used to transform the secret image into a static white noise-like pattern. The non-iterative PPOH approach is applied to transform the image which is encrypted and the cover image into two PPOH, referred to as the secret and cover CGH symbolized by $H_s(x)$ and $H_c(x)$ [60]. Next, a binary-amplitude mask $M(x)$ with random values is generated and utilized to extract the individual pixels, taken from the secret hologram $H_s(x)$. At that point, the binary mask acts as a decryption key as well, with each element separately assigned a value of 0 with a probability of 1/2. The pixels are extracted from the cover CGH $H_c(x)$ using its complimentary pattern,

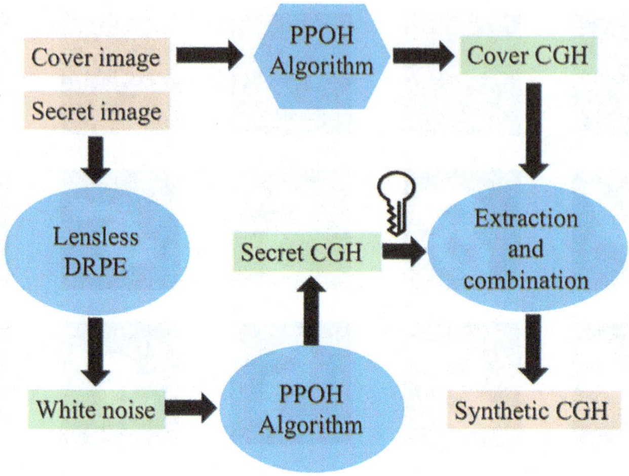

Fig. 13 Schematic illustration of the hybrid non-iterative approach for synthesizing phase CGH. (Figure adapted from [38])

denoted as $M'(x)$. A synthetic phase CGH is produced by combining these two sets of extracted phase pixels, and mathematically it can be expressed as

$$H(x) = H_s(x)M(x) + H_c(x)M'(x) \quad (1)$$

Here, $M(x)$ and $M'(x)$ are binary mask and it's complementary whose sum yields an all-ones matrix. A 50% pixel extraction from the two CGHs is adequate for effectively displaying the cover image details and reconstructing the corresponding secret image. Here, the proposed technique retrieves the secret CGH by using M as the decryption key When the synthetic CGH is illuminated with laser beam, a cover image will be visible on the screen.

- **Dense CNN (DensNet) model architecture**

  The dense CNN (DenseNet) architecture that has been employed is depicted in Fig. 14. The model is trained by employing several cyphertext-plaintext pairings, which are original secret images, and the noise-like patterns observed in the decrypted CGH due to diffraction under fixed optical parameters. The dimensions of each input and output image for the DenseNet model are 128 × 128 pixels. To obtain the feature maps, the input ciphertext is passed through an encoder composed of seven layers, each formed by a "Denseblock + Transtionblock". The Convolution (conv) with tiny filters (size 3 × 3), batch normalization (BN), and rectified linear unit (ReLU) nonlinear activation compose each of the three composite convolution layers that make up each dense block [38]. Each transition block consists of four consecutive operations: BN, ReLU, and a 1 × 1 convolution layer, preceded by average pooling with stride 2 for downsampling. After passing through a dense block, the low-resolution activation maps enter

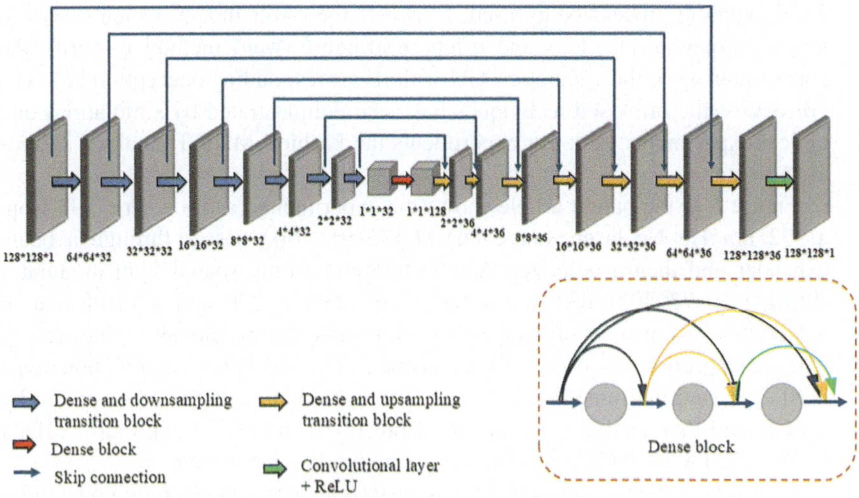

Fig. 14 The proposed DenseNet architecture. (Figure adapted from [38])

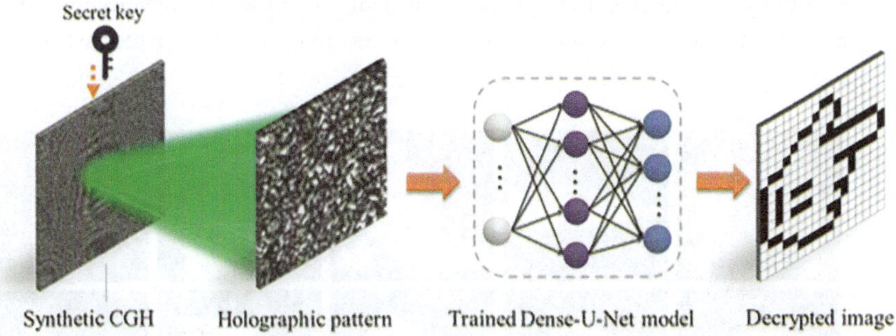

Fig. 15 Deep learning-based speckle pattern decryption. (Figure adapted from [38])

the decoder path and are successively processed by seven 'Upsampling + Dense Block' modules. The plaintext estimate is then generated when the feature maps are passed on to a conventional convolutional layer [38].

Figure 15 depicts the flow diagram of decryption process. A decryption key is multiplied by the CGH that was obtained during the encryption process to generate a speckle pattern. The use of a secret key helps obscure any functional connections between the input and output, making it difficult for a learning-based direct attack model to exploit those relationships. The incorporation of a secret key makes it challenging for models used in direct attacks to recognize or learn the functional relationship between the input and output data [38]. The speckle patterns resembling a purely random noise image, are used as input for a DNN trained in an earlier stage. This network rapidly generates the original secret image. When the same laser beam is directed at the synthetic CGH without the decryption process being used, it reveals the cover image, which serves to mislead potential attackers and offers a straightforward method to verify the correct pairing of the synthetic CGH with its corresponding decryption key. The efficacy of the proposed techniques has been demonstrated by simulations and optical experiments [38]. For experiments the Fashion-MNIST database images are utilized.

Figure 5 in the parent article shows the experimental setup where light from a 532 nm He–Ne laser source DH-SL532-50A first passes through a beam expander and then a polarizer. A reflective phase-only spatial light modulator (SLM, GCI-770402) with a resolution of 1280 × 720 and a pixel size of 6.4 μm × 6.4 μm is utilized as the light-modulating element. The central 720 × 720 portion was used in the experiment. The SLM phase modulation range is [0, 2π], and its frame rate is 60 Hz. The diffraction patterns are captured by a CCD and then analyzed by an Intel Core i7-10750H CPU, with an NVIDIA GeForce GTX 1650 GPU used for neural network computation.

Figure 16 shows five groups for demonstration purposes. Each group includes a secret image, a cover image, the corresponding synthetic hologram and the optically reconstructed image. It demonstrates that when synthetic holograms are directly placed into the optical setup using an SLM, no information about the secret images can be recovered. The cover images and their reconstructions

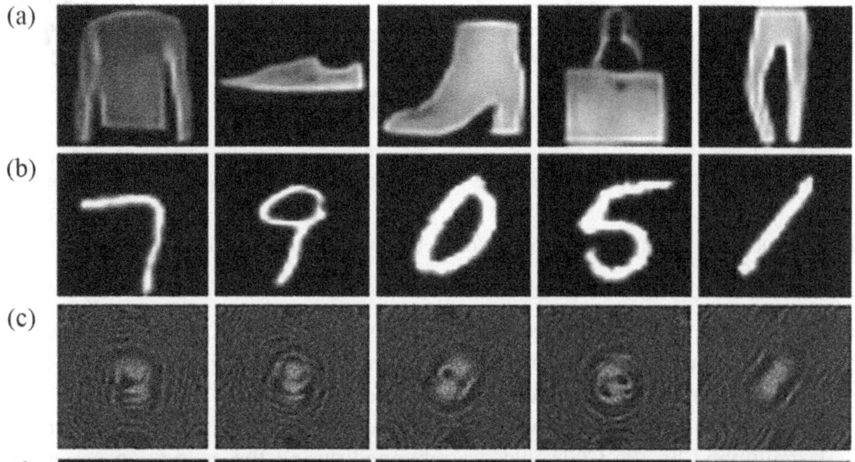

Fig. 16 Optically retrieved images from synthetic CGHs. (**a**) Secret images; (**b**) cover images; (**c**) synthetic CGHs; (**d**) diffraction patterns. (Figure adapted from [38])

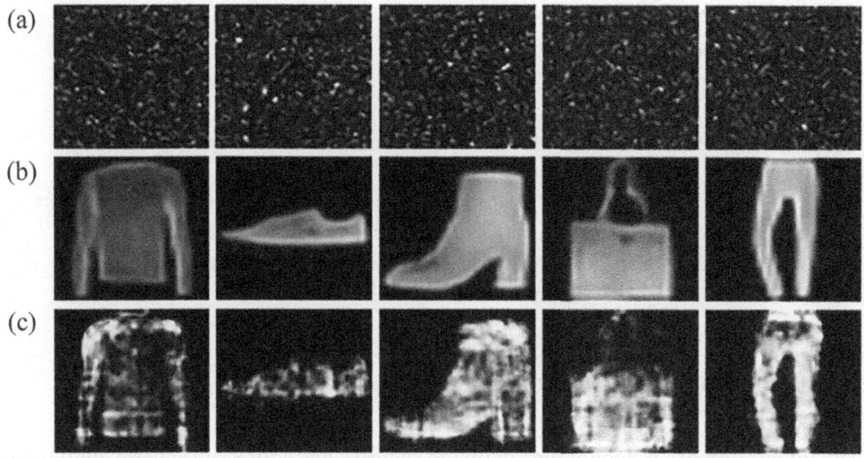

Fig. 17 Decrypted images from speckle patterns. (**a**) Optical speckle patterns; (**b**) Decrypted image with (**b**) DenseNet; (**c**) CNN. (Figure adapted from [38])

have the following correlation coefficients (CC): 0.87, 0.85, 0.81, 0.81, and 0.87, respectively [38].

The optical patterns obtained from correct CGH decryption are displayed in Fig. 17a. These images look like random noise were later analyzed by the trained DenseNet. The results are presented in Fig. 17b, with corresponding CC values of 0.85, 0.93, 0.95, 0.92, and 0.99, respectively. For comparison, a separate network is employed named CNN. Figure 17c shows the decryption results achieved through a CNN learned from same ciphertext and plaintext pairs. The CNN is trained over 100 epochs and consists of three convolutional layers and one fully connected layer. It is evident that during the CNN-based decryption procedure, some information from the original material is lost. The CC value obtained by CNN is depicted for five sets are 0.75, 0.84, 0.89, 0.9 and 0.95, respectively. The DenseNet utilized in proposed approach yields better reconstruction accuracy than the CNN, particularly in the image's periphery. Additionally, decryption has been done using incorrect keys to verify the resilience of the proposed technique, as given in parent article of Fig. 8 [38].

Additionally, Table 1 presents a collection of relevant research papers focused on optical cryptosystems that incorporate DL techniques. These studies explore various approaches where neural networks are applied to enhance encryption, decryption, or cryptanalysis processes within optical systems.

Table 1 Some recent studies on DL based cryptosystem

References	Year	Title
[61, 62]	2019	A deep learning based attack for the chaos-based image encryption, chaos-based image encryption algorithm
[31]	2020	Plaintext attack on joint transform correlation encryption system by convolutional neural network
[63]	2023	Deep learning for medical image cryptography: a comprehensive review
[64]	2023	Towards deep learning for efficient image encryption
[36]	2025	Deep learning-based encryption scheme for medical images using DCGAN and virtual planet domain

5 Future Scope, Challenges and Its Solutions

The integration of DL techniques in image security has addressed several challenges that conventional approaches could not overcome. The characteristics of DL image algorithms include high dependence on large datasets and lengthy pretraining time. Moreover, effectively leveraging these characteristics plays a main role in computing the efficiency of DL models. Future advancements can be highlighted through the points listed below:

1. Designing DL models for image security requires attention to task-specific properties and overall performance balance. For example, DL methods can achieve better robustness and attack resistance capability by applying fixed constraints in the model structure and training approach during design. Due to the weak and limited presence of the stego-signal in the image, enhancing its signal-to-noise ratio is essential for effective image steganalysis. As the difference between the tampered and original image is minimal, further research is needed to effectively leverage DL for forgery detection. A diffusion method is necessary to ensure security in neural network-based end-to-end image encryption because of the low avalanche effect.
2. The lack of theoretical foundations in DL-based image security highlights the urgent need to establish and continuously enhance its theoretical design. Mathematical interpretation is needed for certain DL methods in image security, and their evaluation should also consider diverse perspectives. Improving the theoretical basis will promote more rapid and reliable development in the field of DL-based image security.
3. Expanding DL into other image security domains can help to solve conventional problems. For example, key exchanges in image cryptography using neural networks and neural network-based image homomorphic encryption are both emerging as significant areas of research.
4. Conventional optical encryption techniques have inherent benefits in the field of optical information processing, but there are some issues. For example, the systems require very high alignment requirements for the optical path; the decryption process is susceptible to extra noise during transmission, information

Table 2 Overview of challenges and solutions in DL-based encryption techniques

S. No	Challenges	Solutions
1.	DL-based key generation is computationally expensive	Parallel processing can lower computation cost [65]
2.	Long training duration in most encryption techniques	Can be reduced by transfer learning and advanced computation [66, 67]
3.	The training dataset is extensive	Data augmentation and pretraining model [68, 69]
4.	Limited methods for 3D/4D and multi-resolution images due to complexity and cost	Deep networks such as GAN can be utilized [70]
5.	DL-based encryption is computationally expensive	Hybrid of split learning and federated learning can be employed to make model efficient [71, 72]

is easily lost during transmission, which significantly affects the decryption effect, and so on. To overcome these challenges DL is one of the solutions, however most DL-reliant encryption techniques still need large training datasets and a significant amount of training time. The development of more efficient optical image encryption techniques that are immune to DL-based assaults would be very advantageous.

Additionally, during studies some challenges and their solutions have been identified which are presented in Table 2. Progress in DL is vital for enhancing image security.

6 Conclusion

In this chapter, we have presented a detailed overview of DL-based approaches and how they have been integrated with optical image encryption techniques. It covered the fundamentals of DL, key components and its model architecture. We discussed how DL can be employed not only to enhance encryption methods but also to analyze and potentially break the existing schemes. These findings demonstrate the significant potential of DL in the field of optical security. However, despite these advancements, several challenges remain. Future research should focus on the development of more efficient and robust DL algorithms that can overcome current limitations, such as long training times, high computational costs, complexity to integrate with optical methods for real-time use, etc.

Funding RK would like to acknowledge the support from the Science and Engineering Research Board (SERB), the Government of India, under the SERB SURE research grant (File No. SUR/2022/000910) and SRM University-AP for seed grant SRMAP/URG/SEED/2023-24/044, SRMAP/URG/SEED/2024-25/041.

References

1. Sachin R, Kumar, Sakshi R, Yadav SG, Reddy AK, Yadav, Singh P. Advances in optical visual information security: a comprehensive review. Photonics. 2024;11(1):99. https://doi.org/10.3390/photonics11010099.
2. Biryukov A. The boomerang attack on 5 and 6-round reduced AES. In: International conference on advanced encryption standard. Berlin, Heidelberg: Springer Berlin Heidelberg; 2004. p. 11–5.
3. Al Hasib A, Haque AAMM. A comparative study of the performance and security issues of AES and RSA cryptography. In: Proceedings - 3rd international conference on convergence and hybrid information technology, vol. 2; 2008. p. 505–10.
4. Javidi B, Carnicer A, Yamaguchi M, Nomura T, Pérez-Cabré E, Millán MS, et al. Roadmap on optical security. J Opt (United Kingdom). 2016;18(8):083001. https://doi.org/10.1088/2040-8978/18/8/083001.
5. Liu S, Guo C, Sheridan JT. A review of optical image encryption techniques. Opt Laser Technol. 2014;57:327–42. https://doi.org/10.1016/j.optlastec.2013.05.023.
6. Refregier P, Javidi B. Optical image encryption based on input plane and Fourier plane random encoding. Opt Lett. 1995;20(7):767–9. https://doi.org/10.1364/OL.20.000767.
7. Rodrigo JA, Alieva T, Calvo ML. Experimental implementation of the gyrator transform. J Opt Soc Am. 2007;24(10):3135–9. https://doi.org/10.1364/JOSAA.24.003135.
8. Unnikrishnan G, Joseph J, Singh K. Optical encryption by double-random phase encoding in the fractional Fourier domain. Opt Lett. 2000;25(12):887–9. https://doi.org/10.1364/OL.25.000887.
9. Xin-Xin L, Dao Mu Z. Optical image encryption with simplified fractional Hartley transform. Chin Phys Lett. 2008;25(7):2477. https://doi.org/10.1088/0256-307X/25/7/040.
10. Peng X, Zhang P, Wei H, Yu B. Known-plaintext attack on optical encryption based on double random phase keys. Opt Lett. 2006;31(8):1044–6. https://doi.org/10.1364/OL.31.001044.
11. Carnicer A, Montes-Usategui M, Arcos S, Juvells I. Vulnerability to chosen-cyphertext attacks of optical encryption schemes based on double random phase keys. Opt Lett. 2005;30(13):1644–6. https://doi.org/10.1364/OL.30.001644.
12. Peng X, Wei H, Zhang P. Chosen-plaintext attack on lensless double-random phase encoding in the Fresnel domain. Opt Lett. 2006;31(22):3261–3. https://doi.org/10.1364/OL.31.003261.
13. Kumar R, Bhaduri B. Optical image encryption using Kronecker product and hybrid phase masks. Opt Laser Technol. 2017;95:51–5. https://doi.org/10.1016/J.OPTLASTEC.2017.03.041.
14. Qin W, Peng X. Asymmetric cryptosystem based on phase-truncated Fourier transforms. Opt Lett. 2010;35(2):118–20. https://doi.org/10.1364/OL.35.000118.
15. Wang X, Zhao D. A special attack on the asymmetric cryptosystem based on phase-truncated Fourier transforms. Opt Commun. 2012;285(6):1078–81. https://doi.org/10.1016/j.optcom.2011.12.017.
16. Mehra I, Nishchal NK. Optical asymmetric image encryption using gyrator wavelet transform. Opt Commun. 2015;354:344–52. https://doi.org/10.1016/j.optcom.2015.06.015.
17. Pérez-Cabré E, Abril HC, Millán MS, Javidi B. Photon-counting double-random-phase encoding for secure image verification and retrieval. J Opt. 2012;14(9):094001. https://doi.org/10.1088/2040-8978/14/9/094001.
18. Luan G, Quan C. Optical double-image cryptosystem based on phase truncation in the Fresnel domain. Appl Phys B. 129(8):80, 023. https://doi.org/10.1007/s00340-023-08077-w.
19. Jeon S-H, Gil S-K. Optical asymmetric cryptography modifying the RSA public-key protocol. Curr Opt Photon. 2020;4(2):103–14.
20. Li W, Chang X, Yan A, Zhang H. Asymmetric multiple image elliptic curve cryptography. Opt Lasers Eng. 2021;136:106319. https://doi.org/10.1016/j.optlaseng.2020.106319.
21. Nishchal NK, Joseph J, Singh K. Securing information using fractional Fourier transform in digital holography. Opt Commun. 2004;235(4–6):253–9. https://doi.org/10.1016/j.optcom.2004.02.052.

22. Zhang Y, Wang B. Optical image encryption based on interference. Opt Lett. 2008;33(21):2443–5. https://doi.org/10.1364/OL.33.002443.
23. Liu Q, Liang Y, Liu F, Yang X, Meng X. Optical image encryption based on singular value decomposition ghost imaging and fractional chaotic mapping. Opt Lasers Eng. 2025;186:108779. https://doi.org/10.1016/j.optlaseng.2024.108779.
24. Vardhan H, Sakshi, Reddy SG, Muniraj I, Kumar R. A symmetric optical cryptosystem based on QZ decomposition and Hermite Gaussian beam speckles. Opt Quant Electron. 2024;56:885. https://doi.org/10.1007/s11082-024-06740-w.
25. Sivarajan AK, Vardhan H, Choudhary S, Reddy SG, Kumar R. Asymmetric optical cryptosystem using C-point polarization singularity speckles and polar decomposition in 2D-NS-LCT domain. Phys Scr. 2025;100(5):055504. https://doi.org/10.1088/1402-4896/adc49c.
26. Zhou Q, Wang X, Jin M, Zhang L, Xu B. Optical image encryption based on two-channel detection and deep learning. Opt Lasers Eng. 2023;162:107415. https://doi.org/10.1016/j.optlaseng.2022.107415.
27. Hai H, Pan S, Liao M, Lu D, He W, Peng X. Cryptanalysis of random-phase-encoding-based optical cryptosystem via deep learning. Opt Express. 2019;27(15):21204–13. https://doi.org/10.1364/OE.27.021204.
28. Zhou L, Xiao Y, Chen W. Machine-learning attacks on interference-based optical encryption: experimental demonstration. Opt Express. 2019;27(18):26143–54. https://doi.org/10.1364/OE.27.026143.
29. Qin Y, Wan Y, Gong Q. Learning-based chosen-plaintext attack on diffractive-imaging-based encryption scheme. Opt Lasers Eng. 2020;127:105979. https://doi.org/10.1016/j.optlaseng.2019.105979.
30. Wu C, Chang J, Xu X, Quan C, Zhang X, Zhang Y. Cryptoanalysis of the modified diffractive-imaging-based image encryption by deep learning attack. J Mod Opt. 2020;67(17):1398–409. https://doi.org/10.1080/09500340.2020.1862329.
31. Chen L, Peng B, Gan W, Liu Y. Plaintext attack on joint transform correlation encryption system by convolutional neural network. Opt Express. 2020;28(19):28154–63. https://doi.org/10.1364/OE.402958.
32. Pan S, Liao M, He W, Zhang Y, Peng X. Untrained neural network for cryptanalysis of a phase-truncated-Fourier-transform-based optical cryptosystem. Opt Express. 2021;29(26):42642–9. https://doi.org/10.1364/OE.444126.
33. Wang X, Wei H. Cryptanalysis of compressive interference-based optical encryption using a U-net deep learning network. Opt Commun. 2022;507:127641. https://doi.org/10.1016/j.optcom.2021.127641.
34. Zhou L, Xiao Y, Chen W. Learning-based attacks for detecting the vulnerability of computer-generated hologram based optical encryption. Opt Express. 2020;28(2):2499–510. https://doi.org/10.1364/OE.380004.
35. Montesinos López OA, López AM, Crossa J. Fundamentals of artificial neural networks and deep learning. In: Multivariate statistical machine learning methods for genomic prediction. Cham: Springer; 2022. p. 379–425. https://doi.org/10.1007/978-3-030-89010-0_10.
36. Kumar M, Chivukula AS, Barua G. Deep learning-based encryption scheme for medical images using DCGAN and virtual planet domain. Sci Rep. 2025;15(1):1211. https://doi.org/10.1038/s41598-024-84186-6.
37. Mienye ID, Swart TG, Obaido G. Recurrent neural networks: a comprehensive review of architectures, variants, and applications. Information. 2024;15(9):517. https://doi.org/10.3390/info15090517.
38. Wang X, Wang W, Wei H, Xu B, Dai C. Holographic and speckle encryption using deep learning. Opt Lett. 2021;46(23):5794–7. https://doi.org/10.1364/OL.443398.
39. Zhou L, Xiao Y, Chen W. Learning complex scattering media for optical encryption. Opt Lett. 2020;45(19):5279–82. https://doi.org/10.1364/OL.400174.
40. Zhou L, Xiao Y, Chen W. Learning-based optical authentication in complex scattering media. Opt Lasers Eng. 2021;141:106570. https://doi.org/10.1016/j.optlaseng.2021.106570.

41. Wang X, Wei H, Jin M, Xu B, Chen J. Experimental optical encryption based on random mask encoding and deep learning. Opt Express. 2022;30(7):11165–73. https://doi.org/10.1364/OE.454449.
42. Zhao Q, Li H, Yu Z, Woo CM, Zhong T, Cheng S, et al. Speckle-based optical cryptosystem and its application for human face recognition via deep learning. Adv Sci. 2022;9(27):2202407. https://doi.org/10.1002/advs.202202407.
43. Alzubaidi L, Zhang J, Humaidi AJ, Al-Dujaili A, Duan Y, Al-Shamma O, et al. Review of deep learning: concepts, CNN architectures, challenges, applications, future directions. J Big Data. 2021;8(53):1–74. https://doi.org/10.1186/s40537-021-00444-8.
44. Mienye ID, Swart TG. A comprehensive review of deep learning: architectures, recent advances, and applications. Information. 2024;15(12):755. https://doi.org/10.3390/info15120755.
45. S. S, Chen JIZ, Shakya S. Survey on neural network architectures with deep learning. J Soft Comput Paradigm. 2020;2(3):186–94. https://doi.org/10.36548/jscp.2020.3.007.
46. Gordon J, Hernández-Lobato JM. Combining deep generative and discriminative models for Bayesian semi-supervised learning. Pattern Recogn. 2020;100:107156. https://doi.org/10.1016/j.patcog.2019.107156.
47. Gridach M. Hybrid deep neural networks for recommender systems. Neurocomputing. 2020;413:23–30. https://doi.org/10.1016/j.neucom.2020.06.025.
48. Han X, Zhang K, Jin W, Zhu W, Li Y, Ma L. Deep learning decryption approach for asymmetric computer-generated holography (CGH) cryptosystem. Opt Express. 2024;32(14):22249–62. https://doi.org/10.1364/OE.524044.
49. Ma L, Sun X, Jin W. Symmetric–asymmetric hybrid encryption and decryption system based on chaotic iris phase mask and computer-generated holography. Opt Eng. 2020;59(8):083106. https://doi.org/10.1117/1.OE.59.8.083106.
50. Micikevicius P, Narang S, Alben J, Diamos G, Elsen E, Garcia D, et al. Mixed precision training. arXiv preprint arXiv:1710.03740. 2018.
51. Li S, Zhao Y, Varma R, Salpekar O, Noordhuis P, Li T, et al. PyTorch distributed: experiences on accelerating data parallel training. arXiv preprint arXiv:2006.15704. 2020.
52. Ronneberger O, Fischer P, Brox T. U-net: convolutional networks for biomedical image segmentation. In: Navab N, Hornegger J, Wells WM, Frangi AF, editors. Medical image computing and computer-assisted intervention—MICCAI 2015, vol. 9351. Cham: Springer; 2015. p. 234–41. https://doi.org/10.1007/978-3-319-24574-4_28.
53. Zhou Z, Siddiquee MMR, Tajbakhsh N, Liang J. UNet++: redesigning skip connections to exploit multiscale features in image segmentation. IEEE Trans Med Imaging. 2020;39(6):1856–67. https://doi.org/10.1109/TMI.2019.2959609.
54. Schneier B, Diffie W. Applied cryptography: protocols, algorithms, and source code in C. 2nd ed. Hoboken, NJ: Wiley; 2015.
55. He K, Zhang X, Ren S, Sun J. Deep residual learning for image recognition. In: Proc. IEEE Conf. Comput. Vis. Pattern Recognit. (CVPR), Las Vegas, NV, USA; 2016. p. 770–8. https://doi.org/10.1109/CVPR.2016.90.
56. LeCun Y, Bengio Y, Hinton G. Deep learning. Nature. 2015;521(7553):436–44. https://doi.org/10.1038/nature14539.
57. Badrinarayanan V, Kendall A, Cipolla R. SegNet: a deep convolutional encoder-decoder architecture for image segmentation. IEEE Trans Pattern Anal Mach Intell. 2017;39(12):2481–95. https://doi.org/10.1109/TPAMI.2016.2644615.
58. Zhang Z, Liu Q, Wang Y. Road extraction by deep residual U-net. IEEE Geosci Remote Sens Lett. 2018;15(5):749–53. https://doi.org/10.1109/LGRS.2018.2802944.
59. Xiao H, Rasul K, Vollgraf R. Fashion-MNIST: a novel image dataset for benchmarking machine learning algorithms. arXiv preprint arXiv:1708.07747. 2017.
60. Tsang PWM, Chow YT, Poon T-C. Generation of patterned-phase-only holograms (PPOHs). Opt Express. 2017;25(8):9088–93. https://doi.org/10.1364/oe.25.009088.
61. He C, Ming K, Wang Y, Wang ZJ. A deep learning based attack for the chaos-based image encryption. arXiv preprint arXiv:1902.09351. 2019.

62. Zhang B, Liu L. Chaos-based image encryption: review, application, and challenges. Mathematics. 2023;11(11):2585. https://doi.org/10.3390/math11112585.
63. Lata K, Cenkeramaddi LR. Deep learning for medical image cryptography: a comprehensive review. Appl Sci. 2023;13(14):8295. https://doi.org/10.3390/app13148295.
64. Panwar K, Kukreja S, Singh A, Singh KK. Towards deep learning for efficient image encryption. Procedia Comput Sci. 2023;218:644–50. https://doi.org/10.1016/j.procs.2023.01.046.
65. Fernando E, Murad DF, Wijanarko BD. Classification and advantages of parallel computing in process computation: a systematic literature review. In: 2018 international conference on computing, engineering, and design (ICCED), Bangkok, Thailand; 2018. p. 143–7. https://doi.org/10.1109/ICCED.2018.00036.
66. Shao L, Zhu F, Li X. Transfer learning for visual categorization: a survey. IEEE Trans Neural Netw Learn Syst. 2015;26(5):1019–34. https://doi.org/10.1109/TNNLS.2014.2330900.
67. Zhang X, Reveriano F, Lu J, Fu X, Zhang T. The effect of high-performance computer on deep learning: a face expression recognition case. In: Proc. IEEE Int. Conf. Comput. Sci. Eng. (CSE) and IEEE Int. Conf. Embedded ubiquitous Comput. (EUC), New York, NY, USA; 2019. p. 40–2. https://doi.org/10.1109/CSE/EUC.2019.00017.
68. Mumuni A, Mumuni F. Data augmentation: a comprehensive survey of modern approaches. Array. 2022;16:100258. https://doi.org/10.1016/j.array.2022.100258.
69. Cardenas OA, Nava LMF, Castaneda FG, Cadenas JAM. ECG arrhythmia classification for comparing pre-trained deep learning models. In: Proc. 19th Int. Conf. Electr. Eng., Comput. Sci. Autom. Control (CCE), Mexico City, Mexico; 2022. p. 1–5. https://doi.org/10.1109/CCE56709.2022.9975958.
70. Amrit P, Singh AK, Singh MP, Agrawal AK. EmbedR-net: using CNN to embed mark with recovery through deep convolutional GAN for secure eHealth systems. IEEE Trans Consum Electron. 2023;69(4):1017–22. https://doi.org/10.1109/TCE.2023.3283284.
71. Otoum S, Guizani N, Mouftah H. On the feasibility of split learning, transfer learning and federated learning for preserving security in ITS systems. IEEE Trans Intell Transp Syst. 2023;24(7):7462–70. https://doi.org/10.1109/TITS.2022.3159092.
72. Zhang L, Xu J, Vijayakumar P, Sharma PK, Ghosh U. Homomorphic encryption-based privacy-preserving federated learning in IoT-enabled healthcare system. IEEE Trans Netw Sci Eng. 2023;10(5):2864–80. https://doi.org/10.1109/TNSE.2022.3185327.

SciData-Factory: An Automated Framework for Generating High-Quality Image-Text Datasets

Li Zhu, Shaohua Li, Huan Li, Qianqian Lian, Siao Yu, and Hang Chen

Abstract The scarcity of high-quality image-text paired data severely constrains the development of multimodal large models in complex visual semantic understanding. Existing public datasets commonly suffer from noise introduced by web crawling, lack of domain-specific semantic annotations, and weak logical alignment between images and text, rendering them inadequate for interdisciplinary research. This paper proposes SciData-Factory, an automated framework for generating image-text datasets from academic literature to construct high-quality cross-modal data. The framework integrates a proprietary PDF parsing engine with a YOLO-based layout detection model and a large language model (e.g., Qwen-long) to achieve precise alignment between figures and their corresponding descriptive text in scholarly documents, and implements human-machine collaborative optimization strategies to enhance data quality. Experimental results demonstrate that datasets constructed using this framework significantly improve image understanding performance of multimodal large models in fine-tuning tasks across domains such as satellite imagery and agriculture.

Keywords Multimodal large models · Image-text pair dataset · Automated dataset generation · Academic document processing

1 Introduction

Current mainstream multimodal large models generally possess image recognition capabilities, yet exhibit significant performance disparities, primarily attributable to variations in training data composition-specifically, data scale, quality control

L. Zhu · S. Li · H. Li · Q. Lian · S. Yu
Engineering University of PAP, Xi'an, China

H. Chen (✉)
School of Space Information, Space Engineering University, Beijing, China

H. Chen, Z. Liu (eds.), *Artificial Intelligence in Digital Image Processing: Theories, Methods, and Applications*, Studies in Computational Intelligence 1246,
https://doi.org/10.1007/978-3-032-12891-1_8

strategies, and data diversity [1]. Image-Text dataset, serving as foundational infrastructure for model evolution, have undergone substantial paradigm shifts alongside the advancement of multimodal AI: evolving from early small-scale, task-oriented academic datasets to today's industrial-scale pretraining datasets containing hundreds of billions of samples. The field has now bifurcated into two dominant paradigms: academic research datasets emphasize annotation precision and task-specific adaptability, featuring controlled scale and rigorous curation; industrial pretraining datasets prioritize massive scale and cross-cultural/content heterogeneity, constructed primarily via large-scale web crawling. The systematic differences between these two dataset types-in openness, structural complexity, annotation granularity, and scale-directly determine their suitability for distinct research paradigms and application scenarios.

With the continuous advancement of technology, several large-scale image-text datasets (such as Conceptual Captions, WIT, and OpenAI CLIP) have emerged in recent years, providing a critical data foundation for multimodal pre-training. However, these datasets still commonly suffer from issues such as label noise and image-text mismatches, which affect their quality. Current multimodal large models still face three core bottlenecks in image understanding: (1) Insufficient structured parsing capabilities for complex images. Models struggle to systematically model the internal structure of images, deep logical relationships among elements, and implicit semantics; (2) Weak cognitive ability for interdisciplinary specialized graphics. Significant deficiencies exist in understanding the symbolic systems and semantic structures of specialized graphics such as scientific diagrams, engineering drawings, and medical images; (3) Limited generalization ability for"image-text" logical associations. The multimodal correspondences learned by models are mostly superficial [2], leading to a significant decline in generalization performance when faced with logically complex, novel, or highly implicit semantic scenarios.

The root cause of these three issues lies in the structural limitations at the data level, which can be summarized as the "data bottleneck" in current multimodal research. Firstly, there is a severe lack of deep semantic annotations. Mainstream large-scale datasets (e.g., Conceptual Captions) mostly rely on automated web scraping, resulting in generally shallow textual descriptions that lack fine-grained annotations and cross-modal semantic alignment for structured semantic elements such as spatial relationships, reasoning chains, and implicit semantics, thereby hindering models from acquiring deep structured knowledge. Secondly, high-quality data in specialized domains is scarce. In vertical fields such as medicine, scientific visualization, and engineering drawings, high-quality, semantically aligned image-text pair resources are extremely limited. Moreover, data acquisition costs are high and annotation relies on expert knowledge [3], leading to a severe imbalance in training data distribution. Thirdly, data construction remains highly dependent on manual effort. Currently, constructing image-text pairs with both semantic depth and domain expertise still requires significant human intervention. There is an urgent need to develop standardized, automated, and quality-assured generation pipelines to overcome the dual constraints of efficiency and quality in data construction.

In response to the above issues, this paper proposes the SciData-Factory system framework. Using figures and their corresponding textual descriptions from academic papers as the data source, it constructs an automated processing pipeline for generating high-quality image-text datasets with strong domain specificity, rich semantics, and broad disciplinary coverage, thereby providing high-quality data support for fine-tuning and training multimodal large models.

2 Related Work

2.1 Image Captioning Dataset

Table 1 summarizes the mainstream image-text datasets and their key attributes. From an evolutionary perspective, these datasets exhibit a dual development trajectory characterized by "simultaneous pursuit of scale and precision, and divergence between general-purpose and domain-specific applications." First, data scale expansion has been the most prominent trend in recent years, primarily aimed at providing sufficient training material for foundation models with massive parameter counts. LAION-5B, containing 5.85 billion CLIP-filtered image-text pairs, once served as a benchmark in the open-source community [4], demonstrating the feasibility of constructing ultra-large-scale datasets through automated filtering of web-crawled data using pretrained models [5]. OmniCorpus (86 billion images) and WebLI-100B (100 billion image-text pairs) further push the scale to unprecedented levels. These datasets heavily rely on web crawling, and their annotation paradigms have shifted from costly manual labeling to automated filtering (e.g., via CLIP), weak supervision, and synthetic generation (e.g., PD12M uses Florence-2 to generate captions). Their core characteristic is "high volume, low density," designed to enable models to learn broad, generic associations between visual concepts and linguistic representations-primarily supporting text-to-image models such as Stable Diffusion [6] and multimodal large language models (MLLMs) like GPT-4o [7], endowing them with strong zero-shot and few-shot generalization capabilities.

However, pure scale expansion risks introducing data noise, copyright infringement, and amplification of societal biases. Therefore, alongside scaling, data quality refinement and value-oriented curation constitute a parallel development axis, manifested in three directions: (1) Quality refinement atop large-scale data-e.g., RealSyn achieves higher annotation consistency at 100 M scale via semantic cluster-balanced sampling and rigorous filtering; ImageRewardDB focuses on human aesthetic preferences, providing 137 K expert-annotated pairwise comparisons to optimize generative model alignment. (2) Fine-grained and specialized annotation-though COCO contains only 330 K images, its instance segmentation and keypoint annotations have established it as the authoritative benchmark for object detection and segmentation tasks [8]; CUB-200-2011 targets fine-grained classification, annotating 11 K bird images with 15 part locations and 312 attributes, advancing

Table 1 Mainstream image-text datasets

Dataset name	Data volume	Data source	Annotation method	Key characteristics
COCO	>330K images, >2.5M object instances	Everyday scenes (streets, indoors, parks, etc.)	Manual annotation (bounding boxes, segmentation masks, keypoints, image captions)	High diversity (varied environments, weather, lighting); fine-grained annotations; challenging (occlusion, small objects); fully public
RealSyn	>100M high-quality image-text pairs	Derived from 118M interleaved image-text documents, rigorously filtered and refined	Hybrid of synthetic and real text descriptions, strictly filtered	Large scale, high quality; employs cluster-based semantic-balanced sampling; provides three scale variants (15M, 30M, 100M)
LAION-5B	5.85B image-text pairs	Web-crawled, filtered via CLIP	Automatic filtering using CLIP (similarity threshold); includes watermark and NSFW detection	Largest public multimodal dataset; multiple subsets (LAION-2B-en, LAION-2B-multi, LAION-1B-nolang); includes high-res images
Wukong-CMNER	55K+ annotated image-text pairs	Chinese corpus (annotated from Huawei's Wukong dataset)	Manual annotation (named entity recognition)	Focuses on Chinese multimodal NER; tightly coupled image-text alignment
OBELICS	141M documents, 353M images, 115B text tokens	Web-crawled (processed from raw WARC archives)	Automated processing (HTML extraction, DOM simplification, filtering, deduplication)	Large-scale open-source interleaved document corpus; meticulously filtered and deduplicated; provides interactive visualization tools
ImageRewardDB	137K expert comparisons (from DiffusionDB)	DiffusionDB	Expert preference comparisons and scoring between prompts and generated images	Focuses on human preference in text-to-image generation; includes multiple scoring dimensions (overall, alignment, fidelity); four subset scales (1K, 2K, 4K, 8K)
Text-to-Image-2M	~2M samples	Augmented from high-quality datasets (e.g., LLaVA, ProGamerGov synthetic data); partial text generated by GPT-4o	Text regenerated by Qwen2-VL or retained as original; some images generated by flux-dev	High-quality text-image pairs; available in 512 × 512 and 1024 × 1024 resolutions; designed for fine-tuning text-to-image models

Llama Nemotron VLM v1	~2.864M samples	Re-annotated public image datasets, synthetic/semi-synthetic OCR data (Chinese/English), internal OCR annotations	Refinements and enhancements to original QA or captions	Contains 21 subsets; covers VQA, captioning, and OCR; supports multimodal training and evaluation
MANTA	>1.373M images (86K with anomaly annotations)	Collected from micro-objects in agriculture, pharma, electronics, machinery, groceries	Pixel-level anomaly annotations; text includes declarative knowledge and constructivist learning subsets	Multi-view capture (5 angles per object); focuses on micro-object anomaly detection; rich textual descriptions and MCQs
PD12M (public domain 12M)	12.4M images	Public domain & CC0 sources (GLAM institutions, Wikimedia commons, etc.)	Synthetic captions generated by Florence-2-large	One of the largest public-domain image-text datasets; clear copyright; community-driven governance
OmniCorpus	86B images, 169.6B text tokens	Web-crawled (multilingual websites and video platforms)	Automatic filtering (CLIP aesthetics/NSFW scores, heuristics) + human feedback refinement	Massive scale; supports Chinese & English; high quality (strictly filtered); flexible formats (convertible to text-only or image-text pairs)
Flickr30k	31K images (Flickr30K entities as extension)	Flickr photo-sharing platform	Crowdsourced (5 captions per image); extended version includes coreference chains and bounding boxes	Rich annotations (5 captions/image); diverse everyday scenes; facilitates entity and semantic understanding
WebLI-100B	100B image-text pairs	Web-crawled	Weakly supervised (prioritizes scale expansion over strict filtering; preserves cultural details)	Ultra-large scale; emphasizes cultural and multilingual diversity (including low-resource languages); highly inclusive
CUB-200-2011	11,788 images (200 bird species)	Flickr	Fine-grained annotation (bounding boxes, part keypoints, attribute labels)	Fine-grained visual categorization (birds); extremely rich annotations (15 parts, 312 attributes); high intra-class variation and inter-class similarity

research on highly discriminative visual recognition [9]. (3) Vertical domain or task-specific construction-e.g., Wukong-CMNER focuses on Chinese multimodal named entity recognition; MANTA builds a large-scale dataset for industrial micro-defect anomaly detection; OBELiX serves domain-specific applications in materials science discovery [10].

From a technical application perspective, these datasets are precisely aligned with different stages of model development: ultra-large-scale weakly annotated datasets (LAION, WebLI, OmniCorpus) serve as pretraining foundations, endowing models with general world knowledge; high-quality, densely annotated datasets (COCO, Flickr30k) act as benchmarks for fine-tuning and performance evaluation-Flickr30k's five captions per image make it ideal for image captioning and visual question answering (VQA); highly specialized datasets (CUB, MANTA) are employed to hone cutting-edge capabilities on specific tasks, facilitating real-world industrial deployment.

Looking ahead, multimodal dataset development will continue to balance scale and quality, but future emphasis may shift toward: (1) Synthetic data utilization-as demonstrated by Text-to-Image-2M, generating high-quality, copyright-free data via generative models will become a critical supplement; (2) Complex annotation and reasoning capability construction-extending from static image-text pairs to multimodal structures integrating video, audio, and temporal text, with annotation objectives evolving from object description to dynamic process understanding, causal reasoning, and complex semantic modeling; (3) Enhanced safety and ethical compliance-data cleaning will prioritize bias mitigation, harmful content filtering, and explicit copyright licensing (e.g., PD12M exclusively uses public-domain images), paving the way for legally compliant commercialization. In summary, from macroscopic generality to microscopic specialization, these diverse datasets collectively constitute the ecosystem underpinning multimodal AI research and application.

2.2 *Methodologies for Constructing Image-Text Datasets*

Current approaches to constructing "image-text" datasets primarily rely on four distinct paradigms, each exhibiting inherent systemic limitations that hinder the development of multimodal models towards higher precision, stronger generalization, and reduced bias. The first paradigm involves large-scale web scraping followed by model filtering (e.g., LAION-5B). While these datasets achieve billion-scale size and cover broad visual-semantic spaces, their textual descriptions are often sourced directly from web content, resulting in shallow semantics and high noise levels. Furthermore, filtering models like CLIP introduce their own semantic biases and distribution shifts, leading to widespread "pseudo-alignment" within the data. This significantly impairs the model's ability to learn deep semantic and logical relationships. The second approach is manual re-annotation of existing datasets (e.g., FineHARD, Wukong-CMNER). Although such datasets excel in entity localization and fine-grained semantic alignment, they suffer from high annotation costs and poor scalability. They struggle to incorporate emerging domains or long-tail concepts, and maintaining consistency across human annotators is challenging, often introducing subjective bias. The third paradigm employs synthetic data generated by models (e.g., ImageRewardDB, RealSyn). While this method mitigates copyright concerns and allows for targeted data expansion, it is constrained by the current limitations of text-to-image or image-to-text models in semantic understanding. Generated content often contains factual inaccuracies, attribute mismatches, or semantic ambiguities, creating a negative feedback loop where errors are reinforced through training and regeneration. The fourth category involves expert-curated data in vertical domains (e.g., MANTA, PathVLM). These datasets are characterized by high quality and semantic precision but heavily rely on scarce domain expertise, resulting in long development cycles and non-replicable costs. Consequently, professional domains continue to experience a persistent shortage of high-quality data.

At the annotation mechanism level, automated filtering sacrifices semantic depth for scale, manual annotation sacrifices efficiency for precision, and human preference evaluation (e.g., ImageRewardDB) is susceptible to subjective scoring influences. Hybrid strategies (e.g., Llama Nemotron VLM v1) have yet to establish unified quality evaluation standards. Current methods generally lack semantic consistency validation mechanisms, cross-modal logical reasoning capabilities, and domain-adaptive scalability, resulting in datasets that underperform in complex reasoning, specialized graphic interpretation, and low-resource language scenarios. Particularly in specialized tasks such as agricultural pest and disease identification, remote sensing feature interpretation, and medical imaging report generation, high-quality aligned data is nearly absent. In non-English language systems such as Chinese and Arabic, the semantic gap between images and texts significantly widens.

In summary, current construction paradigms exhibit fundamental bottlenecks across four dimensions: quality controllability, semantic depth, domain scalability, and annotation efficiency. It is therefore imperative to develop an automated framework for generating high-quality image-text datasets. Such a framework should integrate capabilities for controllable generation, semantic rectification, logical consistency verification, and domain knowledge infusion. By minimizing manual intervention, it would enable large-scale production of data with high semantic fidelity, structural integrity, and balanced distribution, thereby establishing a solid, reliable, and generalizable training foundation for next-generation multimodal large models.

3 SciData-Factory Construction

The core challenge in automatically constructing high-quality image-text datasets lies in acquiring semantically coherent, accurate, and sufficiently large-scale sources of paired data. Academic papers, as rigorously structured and peer-reviewed publications, inherently possess a standardized multimodal knowledge representation system. Specifically, the figures, corresponding captions, and related contextual descriptions collectively form semantically consistent, highly professional, and accurately aligned"image–caption–plain text" triplets (as shown in Fig. 1 [11–13]). Additionally, academic papers typically undergo multiple stages of quality control, including author drafting, editorial revision, and peer review, before publication. This process substantially enhances the semantic accuracy and consistency between figures and textual content. Therefore, the precise extraction of such structured multimodal information from large scale academic literature, enabling the construction of a massive and high-fidelity data source, could fundamentally resolve the critical shortage of high-quality training data for multimodal models.

Following this rationale, this paper proposes the "Academic Cross-Modal Data Factory" (SciData-Factory), an end-to-end automated generation framework that takes academic PDFs as input and produces high-quality image-text datasets as output. Inspired by the factory paradigm, the framework comprises four functional modules, as illustrated in Fig. 2.

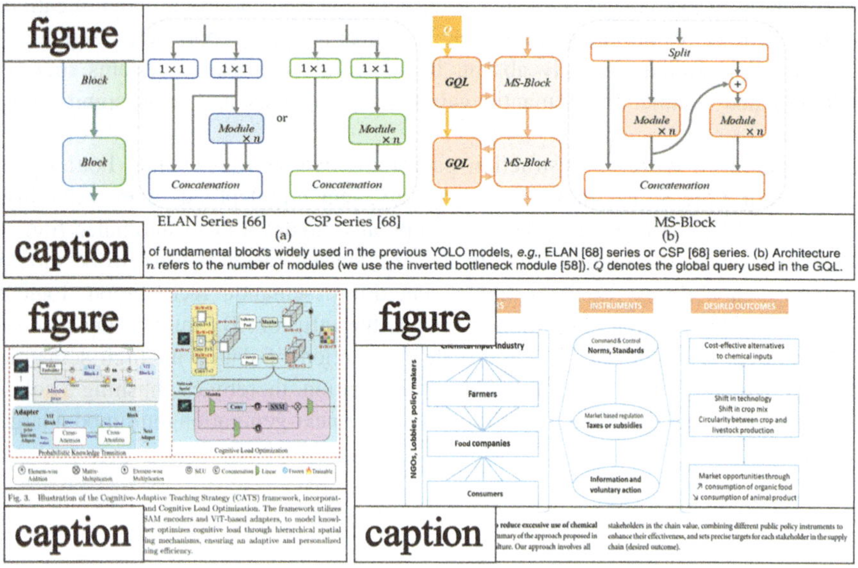

Fig. 1 Example of figures and captions in academic papers

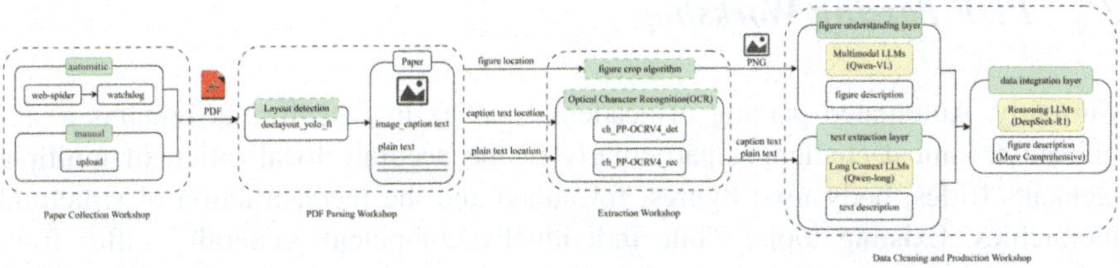

Fig. 2 Overall architecture diagram

3.1 Paper Collection Workshop

As a fundamental module of the data factory, the Paper Collection Workshop is responsible for automatically discovering, ingesting, and performing initial processing of academic literature from external sources. It serves as a core component that ensures data supply for subsequent processes. The module adopts a highly scalable and automated architecture, supporting continuous data inflow and efficient management to meet the demands of interdisciplinary research for highly credible literature resources.

In selecting data sources, this study establishes four criteria: Open Access, Cross-disciplinary Coverage, High Academic Standing, and Stringent Peer Review. Accordingly, core journals including *Nature*, *Science*, *PNAS*, *PLOS ONE*, and leading disciplinary journals (such as *Cell*, *The Lancet*, and the *IEEE Transactionsseries*) are selected as primary literature sources. This approach constructs an authoritative and representative raw corpus, providing a trustworthy data foundation for subsequent cross-disciplinary knowledge mining.

To achieve efficient acquisition of PDF documents, the system integrates two core functionalities: automated crawling with real-time updates and external API import. Firstly, web crawlers are employed to automatically gather publicly available academic resources. A real-time filesystem monitoring layer, built on the watchdog library, continuously watches one or more local directories specified by the BASE_PDF_DIR variable in the configuration file. Upon detection of any new or updated PDF files, the preprocessing pipeline is triggered immediately to enable incremental synchronization and dynamic updating of the database. Secondly, the system provides a manual upload feature via API interfaces (defined in the routers directory), serving as a essential supplement to automated collection. This allows users to upload specific documents on demand, comprehensively addressing database maintenance requirements. Through this design, the paper collection pipeline combines both directory monitoring-based real-time updating capabilities and API-driven flexible upload mechanisms.

3.2 PDF Parsing Workshop

Currently, Automated parsing of academic papers in PDF format remains a significant technical challenge, particularly in the accurate localization of multiple elements (titles, body text, figures, formulas) and the reconstruction of structural hierarchies. Existing tools, while individually competent, generally suffer from functional isolation, modality fragmentation, or computational inefficiency [14]. For instance, PaddleOCR excels in text detection and recognition (PP-OCRv4 supports over 15,000 characters, including Chinese, English, Japanese, and special symbols), yet cannot directly extract image content and exhibits limited capability in understanding complex document layouts. PyMuPDF efficiently extracts text and tabular data but lacks a native OCR module, rendering it ineffective for scanned PDFs. MinerU supports OCR in 84 languages and extracts tables and formulas, but its highly modular architecture incurs high customization and deployment costs. OCRmyPDF relies on CPU-based post-processing, resulting in low efficiency and overly complex output structures, making it unsuitable for high-precision academic parsing. Collectively, these tools fail to achieve end-to-end multimodal co-parsing [15].

To address these issues, this study develops a standardized, configurable academic document parsing pipeline through deep customization and functional extension of multiple open-source PDF parsing projects. The system first employs an enhanced YOLO-v10 model for layout analysis, leveraging diverse pretraining data augmentation and multi-scale contextual perception to accurately localize spatial boundaries and hierarchical relationships of elements such as titles, body text, figures, and formulas, outputting structured JSON metadata [16]. The text extraction module integrates PP-OCRv4_det and PP-OCRv4_rec models, augmented with a text orientation classifier to enable recognition at arbitrary angles, significantly improving accuracy for mixed Chinese-English scenarios while preserving a lightweight architecture. Formula processing adopts a two-stage strategy: Stage one uses a YOLO-based detector to localize inline and display formula regions; stage two employs the UniMERNet model to convert formula images into structured LaTeX expressions, effectively capturing semantic representations of complex mathematical and chemical notations [17]. Through a configuration-driven interface, the system supports dynamic parameter tuning, adapting seamlessly to diverse application scenarios-from fine-grained single-document parsing to large-scale academic corpus construction-achieving optimal balance between parsing accuracy and computational efficiency.

3.3 Extraction Workshop

In a PDF document, the figure, its caption, and the descriptive text within the main body that explains it are typically spatially separated. This is the primary problem that the extraction workshop needs to solve when constructing structured "figure-caption-description" (F-C-D) triplets [18].

The outputs of the PDF document parsing workshop are primarily divided into two categories: Scanned PDFs and Native PDFs. Each category features a unique method for storing document content, which directly impacts subsequent operations such as text search and image extraction [19].

In a Native PDF, elements including text, images, tables, and page layouts are split and stored as individual "objects"-for example, a text block functions as one object, and a table as another. The internal description of an image object contains special tags: either /Subtype or /Image. During actual processing, the system iterates through all descriptive objects in the PDF document. Once an image object is identified, it is decoded and converted into a universal image format [20].

Scanned PDFs are generated by scanning paper documents or capturing images with a camera. In essence, they are PDF files formed by sequentially stitching multiple images together; each page contains only a single image object and no recognizable text code. Clearly, the method used to extract images from Native PDFs is not applicable to Scanned PDFs. Instead, Scanned PDFs require a technology capable of extracting target images from image-based content. Specifically, each page of the Scanned PDF can be treated as a single image. Using the position coordinates obtained from the earlier PDF document parsing process, lossless cropping is performed on this image to isolate the targeted image region [21]. The detailed process of image extraction from Scanned PDFs is presented in Algorithm 1.

Algorithm 1 The detailed pro Incess of image extraction from scanned PDFs
Input: Original PDF file
Output: Extracted image and corresponding title information (stored as JSON file)
FUNCTION process_pdf(pdf_file):
  FOR each page ***IN*** pdf_file:
  // Predicting page layout using layout models
  layout_results = PREDICT_LAYOUT(page)
END FUNCTION
FUNCTION crop_figure(layout_results):
  FOR each image label ***IN*** layout_results:
    // Calculate cropping coordinates based on image coordinates and page image size
    crop_xmin, crop_ymin, crop_xmax, crop_ymax =
      CALCULATE_CROP_COORDINATES(image_coordinates, page_image.size)
END FUNCTION
FUNCTION find_closest_caption(layout_results):
// Initialize image list
  FOR each image In layout_results ***Do***:

```
        // Calculate the distance difference between the image and the annotation
area, and take the minimum value
        distance = abs(cap_y - fig_y)
           If distance < min_distance Then:
              min_distance = distance
              closest_caption = caption
           image_captain_list = APPEND_IMAGE_CAPTION(image,closest_
caption)
     STORAGE_IN_JSON(image_caption_list)
  END FUNCTION
  // Main
  MAIN:
     pdf_file = LOAD_PDF("input.pdf")
     layout_results = process_pdf(pdf_file)
     crop_figure(layout_results)
     find_closest_caption(layout_results)
  END MAIN
```

To acquire the caption information corresponding to images, this paper proposes a graph-text association algorithm that comprehensively considers rules and distances. Specifically, the algorithm proceeds as follows: first, drawing on the typesetting conventions of academic documents, caption texts usually contain index markers such as "Figure X" and "Fig. X". These markers are used to establish a set of caption regions. Second, horizontal and vertical projection operations are performed on each image in the PDF page. By recording the start and end values of continuous value segments, two sets of coordinates are obtained. Then, the distances between the image and all caption regions are calculated, and the minimum distance is identified. The caption region corresponding to this minimum distance is taken as the caption content for the image. Through this approach, batch and accurate extraction of caption content is achieved.

On the basis of the already extracted images and captions, the extraction of image description texts is completed in two steps: "localization" and "screening" [22]. First, the layout is analyzed to lock in the text regions associated with the images. Given the typesetting convention of "proximity-based graph-text association" in academic documents, the spatial search range for candidate texts is set to the three text blocks above and below each image. Subsequently, an OCR tool is used to convert the text blocks within this spatial range into processable text content, while the coordinate information of each text block is retained. Second, core description texts are screened based on semantic features. Such texts usually contain "identifying terms" directly related to the image content; the sentences containing these key terms are marked as "highly relevant texts" and serve as the descriptive texts for the images in the main body. The diagram illustrates the schematic process of extracting these descriptive texts. Through the above steps, the entire process of constructing the "F-C-D triple data" is completed.

3.4 Data Cleaning and Production Workshop

Through the aforementioned processing stages, the acquisition of images and their corresponding textual descriptions has been largely completed. However, the directly extracted "image–caption–plain text" triplets primarily focus on academic semantic content, while overlooking visual elements such as colors, backgrounds, and objects within the images. This leads to issues including semantic fragmentation, informational redundancy, and lack of visual detail in the generated image descriptions. Therefore, how to effectively integrate "textual semantic descriptions of images" and "intrinsic visual feature descriptions of the images" to produce semantically accurate and comprehensively detailed image captions has become a critical challenge addressed in this study. To tackle this issue, we propose a multi-model collaborative approach for cleaning and fusing image-text data, employing a "three-layer architecture + multi-model collaboration" design to achieve efficient alignment and semantic enhancement of multimodal information [23], as illustrated in Fig. 3.

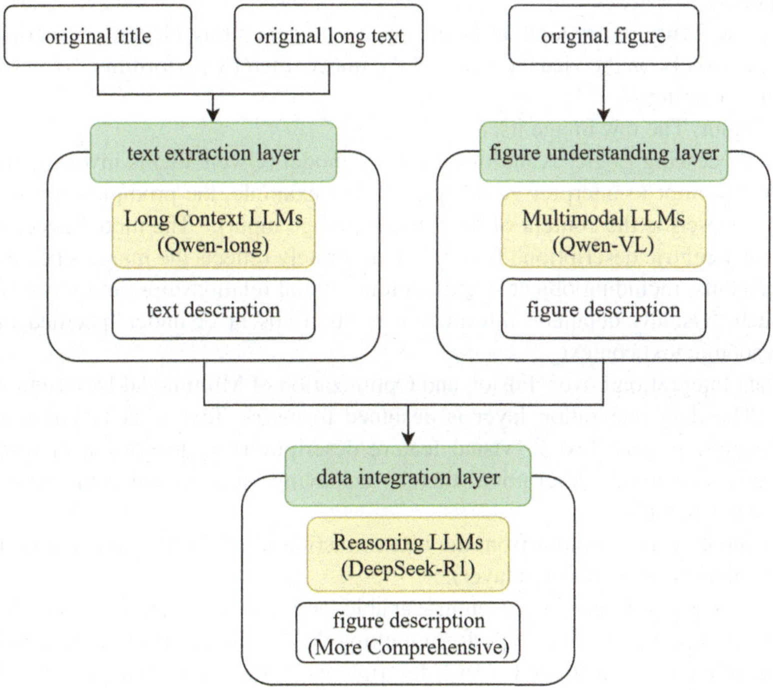

Fig. 3 Data cleaning and production

1. Text Extraction Layer: Structuring Image Descriptions from Long-Context Texts
   The core objective of the text extraction layer is to distill effective image-related descriptions from associated long-form texts (e.g., paper body, figure titles), thereby performing "text-level data cleaning."
   Input: "Original caption" and "original long plain text" semantically linked to the target image. These texts may contain background context, methodological explanations, or result summaries, but often exhibit semantic dispersion or redundancy.
   Processing Logic: A long-context large language model (Qwen-long) is invoked. Through prompt engineering, the model is guided to focus exclusively on image-relevant content. For instance, the prompt is structured as: "Extract descriptive information about the target image from the following text: [original long plain text], in conjunction with the title: [original caption], and generate a concise textual description (Text 1)." The model filters out irrelevant segments and distills core semantic descriptions from dispersed and redundant inputs [24]. Such academically grounded descriptions are inherently inaccessible to purely vision-based models.
2. Figure Understanding Layer: Visual Feature Description from the Image Perspective
   The figure understanding layer focuses on generating objective descriptions based solely on the visual content of the image, thereby performing "visual-level data cleaning."
   Input: The raw image itself.
   Processing Logic: A multimodal large model (Qwen-VL) is invoked, guided by a prompt to interpret visual content. For example, the prompt is formulated as: "Describe the content of this image: [image input]." The model generates a visual-centric description (Text 2) that accurately reflects the image's perceptual elements, including object shapes, colors, spatial relationships, and scene types. Such "visually detailed" information is often absent or under-specified in the academic text context.
3. Data Integration Layer: Fusion and Optimization of Multimodal Descriptions
   The data integration layer is designed to merge Text 1 (academic context description) and Text 2 (visual feature description) to generate a more comprehensive image description, thereby achieving "multimodal data fusion and enhancement."
   Input: Text 1 (output from the text extraction layer) and Text 2 (output from the image comprehension layer).
   Processing Logic: A reasoning-capable large model is employed and guided via prompt engineering to perform multimodal fusion. For example, the following prompt is constructed: "Two descriptions of the same image are provided: Text 1 contains information extracted from the academic paper: [Text 1 content]; Text 2 contains information derived from visual features of the image: [Text 2 content]. Please integrate these two descriptions to generate a more complete and coherent image caption. The output should incorporate both contextual information from the text and visual details from the image, while resolving

any contradictory elements." The model performs deduplication, complementary integration, and linguistic refinement of the two descriptions, ultimately producing a more comprehensive image caption.

This solution offers multiple advantages. First, it possesses multidimensional information fusion capability, simultaneously covering semantic knowledge derived from text (such as academic concepts and contextual logic) and visual details from images (such as color, shape, and spatial layout), thereby achieving deep complementarity between semantic and visual dimensions. Second, it supports data diversity. The strategy of reusing large model capabilities fully leverages the specialized strengths of various models through layered collaboration among long-text comprehension models, cross-modal parsing models, and generative reasoning models, thereby avoiding the limitations of a single model. Third, the process emphasizes both cleansing and enhancement. The first two layers refine and filter textual and visual data respectively, while the integration layer performs creative synthesis, ultimately enabling data enhancement and producing more comprehensive descriptions. Fourth, the architecture exhibits broad scenario applicability. By simply adjusting prompts and model parameters, it can be flexibly adapted to various business needs, such as academic publishing, new media content optimization, and digital asset tag generation.

4 System Design

To achieve end-to-end automation from academic paper acquisition to dataset output, this paper presents a lightweight system designed to construct an efficient and automated data processing pipeline. The system primarily targets PDF documents as data sources and integrates automated extraction, intelligent analysis, structured storage, and visual management into a unified framework.

The system adopts a frontend-backend separation architecture. The backend is built on the Python-based FastAPI framework, while the frontend is a Single Page Application (SPA) developed using the React framework. This architecture clearly delineates the responsibilities of each layer, supports independent development, testing, and deployment, and enhances the system's scalability and maintainability. The core design follows the modular principle: backend services are divided into configuration management, API routing, business logic processing, data model definition, and a tool layer, achieving high cohesion and low coupling. The use of environment variable management isolates configuration data from core code, thereby improving security and flexibility.

The frontend is developed using TypeScript, leveraging the Vite build tool and React framework to implement a responsive user interface. It is primarily responsible for providing users with interactive functionalities for data management and system status monitoring, including file upload and management, visualization of data processing results (charts and tables), triggering of analytical tasks and feedback of results, as well as real-time monitoring of system status and task queues.

The frontend asynchronously calls unified backend API interfaces to retrieve data and execute operations, ensuring real-time consistency in data presentation and processing.

The backend serves as the core of the system, undertaking processing and scheduling tasks. The DirectoryMonitorService monitors specified directories in real time or utilizes an API to receive uploaded PDF files, thereby triggering the processing pipeline. Upon file arrival, the ResourceService employs PyMuPDF to perform document parsing, accurately extracting text, images, and metadata. To enhance data value and usability, the system provides an optional intelligent analysis module, which transmits the extracted text information to a large language model (e.g., DeepSeek) via an API for further optimization and enhancement. All processed, structured, or enriched data records are persistently stored in a MongoDB database through a data access layer built on Motor (an asynchronous MongoDB driver) and Beanie (an asynchronous ODM). MongoDB's flexible schema design is particularly suited for handling semi-structured document data. Finally, the backend exposes the processed structured data to the frontend and application layer through a high-performance FastAPI-based interface layer (located in the routers directory), which offers well-defined RESTful APIs. Uvicorn, as an ASGI server, supports asynchronous processing, effectively handling I/O-intensive requests and ensuring high system throughput and responsiveness.

As shown in Fig. 4, the system achieves a paper download rate of 20 papers per second and a data extraction speed of 10 s per paper. To date, it has successfully generated datasets across multiple domains, including aerospace, agriculture, and mechanical manufacturing.

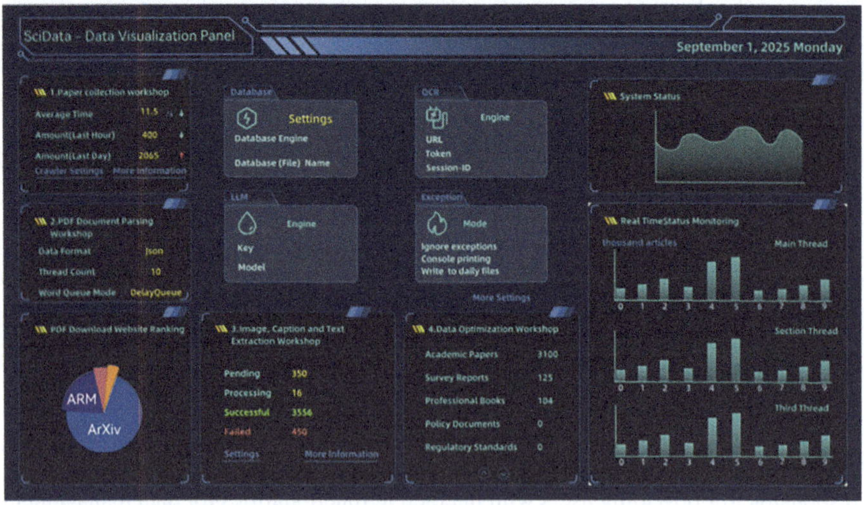

Fig. 4 Interface of the SciData system

5 Fine-Tuning Large Models with the Generated Dataset

To evaluate the quality and practicality of the "image-text" pair dataset constructed in this paper, this study conducted supervised fine-tuning (SFT) on the enhanced version of the large-scale vision-language model Qwen-VL-Plus, using 2297 automatically generated samples.

The fine-tuning experiment was carried out on the Alibaba Cloud BaiLian Large Model Service Platform, which provides elastic computing resources and a standardized training framework to support model development, training, and cloud service deployment. The specific process is as follows: (1) Upload the local dataset to the platform's data management module; (2) Configure training parameters (initial learning rate of 0.0003, training epochs of 5); (3) Perform training using the default AdamW optimizer; (4) Monitor changes in the loss function and learning rate in real time; (5) Deploy the fine-tuned model (SciData-LargeModel) online after training completion.

Experimental results indicate that the model's loss value decreased steadily from an initial 2.1705 to 1.6254. Meanwhile, the learning rate decayed gradually to 0.0001210931609177865 in line with the preset strategy-findings that demonstrate stable convergence of the optimization process. In the validation phase, the time required for a single evaluation remained stable at approximately 260 s, and the evaluation throughput was maintained at 1.7 samples/s and 1.7 steps/s. These metrics reflect the robustness of the training workflow. Figure 5 compares the outputs generated by inputting the same image into Qwen_VL_Max and the fine-tuned SciData-LargeModel, respectively. It is evident that the fine-tuned SciData-LargeModel delivers more professional responses, characterized by greater comprehensiveness and rigor.

It should be noted that due to the constraints of dataset scale, there is an upper limit to the improvement of the model in fine-grained semantic understanding.

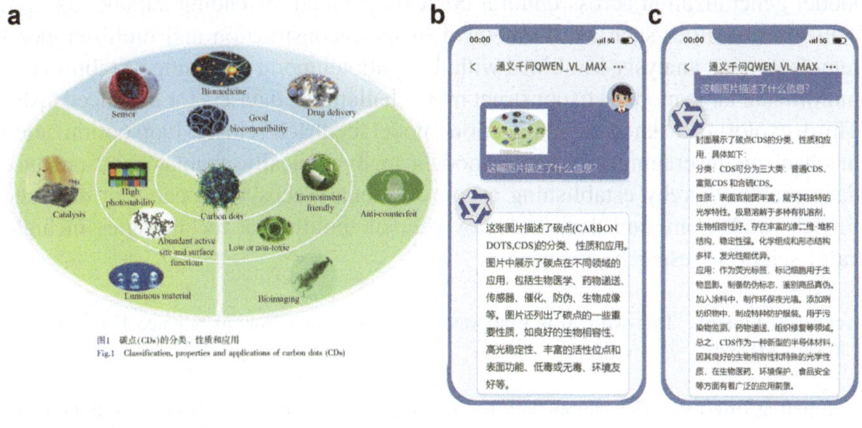

Fig. 5 (**a**) Input image. (**b**) Qwen VL2.5 reply. (**c**) Fine tuning model reply

However, the performance of the fine-tuned model on open-domain visual question answering (VQA) tasks has verified the basic quality and practicality of the dataset, laying a foundation for in-depth optimization in subsequent work when the dataset scale is expanded.

6 Conclusion

To address the scarcity of high-quality image-text data for training multimodal large models, this study proposes SciData-Factory-an end-to-end automated framework that takes academic PDFs as input and outputs high-quality image-text datasets. The system integrates a self-developed, high-performance PDF parsing engine capable of accurately processing complex academic layouts, including two-column formatting, mixed text-figure arrangements, and embedded formulas, enabling fine-grained extraction and semantic alignment of images, captions, and associated contextual text. The full pipeline encompasses core modules including automated literature acquisition, multimodal object detection, cross-page content reassembly, image-text semantic validation, and quality filtering, forming a highly automated, scalable, and minimally human-dependent data production pipeline. This framework not only alleviates the limitations of existing datasets in source authority and image-text logical coherence, but also endows generated data with explicit semantic hierarchies and domain-specific knowledge completeness. Experimental results demonstrate that fine-tuning mainstream multimodal models (e.g., Qwen-VL 2.5) on datasets generated by this framework yields significant performance improvements in scientific visual understanding, fine-grained reasoning, and cross-modal retrieval tasks.

Future work can be extended along three directions: First, enhancing cross-lingual processing capability—the current system focuses on Chinese and English literature; incorporating academic resources in additional languages will improve model generalization across cultural contexts. Second, extending support for specialized modalities-such as 3D medical image reconstruction and multi-temporal remote sensing analysis-combined with the spatiotemporal reasoning capabilities of multimodal large models, to construct more challenging datasets for advanced tasks. Third, exploring a "knowledge injection–model feedback" closed-loop optimization mechanism, leveraging pretrained models to dynamically evaluate and enhance data quality, thereby establishing a co-evolutionary ecosystem between data and models, and promoting deeper deployment of multimodal technologies in high-value scientific research scenarios.

Acknowledgments This work was supported by the National Natural Science Foundation of China (62005320).

Competing Interests The authors have no conflicts of interest to declare that are relevant to the content of this chapter.

References

1. DEEPMIND. Scaling pre-training to one hundred billion data for vision language models. 2025. https://arxiv.org/abs/2502.12597. Accessed Sept 9 2025.
2. Chen L, Wang Z, Li H. The alignment bottleneck in multimodal large language models: a comprehensive survey on text, image, and video integration. Front Comput Sci. 2025;19(3):567–89.
3. Zhang Y, Jiang H, Miao Y, et al. PathVLM: a large vision-language model for histopathology. Nat Commun. 2024;15:3367.
4. Schuhmacher L, Glushkova D, Schuhmacher D. Laion-5B: an open large-scale dataset for training next generation image-text models. 2022. https://laion.ai/blog/laion-5b/. Accessed May 4 2025.
5. Radford A, Kim JW, Halliday J, et al. Learning transferable visual models from natural language supervision. In: Proceedings of the 38th international conference on machine learning. PMLR; 2021. p. 8748–63.
6. Rombach R, Blattmann A, Lorenz D, et al. High-resolution image synthesis with latent diffusion models. In: Proceedings of the IEEE/CVF conference on computer vision and pattern recognition; 2022. p. 10684–95.
7. OPENAI. GPT-4 Technical Report. 2023. https://cdn.openai.com/papers/gpt-4.pdf. Accessed May 4 2025.
8. Lin TY, Maire M, Belongie S, et al. Microsoft COCO: common objects in context. In: European conference on computer vision. Cham: Springer; 2014. p. 740–55.
9. Wah C, Branson S, Welinder P, et al. The Caltech-UCSD Birds-200-2011 dataset. California Institute of Technology; 2011. https://www.vision.caltech.edu/datasets/cub_200_2011/. Accessed Sept 9 2025.
10. Xu H, Huang Y, Chen Z, et al. ImageReward: learning and evaluating human preferences for text-to-image generation. arXiv preprint arXiv:2304.05921. 2023. https://arxiv.org/abs/2304.05921. Accessed Sept 9 2025.
11. Juao J, Wang Z, Yu Z, et al. Exploration of an AIGC-driven intelligent course design and automated teaching resource generation model. Int J High Speed Electron Syst. 2025;2025:2540777.
12. Chen Y, Yuan X, Wang J, et al. YOLO-MS: rethinking multi-scale representation learning for real-time object detection. IEEE Trans Pattern Anal Mach Intell. 2025;47:4240–52.
13. Brunelle T, Chakir R, Carpentier A, et al. Reducing chemical inputs in agriculture requires a system change. Commun Earth Environ. 2024;5(1):369.
14. Hong Z, Ward L, Chard K, et al. Challenges and advances in information extraction from scientific literature: a review. JOM. 2021;73(11):3410–20.
15. Lin J, Wang Y, Li H, et al. Bridging research and readers: a multi-modal automated academic papers interpretation system. 2024. https://arxiv.org/abs/2401.09150. Accessed Sept 9 2025.
16. Wang Z, Zhang Y, Yao R, et al. DocLayout-YOLO: enhancing document layout analysis through diverse synthetic data and global-to-local adaptive perception. 2024. https://arxiv.org/abs/2410.12628. Accessed Sept 9 2025.
17. Gao Y, Li Y, Wu C, et al. UniMERNet: a universal network for real-world mathematical expression recognition. 2024. https://arxiv.org/abs/2404.15254. Accessed Sept 9 2025.
18. Clark C, Divvala S. Looking beyond text: extracting figures, tables and captions from computer science papers. In: Proceedings of the 2015 conference on empirical methods in natural language processing; 2015. p. 1488–98.
19. Arlazarov VV, et al. Document image analysis and recognition: a survey. Comput Opt. 2022;46(4):567–86.
20. Kumar S, Devi M. Identification and extraction of different objects and its location from a pdf file using efficient information retrieval tools. In: 2015 international conference on soft computing and network security (ICSNS); 2015. p. 1–5.

21. Smith, G. R., Stys, J. E., & Price-Whelan, A. M. Figure and figure caption extraction for mixed raster and vector PDFs: digitization of astronomical literature with OCR features. 2022. https://arxiv.org/abs/2209.04460. Accessed Sept 9 2025.
22. Li P, Jiang X, Shatkay H. Figure and caption extraction from biomedical documents. Bioinformatics. 2019;35(7):1238–46.
23. Li Y, Li J, Gan Z, et al. Aligning visual regions and textual concepts for semantic-grounded image representations. Adv Neural Inf Proces Syst. 2019.
24. Yang A, Yu B, Li C, et al. Qwen2.5-1M Technical Report. 2025.

SAR-to-Optical Image Translation Based on Generative Artificial Intelligence

Hang Chen, Jinyu Wang, Haitao Yang, Zhiliang Li, Zhengjun Liu, and Lei Liu

Abstract A cross-modal translation method from SAR images to optical images is provided in this chapter. Based on the latest generative artificial intelligence technologies, the interpretability of SAR images is enhanced. The basic principles and evaluation methods of generative adversarial networks and diffusion models are comprehensively elaborated. Two improved S2O algorithms are introduced, and their effectiveness is demonstrated through experiments. In terms of model design, two algorithm architectures, namely the dual-branch cycle-consistent and the diffusion model, are mainly introduced in this chapter.

Keywords Diffusion model · Generative adversarial network · SAR image augmentation · Image cross modal translation

1 Introduction

SAR is an active microwave remote sensing technology. Its imaging is not affected by time, weather, and illumination. Even when imaging under harsh meteorological conditions, it can still penetrate clouds to obtain surface information [1]. SAR images have extremely high application value in aspects such as ship target detection, land use planning, and natural disaster assessment [2–4]. However, as a

Hang Chen and Jinyu Wang contributed equally to this work.

H. Chen · H. Yang · Z. Li · L. Liu (✉)
School of Space Information, Space Engineering University, Beijing, China

J. Wang
Norinco Group Testing and Research Institute, Xi'an, China

Z. Liu
School of Physics, Harbin Institute of Technology, Harbin, China
e-mail: zjliu@hit.edu.cn

H. Chen, Z. Liu (eds.), *Artificial Intelligence in Digital Image Processing: Theories, Methods, and Applications*, Studies in Computational Intelligence 1246,
https://doi.org/10.1007/978-3-032-12891-1_9

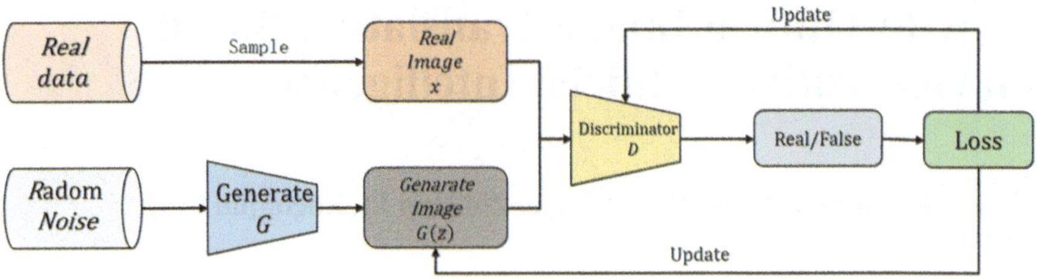

Fig. 1 The algorithm process of GANs

microwave imaging method, SAR images usually have a large number of distortions and shadows, which increases the difficulty of target interpretation. It is very difficult for non-professional interpreters to accurately identify the shape, size, and activity characteristics of aircraft from SAR images [5].

For images of high-value targets, a large amount of visual interpretation work is still required at present [6]. Compared with SAR images, optical images are closer to human visual perception and have richer spatial and color information, which can reduce the workload of interpreters [7, 8]. However, the imaging conditions of optical payloads are relatively harsh and are easily affected by the combined effects of illumination and clouds. In the field of aerospace reconnaissance, the enemy can use camouflage and anti-camouflage means to try to avoid being tracked by optical satellites and target detection algorithms. As shown in Fig. 1b, c, in order to effectively protect high-value targets, means such as installing anti-camouflage measures and constructing false targets can be taken to interfere with the judgment of intelligent detection models. In fact, converting SAR images into optical images for enhanced information has achieved initial results in the task of improving the accuracy of target detection [9–11]. Therefore, the synergy of SAR and optical image information to improve the accuracy of comprehensive target interpretation has become a research hotspot in the field of intelligent interpretation of remote sensing images. This is essentially a cross-modal SAR image data augmentation method [12].

In fact, compared with imaging methods such as infrared and hyperspectral, the cloud and fog penetration characteristics of SAR sensors are not possessed by other payloads. At the same time, as an active imaging method, the image characteristics of SAR are quite different from those of passive imaging means such as optical and infrared. Converting between the SAR and optical modalities can extract more valuable information. In this process, the accurate and efficient conversion effect depends on the modality alignment level of the training data. With the development of GenAI, some scholars have proposed using the cross-modal conversion of S2O as an information complementary strategy to enhance the quality of SAR images and thus improve their readability.

This article aims to achieve the purpose of SAR data augmentation based on GenAI technology. Emphasis is placed on using SAR images for cross-modal conversion of optical images, thereby improving the interpretability of SAR images for downstream tasks.

2 Theories Related to Generative Artificial Intelligence

2.1 Generative Adversarial Network

1. Basic Principles

GAN was initially proposed by Goodfellow. It is a generative model based on the adversarial learning strategy. The core idea of GAN is to combine game theory and deep learning networks to achieve the goal of fitting the real distribution [13]. GAN is usually composed of two neural networks—the Generator and the Discriminator. The two finally reach the Nash equilibrium in the dynamic game. GAN has been widely applied in the field of SAR image data augmentation and plays an important role in solving the problems of insufficient SAR samples and high annotation costs.

(a) Optimization Objective of the Generator

In GANs, the role of the generator is to reduce the gap (distance) between the generated distribution and the real distribution. Therefore, the KL divergence is commonly used to measure the difference between the two. The KL divergence is also known as relative entropy. For two distributions P and Q, the calculation method is as follows:

$$KL\left(P \middle\| Q\right) = \sum\nolimits_{x \in X} P(x) \log \frac{P(x)}{Q(x)} \tag{1}$$

x represents the input sample, and $P(x)$ and $Q(x)$ represent the output and the real distribution, respectively. It can be seen from this that by optimizing the KL divergence and minimizing it to 0, the loss is minimized. However, the drawback of the KL divergence is its asymmetry. When the prior conditions are different, $KL(P\|Q) \neq KL(Q\|P)$. Therefore, in the actual GAN, the JS distance is commonly used as the optimization target, and its expression is as follows:

$$JS\left(P \middle\| Q\right) = \frac{1}{2}\left[KL\left(P \middle\| Q\right) + KL\left(Q \middle\| P\right)\right] \tag{2}$$

The JS distance is symmetric. This characteristic enables the avoidance of the mode collapse problem caused by asymmetric metrics during the training of GAN. In the original GAN proposed by GoodFellow, the JS divergence was used, and when the JS divergence is 0, it indicates that the two distributions have completely overlapped. However, the JS divergence also has the problem of gradient vanishing. Therefore, in subsequent studies, Wasserstein distance has also been used as an alternative to serve as the optimization basic theory for the generator [14].

(b) The discriminator fits the real distribution and calculates the difference.

Due to the powerful nonlinear feature fitting ability of neural networks, neural networks are more applicable to complex distributions compared to the maximum likelihood estimation method. The discriminator, also composed of a deep neural network, discriminates the quality of generation by calculating the difference between the real distribution and the generated distribution. The objective function of the discriminator is expressed as follows:

$$V = \int_x \left[P_{data}(x) \log D(x) + P_G(x) \log (1 - D(x)\right] dx \tag{3}$$

The optimization goal of D is to maximize the objective function. Therefore, by using derivatives, we obtain:

$$D^*(x) = \frac{P_{data}(x)}{P_{data}(x) + P_G(x)} \tag{4}$$

Based on the above equation, substituting the result into the objective function of G gives:

$$V\left(G, D^*\right) = \int_x P_{data}(x) \log \frac{P_{data}(x)}{P_{data}(x) + P_G(x)} + \int_x P_G(x) \log \frac{P_{data}(x)}{P_{data}(x) + P_G(x)} \tag{5}$$

Therefore, from the perspective of optimizing the Jensen-Shannon (JS) divergence, the optimization goal of G is to minimize the JS divergence, while the goal of D is to maximize it. This min-max strategy jointly forms the basis of the game confrontation in the Generative Adversarial Network (GAN), and ultimately leads to the achievement of the Nash equilibrium.

2. Generative adversarial strategy
   After the optimization goals of the generator and the discriminator are determined, the network architecture of GAN is shown in Fig. 1, which consists of two parts: the generator G and the discriminator D. The input of the generator is z sampled from the latent space. After passing through the generator, a preliminary generated image G(z) is obtained. At this time, the gap between the generated image and the real image is relatively large, and it needs to be discriminated for its authenticity by D. The input of D is the image xx sampled from the real data. By discriminating the distribution distance between x and $G(z)$, the probability that $G(z)$ belongs to the real image is output. The range of this probability is (0,1), and 1represents the same as the real image. Therefore, the purpose of D is to make the output of the real data as close to 1 as possible, while the purpose of GG is to "confuse" D so that the output of G can also achieve a score of 1 in D. In this process, with the loss calculation and gradient

update, a Nash equilibrium state is reached in the game between G and D. At this point, the difference between the image generated by the generator and the real image reaches the minimum value. Therefore, the objective function of GAN is as follows:

$$\min_{G}\max_{D} V\,(D, G) = E_{x\sim P_{data}(x)}\,[logD(x)] + E_{z\sim P_Z(z)}\left[\log(\,1 - D\,(G(z))\,\right] \tag{6}$$

$E_{x\sim P_{data}(x)}\,[logD(x)]$ represents the output result of real data in D, and $E_{z\sim P_Z(z)}\left[\log(\,1 - D\,(G(z))\right.$ represents the output result of generated samples in D. In the actual training of GAN, the strategy of first training D while fixing G and then training G while fixing D is adopted.

3. Derivative Networks

According to the current research situation, in the field of SAR image data augmentation, the existing mainstream methods are convolutional, conditional, and cycle-consistent generative adversarial networks. Among them, convolutional GAN has provided a good paradigm for subsequent CNN-based GenAI algorithms. A strategy of using class labels to constrain the generation process was proposed for conditional GAN. The quality of cross-modality image translation has been greatly improved by cycle-consistent GAN, and it has also become the benchmark for subsequent research in the same field. Therefore, in this section, the following three representative derivative networks are analyzed.

(a) Convolutional Generative Adversarial Network

The Deep Convolutional Generative Adversarial Network (DCGAN) is a network that combines CNN and GAN and is specifically used for image generation tasks [15]. The generator architecture in DCGAN is shown in Fig. 2. The principle of DCGAN is the same as that of GAN. The difference is that the generator in DCGAN uses the method of deconvolution for

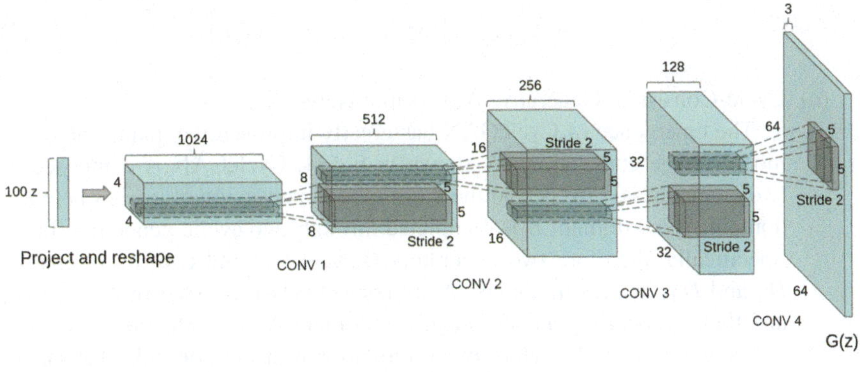

Fig. 2 The network architecture of the DCGAN generator [15]

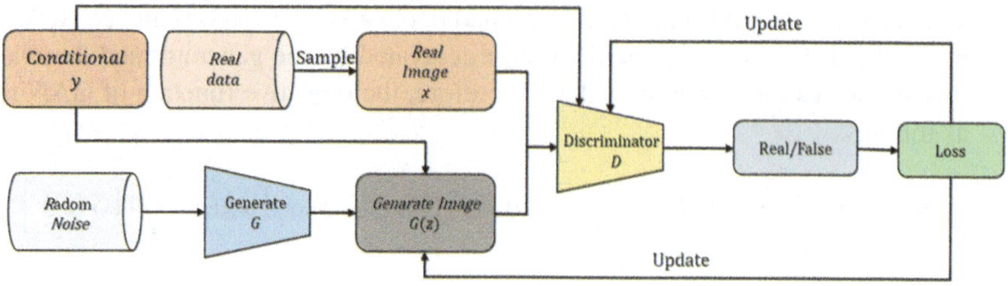

Fig. 3 The network architecture of the CGAN generator [15]

progressive upsampling to complete image output. And the discriminator discriminates between real images and generated images through convolutional layers. The core of DCGAN lies in the use of strided convolutions to replace the pooling layer operation, so that the stability of training is improved.

(b) Conditional Generative Adversarial Network

Generally speaking, the initial random noise of the generator in GAN is usually sampled from a Gaussian distribution, that is, the generator needs to complete the alignment mapping relationship between noise and images. Excessive training degrees of freedom can lead to the problem of uncontrollable generated information during training. For this reason, Van den Oord proposed using the class-conditional synthesis method for image synthesis to improve the quality of generated samples [16]. On this basis, researchers proposed the Conditional Generative Adversarial Nets (CGAN) [17]. The algorithm process of CGAN is shown in Fig. 3. Its core lies in using the conditional information y as additional information, which acts on G and D to constrain the process of adversarial training. Among them, the conditional information can be class labels, or it can be repaired data or multimodal data [18]. The objective function of CGAN is as follows:

$$\min_{G}\max_{D} V(D, G) = E_{x \sim P_{data}(x)}[log D(x|y)] + E_{z \sim P_{Z}(z)}\left[\log(1 - D(G(z|y))\right] \tag{7}$$

(c) Cycle-Consistent Generative Adversarial Network

The emergence of CycleGAN has greatly improved the quality of cross-modal image conversion. As shown in Fig. 4, CycleGAN is composed of two symmetrical GANs, and the cross-modal conversion of images between domain A and domain B is completed through two cyclic generation links. Specifically, there are two generators G_A and G_B, and two discriminators D_A and D_B. Images in domain A are respectively processed by G_A and G_B and then cyclically generate images in domain A. The situation of images in domain B is similar. Thus, two information transmission links that satisfy

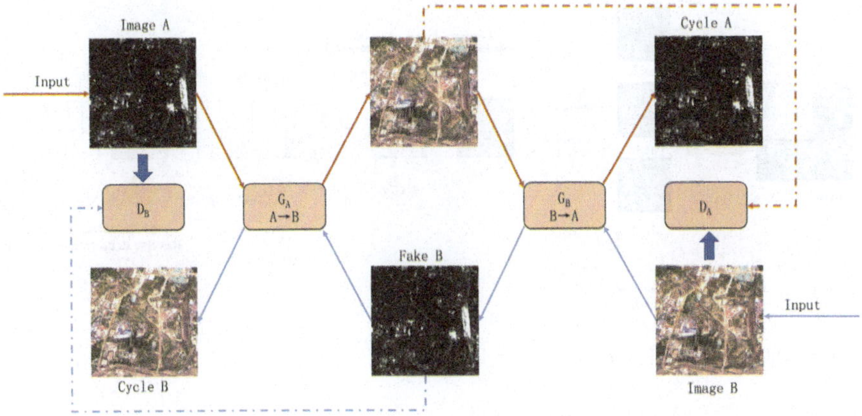

Fig. 4 The network architecture of the CycleGAN generator [15]

cycle consistency are established. Its objective function is as follows:

$$L_{GAN}(G_{AB}, D_B, A, B) = \mathrm{E}_{y \sim p_{data}(y)}\left[\log D_B(y)\right] + \mathrm{E}_{x \sim p_{data}(x)}\left[\log\left(1 - D_B\left(G_{AB}(x)\right)\right)\right] \quad (8)$$

The use of cycle consistency can make the transformation of the feature space more efficient. Therefore, most of the current mainstream S2O cross-modal transformation methods are improved based on this. However, CycleGAN or Pix2Pix can only be trained in a supervised or unsupervised manner, and when the data is complex, simply using cycle consistency loss or L1 loss can no longer meet the requirements of the task.

2.2 *Diffusion Model*

From a training perspective, diffusion models only require noise fitting and prediction, granting them inherent interpretability. Compared to GANs, DMs eliminate the need for multiple competing CNN models, substantially improving training stability. Derived from non-equilibrium thermodynamics, DDPM involves two stages: forward diffusion noise addition and reverse inference (denoising). Figure 5 illustrates its training and inference processes.

1. Forward noise-addition process

   In the forward noise addition process, by continuously adding random Gaussian noise to the original image, the image is gradually transformed into a pure noise sample. This process converts the data distribution into a simple prior

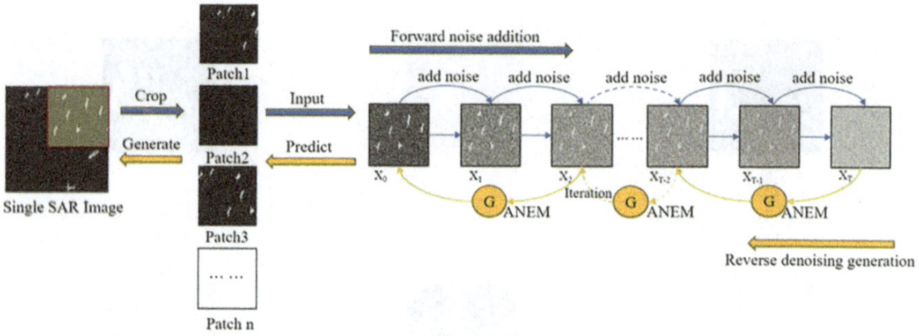

Fig. 5 Diffusion model process

distribution, facilitating fitting through inversion during the reverse inference process. Let the input data sample be $x_0 \sim q(x_0)$, which is sampled from the true data distribution. Given the hyperparameter $\alpha_t = 1 - \beta_t$, when Gaussian noise $x_0, \ldots, x_T$ is incrementally added to the sample, the following Markov process can be obtained:

$$q\left(x_t \mid x_{t-1}\right) = N\left(x_t; \sqrt{1-\beta_t} \cdot x_{t-1}, \beta_t \cdot I\right), \forall t \in \{1, \ldots, T\} \tag{9}$$

In the equation, $t \sim T$ represents the diffusion step count, T denotes the identity matrix. The forward noise addition process is a Markov process, where the probability distribution of the next state depends solely on the current state, meaning the sample distribution at time t is related to that at time x_{t-1}. Therefore, the relationship between the input sample x_0 and the sample distribution at time t can be established through iterative solving:

$$q\left(x_t \mid x_0\right) = N\left(x_t; \sqrt{\hat{\beta}_t} \cdot x_0, \left(1 - \hat{\beta}_t\right) \cdot I\right) \tag{10}$$

In the equation, $\hat{\beta}_t = \prod_{i=1}^{t} (1 - \beta_t)$, thus by applying the reparameterization trick to reformulate Eq. (17), the following result can be obtained:

$$x_t = \sqrt{\hat{\beta}_t} \cdot x_0 + \sqrt{1 - \hat{\beta}_t} \cdot z_t \tag{11}$$

In the equation, $z_t \sim N(0, I)$. Therefore, during the forward noise addition process, the distribution relationship between the input sample and any diffusion step length is established through the Markov chain and reparameterization derivation, ultimately yielding a pure noise image in the final diffusion step.

2. Reverse Inference Process

The reverse inference process is the denoising process, which operates by predicting the noise value at each diffusion step of the forward noise addition and denoising through inverse transformation to ultimately restore the input sample. In reality, $q(x_{T-1} \mid x_T)$ is difficult to obtain directly, necessitating the introduction of the conditional posterior probability $(x_{T-1} \mid x_T, x_0)$, where the mean and variance are $\tilde{\mu}_t$ and $\tilde{\beta}_t$, respectively. Thus, the starting point of the reverse inference process is a sample $x_T \sim N(0, I)$ drawn from Gaussian noise, and according to Bayes' theorem, the following can be derived:

$$q\left(x_{T-1} \mid x_T, x_0\right) = q\left(x_T \mid x_{T-1}, x_0\right) \frac{q\left(x_{T-1}, x_0\right)}{q\left(x_T, x_0\right)} \tag{12}$$

Further, by combining the reparameterization trick and the probability density function of the Gaussian distribution, Eq. (11) can be further derived as:

$$\begin{aligned} q\left(\mathbf{x}_{t-1} \mid \mathbf{x}_t, \mathbf{x}_0\right) &\propto \exp\left(-\frac{1}{2}\left(\frac{\left(\mathbf{x}_t - \sqrt{\alpha_t}\mathbf{x}_{t-1}\right)^2}{\beta_t} + \frac{\left(\mathbf{x}_{t-1} - \sqrt{\hat{\beta}_{t-1}}\mathbf{x}_0\right)^2}{1 - \hat{\beta}_{t-1}} - \frac{\left(\mathbf{x}_t - \sqrt{\hat{\beta}_t}\mathbf{x}_0\right)^2}{1 - \hat{\beta}_t}\right)\right) \\ &= \exp\left(-\frac{1}{2}\left(\left(\frac{\alpha_t}{\beta_t} + \frac{1}{1 - \hat{\beta}_{t-1}}\right)\mathbf{x}_{t-1}^2 - \left(\frac{2\sqrt{\alpha_t}}{\beta_t}\mathbf{x}_t + \frac{2\sqrt{\hat{\beta}_{t-1}}}{1 - \hat{\beta}_{t-1}}\mathbf{x}_0\right)\mathbf{x}_{t-1} + C\left(\mathbf{x}_t, \mathbf{x}_0\right)\right)\right) \end{aligned} \tag{13}$$

The Gaussian probability density function can be expressed as:

$$\exp\left(-\frac{(x - \mu)^2}{2\sigma^2}\right) = \exp\left(-\frac{1}{2}\left(\frac{1}{\sigma^2}x^2 - \frac{2\mu}{\sigma^2}x + \frac{\mu^2}{\sigma^2}\right)\right) \tag{14}$$

Combining Eqs. (13) and (14), the variance and mean of the posterior probability distribution can be derived. Since the variance remains fixed during the noise addition and denoising processes, $\tilde{\beta}_t$ is a known quantity and need not be considered. Therefore, the mean $\tilde{\mu}_t$ is given by:

$$\tilde{\mu}_t = \frac{1}{\sqrt{\alpha_t}}\left(x_t - \frac{1 - \alpha_t}{\sqrt{1 - \hat{\beta}_t}} \cdot z\left(x_t, t\right)\right) \tag{15}$$

From Eq. (15), it can be derived that $\widetilde{\mu}_t(x_t, x_0)$ is only related to x_t and $z(x_t, t)$. Therefore, during the reverse inference process, using Gaussian noise as the training starting point requires only training a noise predictor $z_\theta(x_t, t)$, θ as the neural network model parameter. Based on Eqs. (15) and (16), the original image x_0 can be obtained through reverse iteration, with the final loss function being:

$$\mathcal{L} = \mathbb{E}_{x_0, z \sim \mathcal{N}(0,I)} \left(\| z - z_\theta \left(\sqrt{\hat{\beta}_t} \mathbf{x}_0 + \sqrt{1 - \hat{\beta}_t} z, t \right) \|^2 \right) \tag{16}$$

Here, $\mathbb{E}$ represents the expected value, and x_0 is the original input image. The iterative formula for the final backpropagation process is:

$$x_{t-1} = \frac{1}{\sqrt{\alpha_t}} \left(x_t - \frac{1 - \alpha_t}{\sqrt{1 - \hat{\beta}_t}} \cdot z_\theta (x_t, t) \right) + \sigma_t z \tag{17}$$

2.3 *Image Similarity Evaluation Metrics*

Image similarity is a core method for measuring differences between two or multiple images. With the rapid development of GenAI, the quality of image similarity evaluation metrics directly impacts the final model performance assessment. Due to the slow progress in the field of natural images and S2O modality alignment enhancement, there are limited available evaluation metrics. Thus, this section introduces mainstream metrics in this field. Notably, all similarity evaluation metrics are designed with reference to human visual perception.

1. Traditional Metrics

   (a) Peak Signal-to-Noise Ratio (PSNR)

   The Peak Signal-to-Noise Ratio (PSNR) primarily calculates similarity based on pixel-level mean squared error and maximum error, with its unit typically being decibels (dB). The advantage of PSNR lies in its simplicity and efficiency, focusing on pixel-level differences in images. However, it lacks system perception capabilities akin to human vision. A higher PSNR value indicates greater similarity [19].

$$\text{PSNR} = 10 \times \log_{10} \left(\frac{\text{MAX}^2}{\text{MSE}} \right) \tag{18}$$

   Here, *MSE* represents the mean squared error, and *MAX* denotes the maximum pixel value.

(b) Structural Similarity Index (SSIM)

The Structural Similarity Index (SSIM) is an image similarity evaluation metric based on pixel statistics, incorporating factors such as luminance, contrast, and structure. Thus, it closely aligns with human visual perception [20]. SSIM has comprehensive characteristics, balancing the relationship between objective quantitative metrics and human visual perception to some extent. However, it still cannot fully match human visual judgment. A higher SSIM value indicates greater similarity.

$$\mathrm{SSIM}(x, y) = \frac{\left(2\mu_x\mu_y + C_1\right)\left(2\sigma_{xy} + C_2\right)}{\left(\mu_x^2 + \mu_y^2 + C_1\right)\left(\sigma_x^2 + \sigma_y^2 + C_2\right)} \tag{19}$$

Here, x and y represent the real and generated images, respectively. μ_x and μ_y denote the mean values of the images, σ_x and σ_y represent their standard deviations, and σ_x^2 and σ_y^2 indicate the variances of the images.

2. Intelligent Pre-trained Model Metrics

(a) Learned Perceptual Image Patch Similarity (LPIPS)

Learned Perceptual Image Patch Similarity (LPIPS) is an evaluation metric based on intelligent pre-trained models, with its core referenced from [21]. It works by feeding both real and generated images into a classification model, extracting intermediate feature maps to compute differences, thereby deriving similarity. Compared to traditional methods, LPIPS better aligns with human subjective perception—the lower the score, the higher the similarity. The LPIPS formula is as follows:

$$d(x, x_0) = \sum\nolimits_l \frac{1}{H_l W_l} \sum\nolimits_{h,w} \| \, w_l \odot \left(\hat{y}_{hw}^l - \hat{y}_{0hw}^l\right) \|_2^2 \tag{20}$$

In the formula, $d(_*)$ represents distance, and l denotes the number of extracted feature layers.

(b) Fréchet Inception Distance (FID)

FID is a metric specifically designed to evaluate GenAI algorithms, leveraging the output of a pre-trained InceptionV3 to directly compare the similarity between real and predicted images [22]. The lower the FID, the closer the distributions of the images. Specifically, FID first obtains the pre-trained model outputs for both distributions, then calculates the Fréchet distance to derive the metric value. The FID formula is as follows:

$$d^2 = \| \, u_1 + u_2 \, \|^2 + \mathrm{Tr}\left(C_1 + C_2 - 2 \times \sqrt{(C_1 \times C_2)}\right) \tag{21}$$

Here, the inputs are the output matrices of real and predicted images processed by InceptionV3 [22] to locate the reference source. u_1, u_2 represents the mean, C_1, C_2 the covariance, and Tr the trace of the matrix.

(c) Single Image Fréchet Inception Distance (SIFID)

SIFID is an extension of the FID metric, tailored for evaluating single-sample generation results—i.e., involving only a single image pair in the calculation [23]. Unlike FID, SIFID does not use the activation vector from the final pooling layer of the Inception network but instead extracts convolutions before the second pooling layer, utilizing deeper internal features.

(d) Inception Score (IS)

The IS metric is a parameter-free evaluation method that also employs a pre-trained model [24]. The IS metric assesses a batch of images holistically from two perspectives: single-sample class distribution and diversity distribution. The IS formula is as follows:

$$\mathrm{IS}(G) = \exp\left(E_{x \sim p_g} D_{KL}\left(p\left(y \mid x\right) \parallel \hat{p}(y)\right)\right) \tag{22}$$

Here, $x \sim p_g$ represents the generated sample, $p(y \mid x)$ the output of the InceptionV3 model, and $\hat{p}(y)$ the mean of the probability distributions across all generated samples.

Due to differences in evaluation mechanisms, calculation methods, and interpretability, current image evaluation metrics for GenAI algorithms require a comprehensive analysis approach, and further research is needed on the computation of quality and diversity. Notably, in addition to the mainstream image similarity evaluation metrics mentioned above, this paper also employs other targeted evaluation methods for different algorithms. For instance, the quality of generated images is assessed by comparing the overlap and trends of grayscale distribution curves. In the evaluation of SAR image generation quality, the Signal-to-Noise Ratio (SNR) differences between generated and real images are also compared.

3 SAR-to-Optical Image Translation Model

3.1 Dual-Branch Cycle-Consistent Loss-Based Method

1. Fundamental Principles

The proposed SemiCD algorithm workflow is shown in Fig. 6. Text not specified in document.-1, which consists of two stages: adversarial training and cycle-consistent training. The adversarial training stage employs a semi-supervised learning approach to achieve learning for both supervised and unsupervised modules. This process combines generators and discriminators with parameter-sharing functionality to compute adversarial losses. In the cycle-consistent training stage, a dual-branch cycle-consistent training strategy is introduced, establishing unsupervised and supervised reconstruction losses sep-

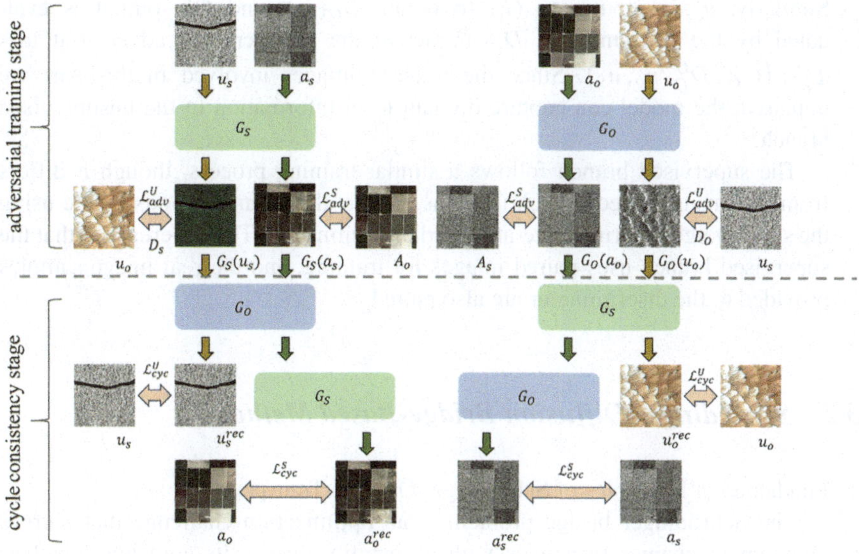

Fig. 6 SemiCD algorithm process

arately. During the transformation processes of $S \rightarrow O$ and $\rightarrow S$, two adversarial losses $\mathcal{L}_{adv}^S\left(G_S, D_S^S, a_o, a_s\right)$ and $\mathcal{L}_{adv}^S\left(G_O, D_O^S, a_o, a_s\right)$ can be computed, enabling the supervised branch to effectively learn pixel-level modality alignment information.

Specifically, the adversarial training phase consists of two parts: an unsupervised branch and a supervised branch. For the unsupervised branch, two generators G_S and G_O are first created, where G_S is used to achieve SAR-to-optical image conversion ($S \rightarrow O$), while G_O is used to achieve optical-to-SAR conversion ($O \rightarrow S$). For simplicity, two distinct image domains $SAR(S)$ and $OPT(0)$ are defined, from which unaligned SAR images $u_s \sim p_{data}(u_s)$ and optical images $u_o \sim p_{data}(u_o)$ are sampled. Similarly, paired SAR and optical image pairs $a_s \sim p_{data}(a_s)$ and $a_o \sim p_{data}(a_o)$ are sampled. Secondly, discriminators D_S^U and D_O^U are designed to evaluate the conversion performance of the unsupervised branch, analogous to the role of the generators, while D_S^S and D_O^S are used to assess the conversion results of the supervised branch.

SemiCD uses unpaired data u_s and u_o to complete unsupervised training. u_s is input into G_S to obtain $G_S(u_s)$, and this output is evaluated by the discriminator D_S^U to derive the unsupervised adversarial loss $\mathcal{L}_{adv}^U\left(G_S, D_S^U, u_s, u_o\right)$. This process can be expressed as follows:

$$\mathcal{L}_{adv}^U\left(G_S, D_S^U, u_s, u_o\right) = E_{u_o}\left[\log D_S^U\left(u_o\right)\right] + E_{u_S}\left[\log\left(1 - D_S^U\left(G_S\left(u_s\right)\right)\right)\right] \tag{23}$$

Similarly, u_o is input into G_O to obtain $G_O(u_o)$, and this output is evaluated by the discriminator D_O^U to derive the unsupervised adversarial loss $\mathcal{L}_{adv}^U(G_O, D_O^U, u_o, u_s)$. Since the training images involved in this part are unpaired, the model can capture domain-level information in the unsupervised branch.

The supervised branch follows a similar training process, though it differs from the unsupervised branch in terms of data alignment methods, while using the same weight-sharing generators and discriminators. The difference is that the supervised branch uses paired images for training, and the real image samples provided to the discriminator are also paired.

3.2 Schrödinger Diffusion Bridge-Based Method

1. Fundamental Principles of Schrödinger Optimal Transport

   The Schrödinger bridge problem is an optimization challenge that merges quantum mechanics principles with probability theory. Its core idea involves identifying a stochastic process (also termed a path) evolving along the time axis, capable of smoothly interpolating from an initial probability distribution to a target as calculated in [25, 26]. This concept aligns closely with the S2O cross-modal image transformation objective. Similar to optimal transport problems, the Schrödinger problem introduces stochasticity, rendering the path connecting the two distribution domains non-deterministic.

   Specifically, given two probability distributions π_0 and π_1 on $\mathbb{R}^d$, the goal is to find a solution $\mathbb{Q}$ connecting these distributions via random samples in the path space Ω. The Schrödinger bridge determines the optimal evolutionary path between these distributions by minimizing an entropy-related cost function. This process accounts not only for transport efficiency but also for the randomness and diversity of paths.

   The Schrödinger bridge problem holds significant importance in numerous applications, particularly in data alignment and enhancement. It identifies natural transformation paths—acting as optimal "bridges"—between different distribution domains while preserving statistical properties of the data. This approach captures complex inter-data relationships, offering a more robust and flexible modeling framework. The mathematical formulation of the Schrödinger bridge is as follows:

$$\mathbb{Q}^{SB} = \underset{\mathbb{Q}\in\mathcal{P}(\Omega)}{\operatorname{argmin}} D_{KL}\left(\mathbb{Q} \parallel \mathbb{W}^{\tau}\right) \text{ s.t.} \mathbb{Q}_0 = \pi_0, \mathbb{Q}_1 = \pi_1 \tag{24}$$

$\mathbb{W}^{\tau}$ represents the Wiener measure with variance τ, while $\mathcal{P}(\Omega)$ is the probability measure space on the path space Ω. Therefore, by minimizing the KL divergence, $\mathbb{Q}$ is made to infinitely approach the optimal solution $\mathbb{Q}^{SB}$, which forms the fundamental framework of the Schrödinger bridge for solving the optimal transport path. Given $\{\mathbf{x}_t\}$ represents the distribution of time t sampled on $\mathbb{Q}^{SB}$ its dynamic process in the form of stochastic control follows the following stochastic differential equation:

$$d\mathbf{x}_t = \mathbf{u}_t^{SB}dt + \sqrt{\tau}d\mathbf{w}_t \tag{25}$$

Here, $\{w_t\} \sim \mathbb{W}^\tau$, u_t^{SB} denotes the time-varying drift at time. It can be observed that when $\rightarrow 0$, $\{\mathbf{x}_t\}$ converges to the optimal ordinary differential trajectory. This formula proves that $\{\mathbf{x}_t\} \in [0, 1]$ is a strict dynamic Markov chain. Thus, it can be obtained by using the static formula (26):

$$\mathbb{Q}_{01}^{SB} = \underset{\gamma \in \Pi(\pi_0, \pi_1)}{\operatorname{argmin}} \ \mathbb{E}_{(x_0, x_1) \sim \gamma} \left[\| \ \mathbf{x}_0 - \mathbf{x}_1 \ \|^2 \right] - 2\tau H(\gamma) \tag{26}$$

$\mathbb{Q}_{01}^{SB}$ represents the joint distribution over time $t \in [0, 1]$, π_0 and π_1 are the two endpoint distributions, and H denotes the entropy function. This indicates that the Schrödinger bridge problem can be solved through entropy regularization. There exists a time sequence $\{t_i\}_{i=0}^N$, where $t_0 = 0$, $t_N = 1$ In the diffusion Schrödinger bridge process, there is a Markov process p used for the transition from $\pi_0 \rightarrow \pi_1$:

$$p\left(\{\mathbf{x}_{t_n}\}\right) = p\left(\mathbf{x}_{t_N} \mid \mathbf{x}_{t_{N-1}}\right) p\left(\mathbf{x}_{t_{N-1}} \mid \mathbf{x}_{t_{N-2}}\right) \cdots p\left(\mathbf{x}_{t_1} \mid \mathbf{x}_{t_0}\right) p\left(\mathbf{x}_{t_0}\right) \tag{27}$$

Here, $p\left(\{x_{t_n}\}\right)$ represents the generated result at time t_i. Through the Markov propagation method, the transition probability $p\left(\mathbf{x}_{t_{i+1}} \mid \mathbf{x}_{t_i}\right)$ can be easily computed. By learning this parameter, the optimized path from $p(\{\mathbf{x}_0\}) \rightarrow p\left(\{\mathbf{x}_{t_i}\}\right) \rightarrow p(\{\mathbf{x}_1\})$ can be obtained.

It is worth noting that the above derivation assumes the existence of an optimal diffusion Schrödinger bridge from $0 \rightarrow \pi 1$, meaning there is a direct interpolative mapping relationship between these two domains. Therefore, another process q is defined, which fits the process p by learning the parameter. According to Schrödinger bridge theory, the objective function can be expressed as:

$$\begin{cases} \min\limits_{\phi_i} \mathcal{L}_{SB}(\phi_i, t_i) := \mathbb{E}_{q_{\phi_i}(x_{t_i}, x_1)} \left[\| \ \mathbf{x}_{t_i} - \mathbf{x}_1 \ \|^2 \right] - 2\tau(1 - t_i) H\left(q_{\phi_i}\left(\mathbf{x}_{t_i}, \mathbf{x}_1\right)\right) \\ \qquad q_{\phi_i}\left(\mathbf{x}_{t_i}, \mathbf{x}_1\right) := q_{\phi_i}\left(\mathbf{x}_1 \mid \mathbf{x}_{t_i}\right) p\left(\mathbf{x}_{t_i}\right) \end{cases} \tag{28}$$

Here, $q_{\phi_i}\left(\mathbf{x}_1 \mid \mathbf{x}_{t_i}\right)$ represents a generator whose input is the sampled image at time t_i, and the output is the predicted $\mathbf{x}_1$. ϕ denotes the model parameters, typically fitted by a deep neural network. Since $\mathbf{x}_1$ is uniformly given, an iterative method from $x_{t_i} \rightarrow x_{t_{i+1}}$ is established reparameterization techniques based on the static control equation, with the specific calculations as follows:

$$\begin{cases} p\left(\mathbf{x}_{t_{i+1}} \mid \mathbf{x}_1, \mathbf{x}_{t_i}\right) := \mathcal{N}\left(\mathbf{x}_{t_{i+1}} \mid s_{i+1}\mathbf{x}_1 + (1-s_{i+1})\,\mathbf{x}_{t_i},\, s_{i+1}\,(1-s_{i+1})\,\tau\,(1-t_i)\,\mathbf{I}\right) \\ s_{i+1} := (t_{i+1} - t_i)\,/\,(1 - t_i) \end{cases} \tag{29}$$

2. SBS2O Algorithm

SBS2O focuses on resolving cross-domain conversion modal mismatch issues under unsupervised training conditions, utilizing the random optimal transport process of the Schrödinger bridge to fit the interpolation path between the SAR domain and the optical domain. Grounded in the theoretical framework of the Schrödinger optimal transport problem, SBS2O employs an end-to-end domain transformation method that utilizes a noise interpolation technique to convert between SAR and optical images. Its algorithmic workflow is illustrated in Fig. 7.

SBS2O combines diffusion bridge theory with adversarial generation strategies, incorporating both forward training and backward inference generation processes. It should be noted that SBS2O differs fundamentally from GANs. GANs directly connect two distinct domains, with the generator's inputs and outputs during training being $\boldsymbol{x}_0$ and $\boldsymbol{x}_1$. In contrast, SBS2O generates images by predicting diffusion bridge interpolation, where the inputs and outputs are the interpolated $\boldsymbol{x}_{t_i}$ and $\boldsymbol{x}_1\left(\boldsymbol{x}_{t_i}\right)$, respectively.

Specifically, as a generative model, the SBS2O-based S2O image cross-modal transformation process also incorporates a generator G and a discriminator

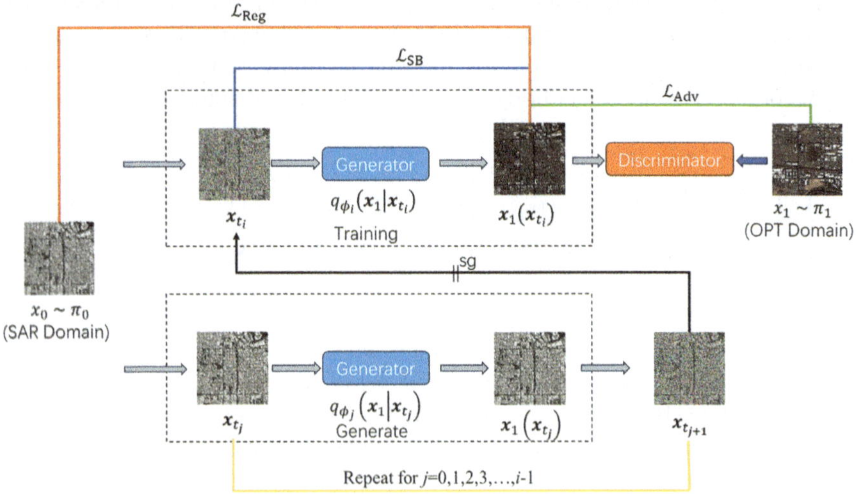

Fig. 7 SemiCD algorithm process

D. During the forward process (training phase), G receives an intermediate image $\boldsymbol{x}_{t_i}$ at a random time t_i—that is, a random image within the optimal transport path. Inputting $\boldsymbol{x}_{t_i}$ into G yields an estimated target domain sample $\boldsymbol{x}_1\left(\boldsymbol{x}_{t_i}\right)$. Subsequently, the Schrödinger loss distance $\mathcal{L}_{SB}\left(\phi, t_i\right)$ between them can be computed, enabling optimization of G's parameters to achieve the theoretical optimal transport (OT). Next, D is designed to assess the discrepancy between the generated inference image $\boldsymbol{x}_1\left(\boldsymbol{x}_{t_i}\right)$ and the ground truth image $\boldsymbol{x}_1$. In this chapter's methodology, unsupervised training is employed for model optimization. Given the presence of adversarial training strategies, the adversarial loss is defined as follows:

$$\mathcal{L}_{Adv}\left(\phi_i, t_i\right) := D_{KL}\left(q_{\phi_i}\left(\boldsymbol{x}_1|\boldsymbol{x}_{t_i}\right) \parallel p\left(\boldsymbol{x}_1\right)\right) \tag{30}$$

During the inference process (generation phase), first select the sequence $\left\{t_j\right\}_{j=0}^{i-1}$, then compute the transition probability $q_\phi\left(\boldsymbol{x}_{t_i}|\boldsymbol{x}_{t_{i-1}}\right)$. For j = 0, 1, 2, 3, ..., i-1, this process is repeated, ultimately yielding $q_\phi\left(\boldsymbol{x}_{t_i}|\boldsymbol{x}_{t_{i-1}}\right)$. In other words, as G completes its training optimization, $\boldsymbol{x}_1 \sim q_{\phi_j}\left(\boldsymbol{x}_1|\boldsymbol{x}_{t_j}\right) = p\left(\boldsymbol{x}_{t_i}\right)$. Thus, a sampling chain from $\boldsymbol{x}_0 \rightarrow \boldsymbol{x}_{t_i} \rightarrow \boldsymbol{x}_1\left(\boldsymbol{x}_{t_i}\right)$ is established, ultimately enabling the completion of the S2O task. The objective function for the entire process is as follows:

$$\mathcal{L}\left(\phi, t_i\right) := \mathcal{L}_{Adv}\left(\phi, t_i\right) + \alpha\mathcal{L}_{SB}\left(\phi, t_i\right) + \beta\mathcal{L}_{Reg}\left(\phi, t_i\right) \tag{31}$$

α and β are proportionality factor hyperparameters used to determine the weighting of specific losses during the objective function optimization process.

SBS2O also incorporates a regularization loss $\mathcal{L}_{Reg}$ to constrain the mapping relationship between $\boldsymbol{x}_1\left(\boldsymbol{x}_{t_i}\right)$ and $\boldsymbol{x}_0$. This regularization loss is designed to ensure that generated images not only perform well during adversarial training of the discriminator but also retain certain features of the original input image. This helps preserve key information from the original image while maintaining generation quality. By adjusting the values of α and β, a balance can be achieved between preserving image features and generation quality, thereby optimizing overall model performance. $\mathcal{L}_{Reg}$ is expressed as follows:

$$\mathcal{L}_{Reg}\left(\phi, t_i\right) := \mathbb{E}_{p\left(\boldsymbol{x}_0, \boldsymbol{x}_{t_i}\right)}\mathbb{E}_{q_\phi\left(\boldsymbol{x}_1|\boldsymbol{x}_{t_i}\right)}\left[\mathcal{R}\left(\boldsymbol{x}_0, \boldsymbol{x}_1\right)\right] \tag{32}$$

4 S2O Image Translation Results

4.1 SemiCD Method

1. Dataset Configuration

   (a) Experimental Environment Setup

   The software and hardware environment for the experiment is detailed in Sect. 4.1 of chapter "Adversarial Attacks and Defenses in Computer Vision." Training data is uniformly resized to $3 \times 256 \times 256$ dimensions. For hyperparameter settings, λ_1 and λ_2 are set to 10, λ_3 at 50. During training, the number of epochs is set to 400, using the Adam optimizer with a hyperparameter value of 0.999. Additionally, the learning rate is set to 0.0002, employing a learning rate decay strategy where it linearly decreases to 0 after 350 epochs. To achieve optimal performance, the batch size is set to 4.

   (b) Dataset Preparation

   The dataset used is the SEN1-2 dataset proposed by Schmitt M. These datasets originate from Sentinel 1 and Sentinel 2 satellites [27]. As shown in Fig. 8 this chapter selects 1153 image pairs from the SEN1-2 dataset for training, which have been carefully evaluated to remove unqualified pairs. The selected dataset includes five geographical classifications: residential areas, farmland, deserts, mountains, and rainforests. Additionally, 1100 pairs are chosen as the model training set, with the remaining 53 pairs serving as the test set to evaluate translation accuracy. The quantity of each land type in the training samples is listed in Table 1. Text not styled in document.-1, and the selected dataset contains images of both natural and artificial terrains.

Fig. 8 Introduction to the SemiCD dataset

Table 1 Distribution of the SemiCD dataset

Land	Train (pairs)	Test (pairs)
Residence	326	11
Cropland	267	9
Desert	190	14
Mountain	98	10
Vegetation	219	9
Total	1100	53

Since SeimiCD employs a semi-supervised training strategy, in the supervised training branch, the dataset is directly fed into the model, while in the unsupervised training branch, the training data is shuffled using random mixing to simulate unpaired scenarios. Notably, we conducts mixed-scenario training on the dataset rather than modeling based on terrain types separately. A dataset with imbalanced quantity and inter-class distribution effectively tests the model's transformation capability. Due to differences in texture and feature quantity, this poses a challenge for the model to distinguish varying feature patterns. Such a strategy can test the model's robustness, enabling it to better adapt to complex scenes and diverse image features.

2. Transformation Results

As is shown in Fig. 9, SemiCD outperforms comparative algorithms in the realism of generated images, achieving high standards in both texture details and color reproduction. Specifically, regarding texture details, the four methods—CycleGAN, UGATIT, NICEGAN, and UNSB—all exhibit some degree of texture distortion, and their performance in generating artificial terrain is significantly inferior to that of natural terrain. For instance, CycleGAN fails to accurately restore the course of rivers in rainforest recognition. UGATIT lacks effective reconstruction of straight roads in residential and farmland conversions, particularly failing to restore main roads in residential area images. Although NICEGAN restores main roads in residential area images, it misses the modeling of internal roads. Compared to the three aforementioned methods, UNSB performs better in restoring artificial terrain than natural terrain, such as exhibiting texture distortion in desert scene conversions.

Additionally, the same issues exist across the four supervised models: Pix2Pix, Conditional Diffusion, BBDM, and DDColor. For instance, while Pix2Pix accurately models texture details, it exhibits noticeable checkerboard artifacts in desert and residential area scenarios. Conditional Diffusion demonstrates unique conversion performance, effectively restoring desert texture details with visually appealing results, but fails in other scene conversions. DDColor achieves the best texture detail performance among all supervised algorithms, yet suffers from severe color shift issues, failing to learn tonal information between SAR and optical modalities. BBDM excels in restoring natural terrain but lacks precise understanding of linear textures in residential and farmland areas, leading to noticeable distortions despite correctly reflecting the positions of urban roads and field paths. SemiD2, as a semi-supervised architecture, performs well in texture detail generation but shows obvious checkerboard artifacts across all four land cover types except rainforest.

Further analysis of the underlying reasons reveals that unsupervised algorithms, due to the lack of pixel-level alignment information, can only perform image conversion through domain-level mapping. When algorithm performance is limited, the model fails to interpret detailed markers in SAR images, such as straight roads or winding rivers, resulting in texture-level training failure. With supervised architectures, the model leverages pixel-level information for

Fig. 9 Subjective generation results of SemiCD versus comparative algorithms

texture conversion, but this necessitates more complex pixel-aligned mappings, significantly increasing data requirements. Consequently, texture conversion performance in purely supervised learning architectures remains mediocre. Semi-supervised architectures, however, enable collaborative learning using both domain-level and pixel-level information, facilitating more accurate optical image conversion.

3. Color restoration accuracy

Color is a crucial interpretive marker in optical image analysis, and accurately restoring the colors of corresponding targets in SAR images is essential for enhancing interpretation capabilities. Among all the algorithms, SemiCD can accurately reconstruct colors, not only effectively preserving style but also performing well on targets such as residential areas, roads, deserts, mountains, and rivers. In comparison, Conditional Diffusion and DDcolor exhibit noticeable color distortion. Specifically, Conditional Diffusion only learned the color features of deserts and residential areas, mistakenly transferring the characteristics of residential areas to rainforests. The images generated by the DDcolor algorithm still retain the SAR style, particularly maintaining pure black-and-white tones in mountainous scenes. As for the remaining algorithms, CycleGAN failed to learn the color information of mountains, while UGATIT and NICEGAN did not capture the features of rainforests. SemiD2 and BBDM perform relatively well in color restoration.

Further analyzing the reasons behind the differences in color restoration, it is evident that Conditional Diffusion and DDcolor exhibit more color shifts. This is because both algorithms focus on model improvements at the algorithmic level but overlook the importance of loss function constraints in addressing color shifts [28]. Meanwhile, insufficient data volume is also a reason for the failure of these two algorithms. Particularly for DDcolor, subjective generation results indicate that Transformer is highly data-dependent and not entirely suitable for cross-modal S2O image conversion. In contrast, the other methods employ traditional cycle-consistent loss or end-to-end regularization constraints, ensuring that the models do not lose existing color information while completing texture conversion, thereby achieving precise color modeling. CycleGAN deviated in desert color conversion, incorrectly transferring rainforest color information to deserts. UGATIT and NICEGAN also exhibit similar issues. Therefore, a comprehensive consideration of texture and color restoration factors is necessary to judge the quality of image conversion.

4. Inter-class discrimination capability

As shown in Fig. 10, SemiCD adopts a hybrid training approach, i.e., training only one GenAI model to verify generation performance. Therefore, distinguishing between different land categories in terms of texture and color is crucial, and class imbalance in the dataset can lead to conversion failures. Specifically, Conditional Diffusion and DDcolor learned only one category (desert), which is due to the high data requirements of DDPM and Transformer. CycleGAN, UGATIT, and NICEGAN learned three categories, while DDcolor captured style information for four categories, indicating limited inter-class discrimination

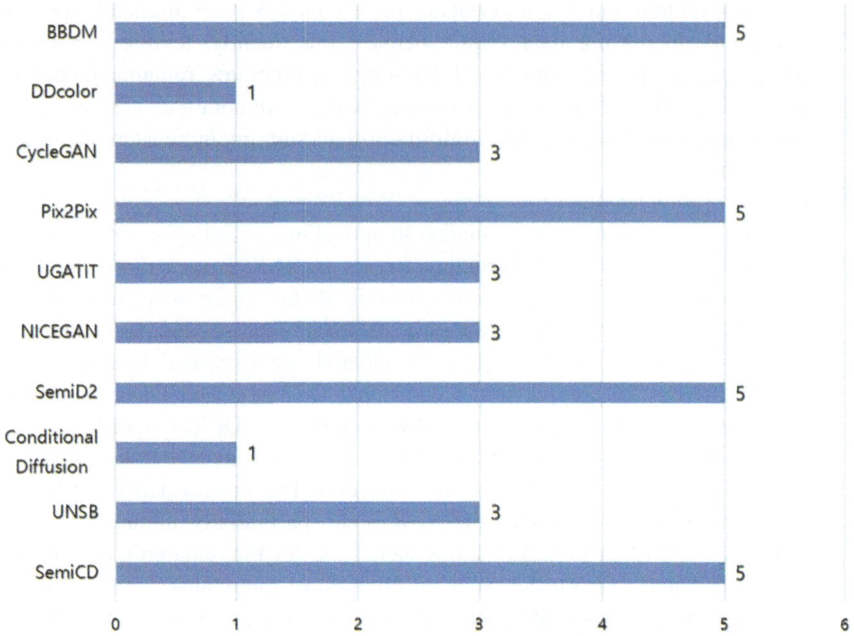

Fig. 10 Number of correctly converted land categories by different algorithms

capabilities of these methods. Although BBDM, Pix2Pix, and SemiD2 restored the style of all land categories in color, their texture modeling capabilities were problematic. In contrast, SemiCD accurately accomplished the S2O cross-modal image conversion task in both texture and color. The reason lies in its dual-branch architecture, which efficiently complements domain-level and pixel-level information simultaneously, ensuring superior texture restoration. Additionally, the dual-branch cycle-consistency loss guarantees accurate generation in style and color. Overall, this further validates the effectiveness of the proposed method in this chapter.

Furthermore, based on subjective evaluation and inter-class discrimination assessment, the following patterns can be observed:

(a) The architecture of the algorithm model is critical for texture detail conversion, which depends on the data volume and training strategy. When training data is limited, semi-supervised models outperform supervised models, while unsupervised models perform poorly. SemiCD excels in practical conversion by focusing on texture details due to its dual-branch training strategy.
(b) The objective function or detailed loss functions determine color conversion or style transformation results. Generally, algorithms under a cycle-consistency architecture can achieve accurate color restoration. Without

effective regularization constraints, even the latest diffusion models or Transformer methods may still suffer from color deviation. The dual-branch cycle-consistency loss has advantages in addressing these issues.

5. Evaluation of generation details

   In terms of detail generation, the SemiD2 and Pix2Pix algorithms were selected to compare cross-modal conversion performance. Since both SemiD2 and Pix2Pix involve supervised learning, they demonstrate strong capabilities in transforming image styles and textures. As shown in Fig. 11, the proposed method in this chapter excels in detail generation, particularly in producing optical image edges with sharper textures. Enlarged local features from the figure reveal that SemiD2 exhibits noticeable jagged blocks and checkerboard patterns, indicating overfitting issues in the model. This occurs because the method fails to account for the importance of feature extractor information in the generator's design. Pix2Pix, on the other hand,

   To some extent, this phenomenon is avoided, indicating that the model's complexity is better aligned with the dataset. However, in local details, the results generated by Pix2Pix appear relatively blurry. For example, the generated residential areas fail to effectively distinguish buildings, and the river bends in vegetation are not smoothly rendered. In contrast, SemiCD's output is more refined, with magnified details appearing smoother and clearer than in other images.

Additionally, Fig. 12 uses farmland as an example to conduct a comparative analysis of the SemiD2 translation results. The findings indicate that SemiCD excels in reducing speckle noise and accurately restoring scene details. The algorithm proposed in this chapter effectively preserves more diverse ground colors in farmland grayscale features, thereby enhancing image details and improving translation accuracy. For instance, when processing multiple farmland images, SemiCD did not generate checkerboard patterns, suggesting that the current data volume is sufficient to support model training. Secondly, for small targets like simple houses in SAR images, SemiCD can also accurately capture and restore their details. Furthermore, the method in this chapter effectively converts straight linear textures in farmland and accurately restores brightness and color information. This further demonstrates the generative capability of SemiCD.

4.2 SBS2O Method

1. Dataset Settings

   The dataset used by SBS2O is the SEN1-2 data proposed by M. Schmitt, with images acquired by Sentinel-1 and Sentinel-2 satellites [27]. This dataset was initially applied in S2O image fusion. The Sentinel-1 series consists of two polar-orbiting satellites equipped with C-band SAR sensors, enabling ground image acquisition under any weather conditions. Sentinel-2 comprises two polar-

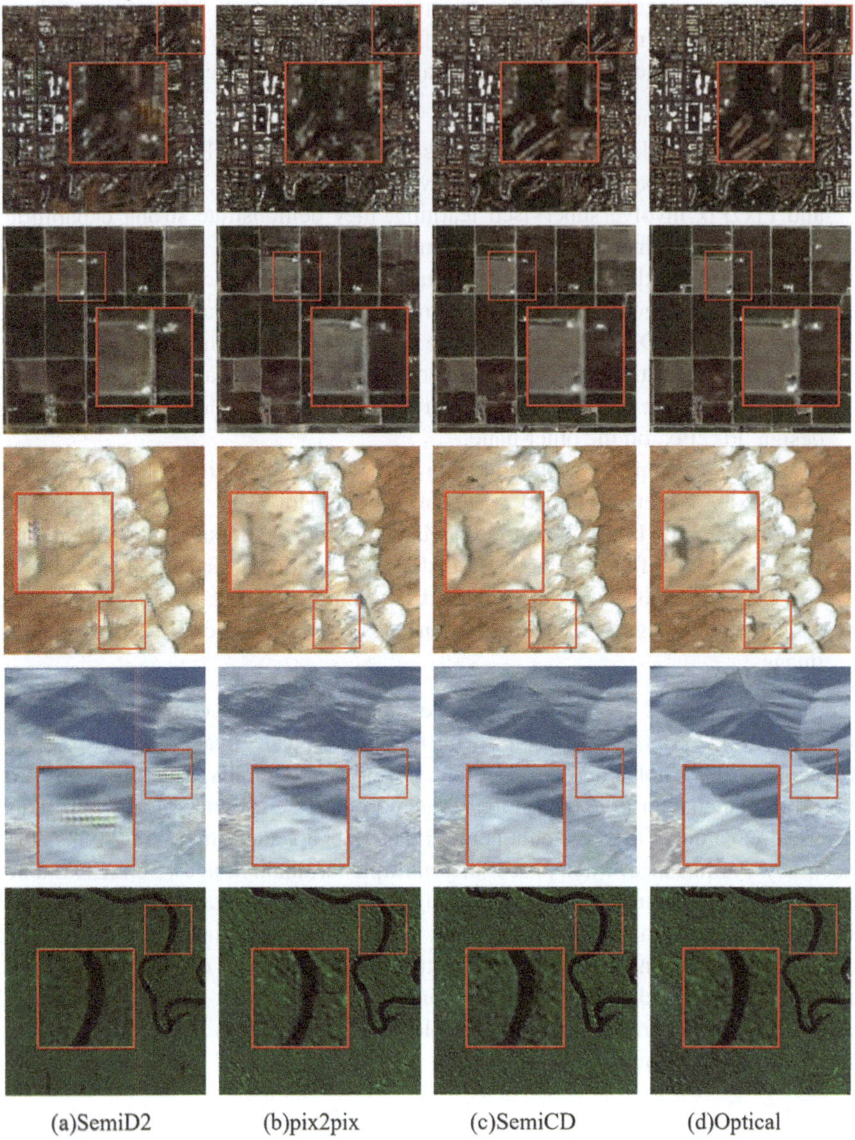

Fig. 11 Comparison of conversion details between SemiCD, SemiD2, and Pix2Pix, with the red areas indicating enlarged detail regions. (**a**) SemiD2; (**b**) Pix2Pix; (**c**) SemiCD; (**d**) Optical

orbiting satellites on the same orbit with a 180° phase difference. Therefore, the SEN1-2 dataset contains 282,384 filtered patches of SAR and optical images from different seasons.

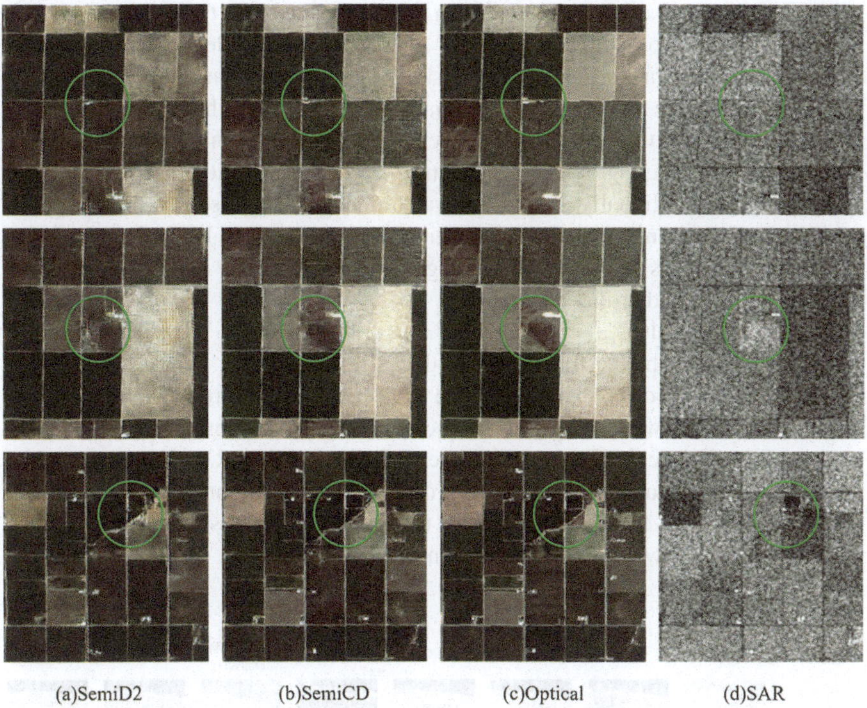

Fig. 12 SemiCD and similar semi-supervised algorithm SemiD2 details. Section comparison: (**a**) SemiD2; (**b**) SemiCD; (**c**) Optical; (**d**) SAR

Fig. 13 Example dataset used by SBS2O

Table 2 Land class distribution of the SBS2O dataset

Land	Island	Residence	Bare-soil	Cropland	Desert	Mountain	Vegetataion	Total
Numbers	202	337	204	276	204	208	228	1659

Furthermore, 1659 image pairs were selected from the SEN1-2 dataset for research, covering diverse natural terrains and primarily sourced from summer and autumn images in the original dataset. Poor-quality data were removed through manual evaluation, resulting in seven land-use types. As shown in Fig. 13, each type contains 202–337 data pairs (see Table 2). In the training set, the dataset was further randomly shuffled to simulate unsupervised training. Finally, 25 image pairs from each domain were selected as the test set, which did not participate in training. This approach aimed to reduce computational pressure while better reflecting the model's feature-learning capability under limited data conditions. Additionally, the model was trained on single scenes, meaning that for each algorithm, seven transformation models were established for different land covers. This helps evaluate the generalization ability of various methods across natural and artificial terrains.

2. Conversion Results

To validate the performance of the SBS2O method in the cross-modal conversion task of S2O images, one conversion result from each method is selected, as shown in Fig. 13. These images are all from the test set. The first and second columns represent the corresponding SAR and optical images, respectively. Columns three to seven display the generation results of baseline algorithms. The last column shows the experimental results of the SBS2O method. Each row represents a different land cover type.

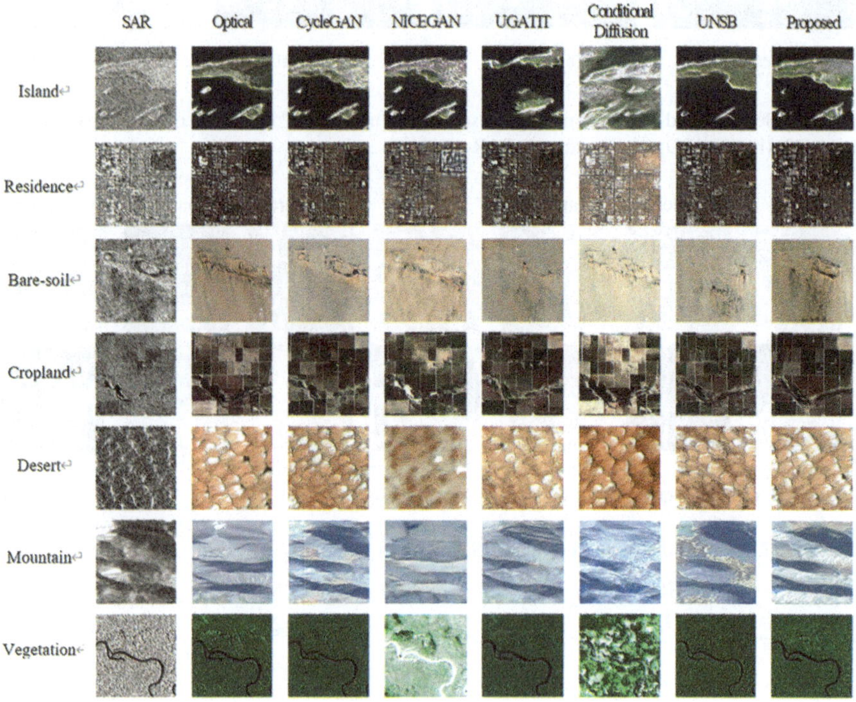

Fig. 14 Conversion results of SBS2O and comparison algorithms

From Fig. 14, it can be observed that the SBS2O method subjectively achieves better experimental results. This model accurately distinguishes the boundary contours of natural and artificial terrains, such as island coastlines, rivers in vegetation, roads in residential and farmland areas, and mountain shadows. Additionally, the images generated by SBS2O exhibit color accuracy and saturation closer to real optical images. Although some details are missing, this already represents optimal performance for unsupervised image conversion tasks.

In comparison, UNSB yields results relatively close to the proposed method in all cases but performs poorly in generating scenes with noisy distributions (e.g., deserts), failing to complete the image conversion task effectively. GAN-based architectures like CycleGAN, NICEGAN, and UGATIT exhibit some information errors in image translation. Among them, CycleGAN effectively extracts contour information but shows deviations in color learning. Notably, CycleGAN performs better on natural terrains with simple textures than on artificial terrains, as its lower model complexity prevents effective information mapping between SAR and optical images for complex textures.

UGATIT exhibits checkerboard artifacts in the conversion of bare soil and mountains due to its high model complexity, which leads to incomplete feature learning due to insufficient data. ConditionalDiffusion performs better in texture conversion for certain features but suffers from color distortion and imbalance, such as generating incorrect rainforest representations.

In contrast, the SBS2O method employs a novel generator architecture to comprehensively extract multi-dimensional feature map information, thereby minimizing the aforementioned issues. Further analysis of results from similar algorithms reveals that both UNSB and SBS2O models adopt a Schrödinger generator architecture, and their generated images exhibit highly consistent tones with real images. This demonstrates that the interpolation images predicted by stochastic optimal transport methods can achieve richer image details. Notably, UNSB achieves good results in natural image processing, but its high data requirements cause it to lag behind SBS2O in texture detail generation.

3. Qualitative Evaluation of Details

To observe the details of image generation more clearly, farmland is used as an example, as shown in Fig. 15. Farmland has a more regular texture structure, containing numerous straight roads and field textures. For unsupervised domain-level modality alignment tasks, generating such artificial terrain is inherently challenging. The red and yellow boxes in the figure highlight the advantages in generating field and riverbed details, clearly demonstrating that the SBS2O method outperforms other methods in terms of texture fineness and structural clarity. It particularly excels in capturing the complex features of farmland, better reflecting the authentic characteristics of natural environments.

Figure 16 displays the translation results of desert in natural terrain. Due to the presence of significant noise and complex feature contours in this scenario, the image conversion task is challenging. The red circles mark the regions of interest,

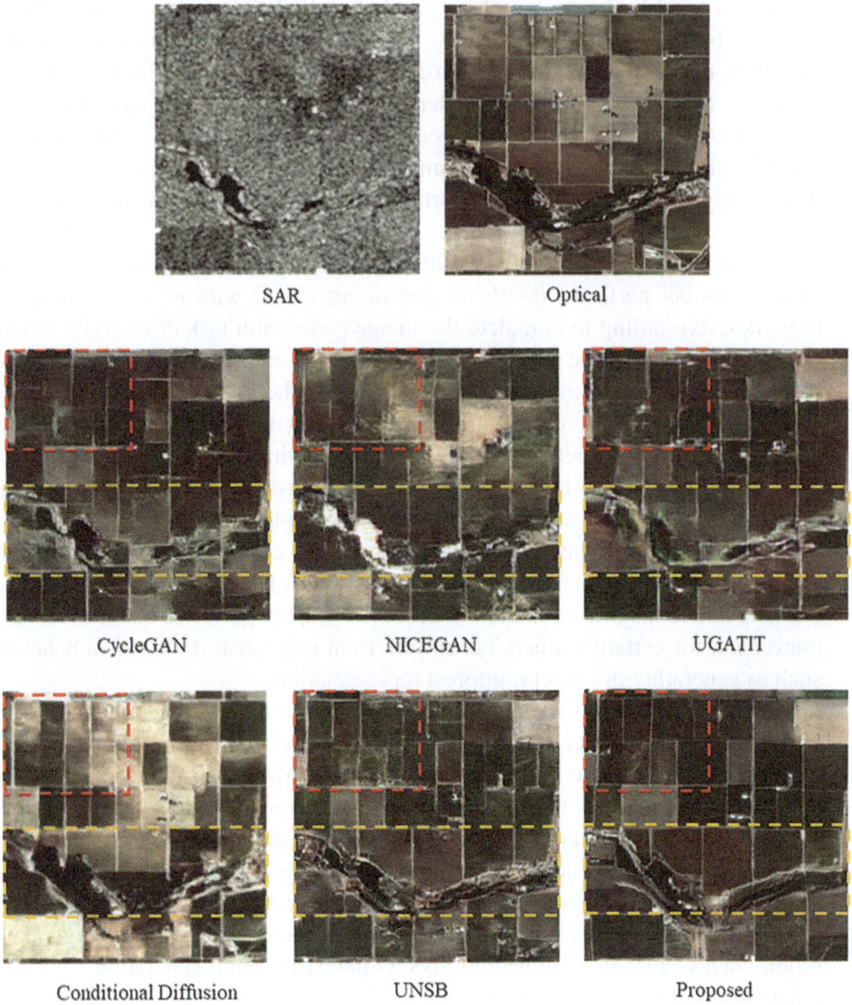

Fig. 15 The farmland category conversion results, with the first row being real images and the second and third rows being converted images

showing that the SBS2O method effectively establishes a mapping relationship between SAR and optical information in both texture and color generation.

5 Conclusion

In this chapter, the SAR-to-optical image modality conversion method based on generative artificial intelligence is systematically elaborated. From the perspective

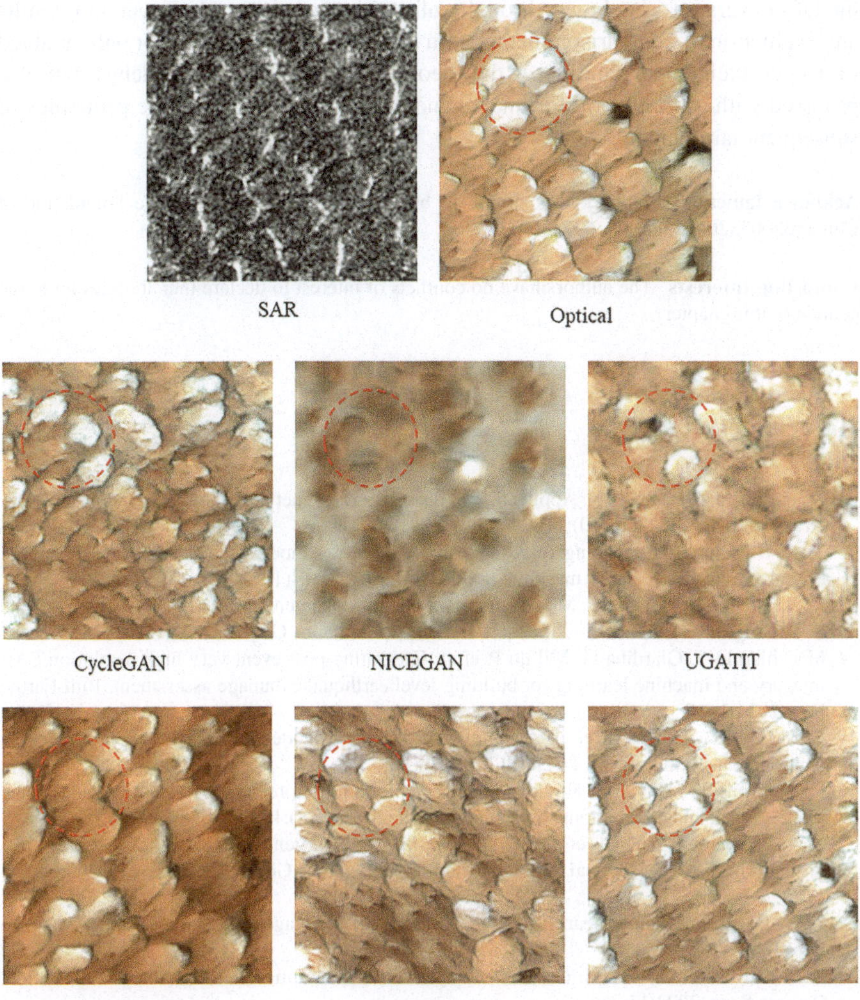

Fig. 16 The desert category conversion results, with the first row being real images, and the second and third rows being converted images

of GenAI theory, the basic principles of generative adversarial networks and diffusion models are deeply explained. The important roles of these two methods in the image cross-modal conversion task are demonstrated. On this basis, two advanced S2O image cross-modal translation methods are introduced. SemiCD adopts a semi-supervised strategy, and the SBS2O method uses an unsupervised strategy. The mathematical principles, network architectures, and loss function designs of the models are specifically introduced. Furthermore, a sub-dataset is extracted from the SEN1-2 dataset for experiments. The advantages of GenAI in image cross-modal conversion are shown through subjective evaluation of conversion results and evaluation of comparison methods. In this chapter, readers are not only enabled to deepen their understanding of the theories of deep learning branches, but also provided with a foundation for understanding and applying the basic principles of subsequent large models.

Acknowledgments This work was supported by the National Natural Science Foundation of China (62005320, 62375065, 62575091).

Competing Interests The authors have no conflicts of interest to declare that are relevant to the content of this chapter.

References

1. Zhang J, Liu Z, Jiang XLX. Application of deep generative networks for SAR/ISAR: a review. Artif Intell Rev. 2023;56(10):11905–83.
2. Zhao B, Song R. Enhancing two-stage object detection models via data-driven anchorbox optimization in UAV-based maritime SAR. Sci Rep. 2024;14(1).
3. Li X, Zhang G, Cui H, et al. MCANet: a joint semantic segmentation framework of optical and SAR images for land use classification. Int J Appl Earth Obs Geoinf. 2022;106:102638.
4. Macchiarulo V, Giardina G, Milillo P, et al. Integrating post-event very high resolution SAR imagery and machine learning for building-level earthquake damage assessment. Bull Earthq Eng. 2024:1–27.
5. Li M, Wen G, Huang X, et al. A lightweight detection model for Sar aircraft in a complex environment. Remote Sens. 2021;13(24):5020.
6. Fu K, Wang P, Feng Y, et al. Cross-modal intelligent interpretation of remote sensing data: models, data, and applications. Sci China Inf Sci. 2023;53(8):1529–59.
7. Yang X, Wang Z, Zhao J, et al. FG-GAN: a fine-grained generative adversarial network for unsupervised SAR-to-optical image translation. IEEE Trans Geosci Remote Sens. 2022;60:1–11.
8. Li Y. Research on deep learning-based SAR-to-optical image translation methods. Ningbo University; 2025.
9. Sommervold O, Gazzea M, et al. A survey on SAR and optical satellite image registration. Remote Sens. 2023;15(3):850.
10. Yuan Y, Rao Z, Lin C, et al. Adaptive ship detection from optical to SAR images. IEEE Geosci Remote Sens Lett. 2023;20:1–5.
11. Lee IH, Park CG. SAR-to-virtual optical image translation for improving SAR automatic target recognition. IEEE Geosci Remote Sens Lett. 2023;20:1–5.
12. Kulkarni SC, Rege PP. Pixel level fusion techniques for SAR and optical images: a review. Inf Fusion. 2020;59:13–29.
13. Goodfellow I, Pouget-Abadie J, Mirza M, et al. Generative adversarial networks. Commun ACM. 2020;63(11):139–44.
14. Adler J, Lunz S. Banach wasserstein gan. Adv Neural Inf Process Syst. 2018:31.
15. Radford A, Metz L, Chintala S. Unsupervised representation learning with deep convolutional generative adversarial networks. arXiv preprint arXiv:1511.06434. 2015.
16. Van den Oord A, Kalchbrenner N, Espeholt L, et al. Conditional image generation with pixelcnn decoders. Adv Neural Inf Process Syst. 2016:29.
17. Mirza M, Osindero S. Conditional generative adversarial nets. arXiv preprint arXiv:1411.1784. 2014.

18. Goodfellow I, Mirza M, Courville A, et al. Multi-prediction deep Boltzmann machines. Adv Neural Inf Process Syst. 2013:26.
19. Huang Y, Niu B, Guan H, et al. Enhancing image watermarking with adaptive embedding parameter and PSNR guarantee. IEEE Trans Multimedia. 2019;21(10):2447–60.
20. Wang Z, Bovik AC, Sheikh HR, et al. Image quality assessment: from error visibility to structural similarity. IEEE Trans Image Process. 2004;13(4):600–12.
21. Zhang R, Isola P, Efros AA, et al. The unreasonable effectiveness of deep features as a perceptual metric. In: Proceedings of the IEEE conference on computer vision and pattern recognition; 2018. p. 586–95.
22. Xia X, Xu C, Nan B. Inception-v3 for flower classification. In: 2017 2nd international conference on image, vision and computing (ICIVC), vol. 2017. IEEE. p. 783–7.
23. Shaham TR, Dekel T, Michaeli T. Singan: learning a generative model from a single natural image. In: Proceedings of the IEEE/CVF international conference on computer vision; 2019. p. 4570–80.
24. Barratt S, Sharma R. A note on the inception score. arXiv preprint arXiv:1801.01973. 2018.
25. Zhou X, Zheng F, Wang H, et al. Towards real-time detection of wakes for various sea states with lightweight deep learning model in synthetic aperture radar images. Remote Sens. 2024;16(24):4798.
26. Liu GH, Vahdat A, Huang DA, et al. I2SB: image-to-image Schrodinger bridge. arXiv preprint arXiv:2302.05872. 2023.
27. Schmitt M, Hughes LH, Zhu XX. The SEN1–2 dataset for deep learning in SAR-optical data fusion. arXiv preprint arXiv:1807.01569. 2018.
28. Bai X, Pu X, Xu F. Conditional diffusion for SAR to optical image translation. IEEE Geosci Remote Sens Lett. 2023;21:1–5.

Q-Learning Implementation for Decision Problems in Digital Image Processing: Application to Adaptive Image Segmentation in Histological Cancer Detection

Camel Tanougast, Hang Chen, Marouane Ben-Akka, and Ahmad Chaddad

Abstract Digital image processing is fundamental to modern medical imaging, enabling feature extraction, visual structure enhancement, and computer-aided diagnosis. With the development of computational power and AI, traditional image analysis techniques have evolved into hybrid systems combining classical signal processing with machine learning, especially in histology for cancer cell detection and classification. AI, particularly machine learning, has brought new opportunities in digital pathology, like CNNs showing high accuracy in tissue differentiation. However, they rely on large annotated datasets and lack adaptability. This has led to the exploration of Reinforcement Learning (RL), which formulates image analysis as a sequential decision-making process. In histopathology, segmentation is crucial for quantitative analysis, and Q-learning, a model-free RL algorithm, can iteratively adjust segmentation parameters without pre-annotated data. Q-learning in medical imaging offers adaptivity to image-specific characteristics and robustness against suboptimal segmentation. This chapter covers image processing and histological segmentation fundamentals, RL framework introduction, reformulating adaptive thresholding as a Q-learning problem, implementation details, case studies on real histological images, and a discussion of the approach's benefits, limitations, challenges, and future perspectives.

C. Tanougast (✉) · M. Ben-Akka
Université de Lorraine, LGIPM, Metz, France
e-mail: camel.tanougast@univ-lorraine.fr

H. Chen
Université de Lorraine, LGIPM, Metz, France

School of Space Information, Space Engineering University, Beijing, China

A. Chaddad
School of Artificial Intelligence, Laboratory for Artificial Intelligence of Personalized Medicine, Guilin University of Electronic Technology, Guilin, China

H. Chen, Z. Liu (eds.), *Artificial Intelligence in Digital Image Processing: Theories, Methods, and Applications*, Studies in Computational Intelligence 1246,
https://doi.org/10.1007/978-3-032-12891-1_10

Keywords Digital image processing · Reinforcement learning · Q-learning · Medical imaging · Histological segmentation

1 Introduction

Digital image processing has become a cornerstone of modern medical imaging, enabling the extraction of relevant features, the enhancement of visual structures, and the support of computer-aided diagnosis. With the exponential growth of computational power and artificial intelligence (AI), traditional image analysis techniques—such as thresholding, edge detection, and morphological filtering—have progressively evolved into hybrid systems combining classical signal processing with machine learning. This shift has profoundly impacted medical imaging, particularly histology, where the detection and classification of cancerous cells rely heavily on robust segmentation strategies [1–7]. Artificial Intelligence (AI), and more specifically machine learning, has opened new opportunities in digital pathology. Convolutional Neural Networks (CNNs) and supervised classification techniques have shown remarkable accuracy in differentiating normal and abnormal tissues [8–9]. However, such approaches depend strongly on large, annotated datasets and lack adaptability when facing variations in staining, acquisition conditions, or tissue morphology. This limitation has motivated the exploration of Reinforcement Learning (RL), a branch of AI that formulates image analysis as a sequential decision-making process, allowing adaptive optimization of segmentation criteria. The general principle of reinforcement learning is illustrated in Fig. 1, where an agent interacts with its environment by selecting actions, observing states, and receiving rewards. This trial-and-error process allows the agent to progressively learn an optimal policy without requiring explicit supervision. In the context of image processing, the environment corresponds to the image itself, while the agent's actions involve modifying parameters such as thresholds or segmentation masks [9–13].

In histopathology, segmentation is an essential first step for quantitative analysis. An example of a histological tissue image is shown in Fig. 2, where the complexity

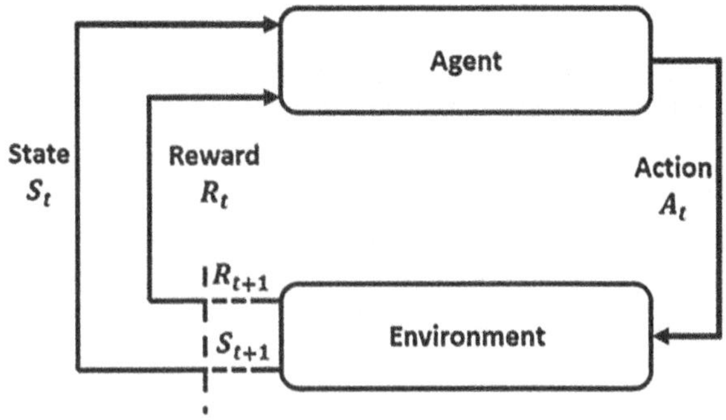

Fig. 1 Reinforcement learning schema illustrating the interaction between the agent and the environment

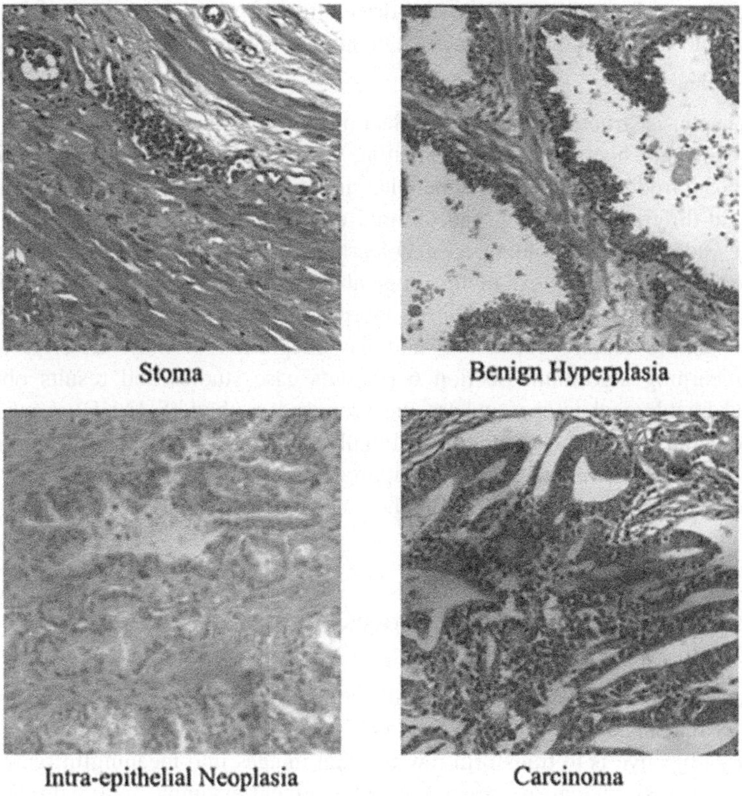

Fig. 2 Example of histological cancer tissue image (colon biopsy, H&E staining)

of cellular structures highlights the difficulty of distinguishing cancerous regions from normal ones using fixed thresholding approaches. By isolating nuclei, cytoplasm, and tissue regions, it becomes possible to extract texture features-such as Haralick descriptors-which play a crucial role in the classification of cancerous versus non-cancerous tissues.

Within RL, Q-learning has emerged as a powerful model-free algorithm that learns optimal decision policies through interactions with an environment. Instead of relying on pre-annotated data, Q-learning agents iteratively improve their performance by receiving rewards or penalties for their actions. Applied to histological images, this paradigm translates into iteratively adjusting segmentation parameters—such as threshold values—until the optimal separation of structures is achieved.

The relevance of Q-learning for medical imaging is twofold:

1. It provides adaptivity, allowing algorithms to dynamically adjust to image-specific characteristics rather than applying fixed global parameters.

2. It ensures robustness, as the reward-driven framework encourages exploration of multiple solutions, reducing the risk of suboptimal segmentation caused by local minima.

This chapter is structured as follows: Section 2 reviews the fundamentals of image processing and histological segmentation, with emphasis on texture descriptors. Section 3 introduces the reinforcement learning framework, focusing on Q-learning, temporal-difference learning, and tailored exploration strategies. Section 4 reformulates adaptive thresholding as a Q-learning problem, defining states, actions, and rewards in the context of histological image analysis. Section 5 details the implementation framework, including morphological preprocessing, Haralick and related texture feature extraction, and the integration of these descriptors into the Q-learning algorithm. Section 6 presents case studies and results obtained on real histological images, where the proposed method (SHAKE) is evaluated both qualitatively and quantitatively. Finally, Section 7 provides a discussion of the benefits and limitations of the approach, highlights challenges and future perspectives, and concludes the chapter.

2 Fundamentals of Image Processing

Medical image processing has emerged as one of the most dynamic and impactful fields at the intersection of artificial intelligence and biomedical engineering. Its primary objective is to transform raw medical images into meaningful representations that support diagnosis, prognosis, and treatment planning. Within this domain, segmentation occupies a central role. By partitioning an image into regions of interest, segmentation allows researchers and clinicians to isolate nuclei, cytoplasm, and tissue structures from the complex background of histological slides. Such delineation is particularly important in cancer detection, where the identification of subtle cellular abnormalities can determine diagnostic accuracy.

2.1 Principles of Medical Image Segmentation

Segmentation can be defined as the process of dividing an image into non-overlapping regions that correspond to specific anatomical or pathological structures. In histopathology, this task typically involves distinguishing between healthy and malignant cells, detecting nuclei boundaries, or separating tissue compartments. Unlike segmentation in natural images-where uniform textures or color differences often simplify the task—histological images are characterized by substantial variability. Differences in staining protocols, acquisition devices, and biological heterogeneity result in highly complex visual patterns. Preparation artifacts such as tissue folds or uneven staining further complicate the analysis, often leading to irregular and ambiguous structures. As a result, simple rule-based methods are insufficient, and designing segmentation algorithms that are both accurate and robust across diverse cases remains a central challenge in medical image analysis.

2.2 *Histological Slides: Complexity and Challenges*

Histological slides stained with Hematoxylin and Eosin (H&E) vividly illustrate the complexity of medical image segmentation. Hematoxylin stains cell nuclei in shades of blue or purple, while eosin highlights cytoplasmic and extracellular matrix regions in pink. This dual staining creates a visually rich landscape but also introduces significant analytical challenges. In practice, nuclei often overlap, cell morphology is irregular, and staining intensity varies both within and across samples. Preparation artifacts such as tissue folds, air bubbles, or uneven dye penetration further complicate segmentation. Figure 2, introduced in the previous section, provides an example of a histological cancer tissue image where pathological and healthy structures blend together, making automatic separation particularly difficult. These difficulties demonstrate the inadequacy of simplistic global segmentation methods and emphasize the need for adaptive approaches that can account for local variations, noise, and the complex morphology of biological tissues.

2.3 *Classical Segmentation Approaches*

Over the past decades, several classical segmentation techniques have been applied to histological image analysis. Although often limited in adaptability, they form the foundation upon which more advanced AI-driven methods such as reinforcement learning are built. Thresholding methods represent the most intuitive approach. A global threshold, for instance using Otsu's method, separates foreground (cells or nuclei) from background by optimizing intra-class variance. This approach works well when there is a clear intensity difference between structures but fails when staining variations or heterogeneous tissue distributions exist. Figure 3 illustrates the effect of global thresholding applied to a histological image. While some nuclei are correctly segmented, others are either missed or incorrectly merged with the background due to uneven staining.

Edge-based methods, such as the Sobel operator, detect boundaries by highlighting sharp intensity changes between neighboring pixels. As illustrated in Fig. 4, the Sobel filter can effectively outline nuclei contours. However, its sensitivity to noise and irregular staining often results in discontinuous or fragmented edges, which are not sufficient for reliable cancer detection.

Region growing techniques represent another classical approach, starting from initial seed points and expanding regions until homogeneity criteria are no longer satisfied. For example, in histopathology, region growing can be used to capture

Fig. 3 Example of global thresholding applied to a histological image

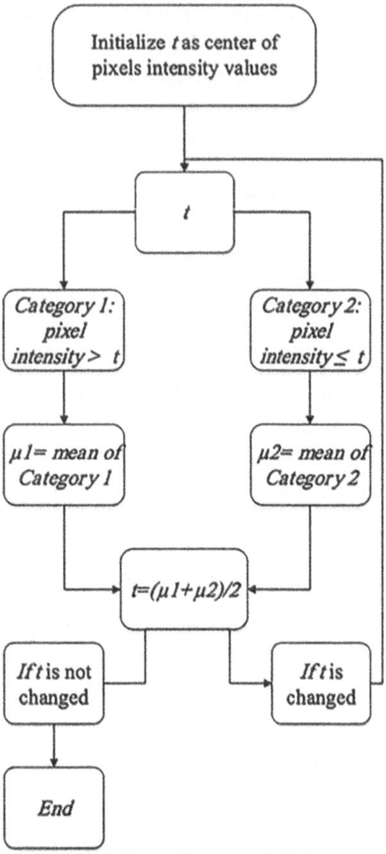

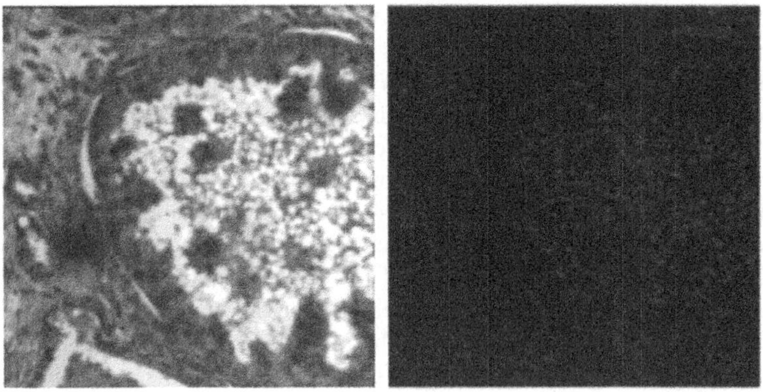

Fig. 4 Comparison of edge-based segmentation using the Sobel operator

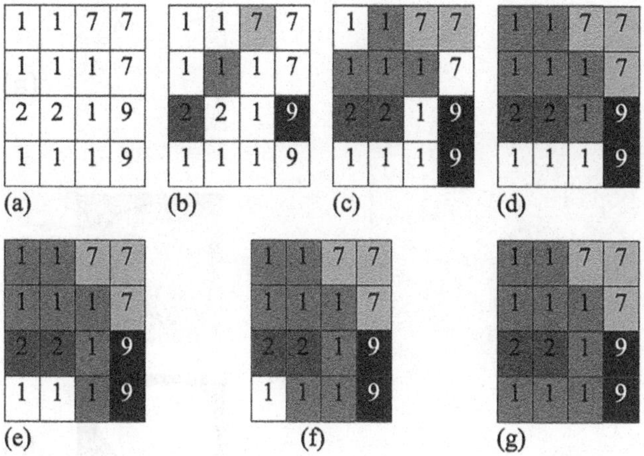

Fig. 5 Illustration of region growing applied to nuclei detection. An example of seeded region method, (**a**) Original image, (**b**) first step, (**c**) second step, (**d**) third step, (**e**) fourth step, (**f**) fifth step, (**g**) sixth step

nuclei regions by initializing seeds within high-intensity zones. The algorithm for seeded region based method is:

(a) Select the initial seed points say, {p1, p2, ..., pn},
(b) Compute difference between the pixel values of initial seed points and its neighboring points,
(c) If the computed difference is less than the threshold, add the neighboring point to the cluster where seed point belonged,
(d) Re-compute the boundary of the cluster and set the boundary points as new seed points,
(e) Re-compute the mean pixel values of the clusters,
(f) Repeat the steps (b) to (e) for all pixels until they are allocated to a cluster.

An example about the seeded region is represented in Fig. 5, which shows how to add each pixel to its region according to the nearest. As shown in Fig. 5, this approach can yield relatively homogeneous segmentation, but its performance heavily depends on the initial seed placement and the predefined similarity criteria. Incorrect seeds often lead to over-segmentation or leakage into neighboring regions.

A more advanced classical method is active contour modeling, also known as snakes. This approach involves initializing a curve around a potential structure of interest, which then evolves iteratively to fit object boundaries based on internal and external energy terms. In histological images, active contours are particularly useful for capturing irregularly shaped nuclei. Figure. 6 illustrates how active contours adapt to cell boundaries, providing more precise segmentation than thresholding or Sobel filters. Nonetheless, this method requires careful initialization and remains computationally expensive for large images.

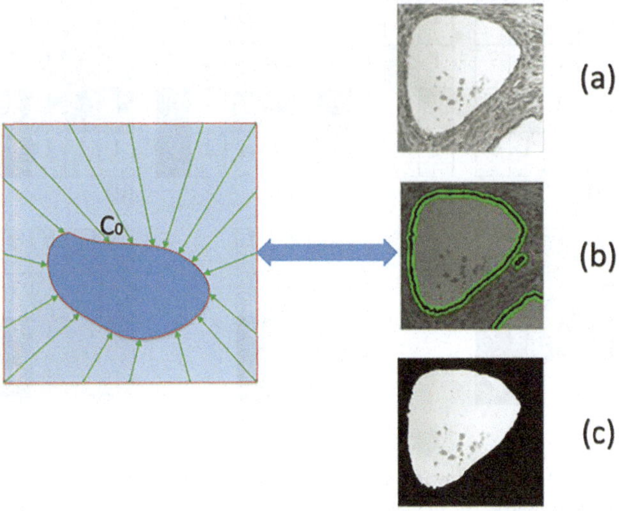

Fig. 6 Active contour (snake) segmentation in histopathology. Automatic and manual segmentation of carcinoma cell type. Example of snake algorithm, (**a**) initial image, (**b**) contour after many iterations; and (**c**) segmented image using snake method

2.4 Principles of Medical Image Segmentation

Segmentation is not an end in itself, but a means to enable quantitative analysis and classification. Once structures such as nuclei are isolated, the next step is to characterize them using discriminative features. Among these, texture features have demonstrated strong relevance in histopathology. Texture analysis aims to capture spatial variations in pixel intensities that reflect the underlying tissue organization. For example, malignant tissues often display more irregular, heterogeneous textures than normal tissues. Among texture descriptors, Haralick features, derived from the Gray-Level Co-occurrence Matrix (GLCM), are widely used. These include:

- Contrast, which measures local intensity variations,
- Correlation, which captures the linear dependency between neighboring pixels,
- Energy, reflecting textural uniformity, and
- Homogeneity, assessing the closeness of distribution of elements in the GLCM.

These descriptors, once extracted from segmented nuclei or regions of interest, provide crucial input for classification models. They allow reinforcement learning agents not only to optimize segmentation parameters but also to evaluate how well segmentation outcomes contribute to distinguishing between cancerous and non-cancerous tissues. In summary, classical segmentation methods provide the basis for histological image analysis, but their limitations motivate the integration of adaptive approaches. Reinforcement learning, and specifically Q-learning, offers a promising alternative by treating segmentation as a dynamic optimization process, where thresholding and morphological operations are adapted iteratively based on feedback. This perspective will be further developed in Sect. 3.

3 Reinforcement Learning Foundations

Reinforcement Learning (RL) represents one of the most active and influential fields of artificial intelligence. Its importance lies in its ability to model situations where an agent must learn to act optimally in an uncertain and dynamic environment by trial and error. In contrast to supervised learning, which requires large sets of labeled examples, RL agents rely on interactions with their environment to improve their decision-making strategies. This difference is crucial in image processing, and even more so in medical imaging, where annotated data are scarce, heterogeneous, and sometimes subjective. RL thus provides a flexible framework for developing adaptive algorithms that can refine their parameters iteratively until optimal performance is reached.

3.1 The Markov Decision Process

The theoretical basis of RL is the Markov Decision Process (MDP). An MDP is defined by five elements: a set of states S, a set of actions A, a transition function P(s′|s,a), a reward function R(s, a), and a discount factor γ. At each time step, the agent observes its current state st ∈ S, selects an action a_t ∈ A, and as a result the environment transitions to a new state s_{t+1} with probability P(s′|s, a). The agent then receives a reward r_{t+1}, which indicates the quality of the chosen action. The process repeats, generating a trajectory of states, actions, and rewards. The objective of the agent is to learn a policy π: S → A that maximizes the expected cumulative discounted reward, defined as:

$$G_t = E\left[\sum_{k=0}^{\infty} \gamma^k r_{t+k+1} \mid s_t\right] \tag{1}$$

The agent–environment loop has already been introduced in Fig. 1 of chapter "Fundamentals of Deep Learning." This figure illustrates how the agent continuously updates its strategy by observing the effects of its actions on the environment and adjusting future decisions accordingly. In the context of image processing, the state may correspond to the current segmentation mask, the action to a modification of the threshold value, and the reward to a quality index measuring the similarity between the obtained segmentation and the desired outcome. This closed-loop nature is fundamental: the algorithm is not applied once but interacts iteratively with the image until it reaches an optimal segmentation solution.

3.2 Q-Learning Algorithm

Within the family of RL algorithms, Q-learning stands out for its simplicity and efficiency. Introduced by Watkins in 1989, Q-learning is a model-free algorithm, meaning that it does not require prior knowledge of the transition probabilities P(s′ıs,a). Instead, it directly estimates the value of each state–action pair. This makes it particularly suitable for problems like image segmentation, where the dynamics of the environment (i.e., the relationships between thresholds, features, and segmentation results) are unknown or highly complex. The Q-function Q(s,a) represents the expected cumulative discounted reward when taking action a in state s and then following the optimal policy. It is formally defined as:

$$Q^{*}(s, a) = \max_{\pi} E\left[\sum_{k=0}^{\infty} \gamma^{k} r_{t+k+1} \mid s_t = s, a_t = a, \pi\right] \tag{2}$$

Where π is the strategy being followed. The algorithm updates its estimate of Q(s,a) iteratively using the temporal-difference update rule:

$$Q_{t+1}(s_t, a_t) = Q_t(s_t, a_t) + \alpha\left[r_{t+1} + \gamma \max_{a'} Q_t\left(s_{t+1}, a'\right) - Q_t(s_t, a_t)\right] \tag{3}$$

Where α is the learning rate, γ the discount factor, and the term in brackets is the temporal-difference (TD) error. Equations (2) and (3) jointly capture the essence of Q-learning: the first defines the optimal action–value function, while the second describes the iterative update mechanism that progressively corrects the estimates. Q-learning belongs to the family of Temporal Difference (TD) learning methods. The term temporal difference refers to the fact that learning is driven by the difference between two consecutive value estimates at successive time steps. This difference, known as the TD error, is defined as:

$$\delta_t = r_{t+1} + \gamma \max_{a'} Q\left(s_{t+1}, a'\right) - Q(s_t, a_t) \tag{4}$$

The TD error δt quantifies how much the current prediction Q(st,at) diverges from a better-informed estimate that incorporates the immediate reward and the next state. A positive error indicates that the action performed better than expected, while a negative error means it was worse. The update in Eq. (3) of chapter "Adversarial Attacks and Defenses in Computer Vision" can therefore be seen as an incremental correction proportional to the TD error. To complement these equations, Algorithm 1 summarizes the standard Q-learning procedure in a general form, independent of any particular domain.

Algorithm 1 Standard Q-Learning Procedure (Adapted for This Chapter)

```
Initialize Q(s,a) arbitrarily
Repeat for each episode:
   Initialize state s
   Repeat for each step of the episode:
```

Choose action a from state s using ε-greedy policy
Execute action a, observe reward r and next state s′
Update:
$Q(s,a) \leftarrow Q(s,a) + \alpha\, [r + \gamma$ max_a′ $Q(s',a') - Q(s,a)]$ (Eq. 3.3)
Set $s \leftarrow s'$
until s is terminal

Figure 7 shows how Q-values are updated across iterations. At each step, the agent compares the observed reward plus the estimated value of the next state to the current estimate. This difference, known as the temporal-difference error, drives the learning process. Over time, Q-values converge to the optimal values, provided the agent continues to explore all state–action pairs sufficiently. The figure highlights the iterative improvement of Q-values and their eventual convergence, which is essential for understanding why Q-learning can be applied effectively to image segmentation tasks. Figure 7 (constructed from Eqs. (2)–(4) of chapter "Adversarial Attacks and Defenses in Computer Vision" and Algorithm 1) visualizes the TD mechanism: at each interaction the agent observes a state, selects an action, receives a reward, and updates the corresponding Q-value according to Eq. (3) of chapter "Adversarial Attacks and Defenses in Computer Vision." Over time, the Q-values converge, illustrating how learning progresses step by step until the optimal decision policy emerges. This figure highlights why Q-learning is well suited for iterative problems such as adaptive thresholding in image segmentation.

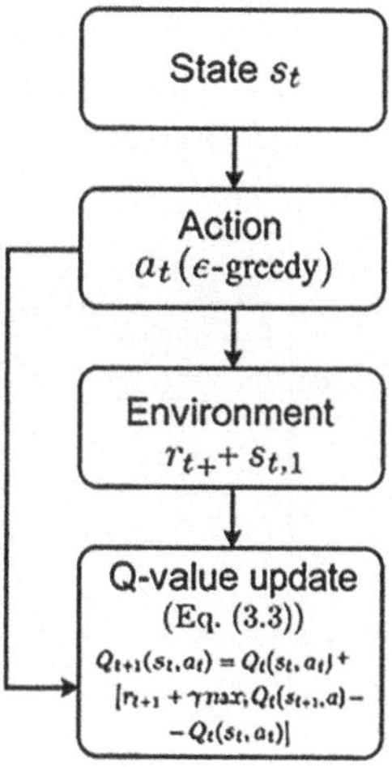

Fig. 7 Q-learning update mechanism (TD rule)

3.3 Exploration Versus Exploitation

A key challenge in Q-learning is balancing exploration and exploitation. Exploitation means choosing the action with the highest current Q-value, while exploration means trying alternative actions that may yield higher rewards in the long run. Too much exploitation can trap the agent in a suboptimal solution, while too much exploration slows down learning. The most widely used strategy is the ε-greedy policy:

$$a_t = \begin{cases} \text{argmax}_a Q(s_t, a), & withprobability 1 - \epsilon, \\ randomaction, & withprobability \epsilon \end{cases}$$

Initially, ϵ is set high to encourage exploration. As learning progresses, ϵ is gradually reduced (a process known as ε-decay), so that the agent increasingly exploits what it has learned. In our recent research, we proposed an improved approach named TRE-QL (Tailored Reward and Epsilon-Greedy Decay Q-learning). In this contribution, ε does not decrease according to a fixed schedule but adapts dynamically to the agent's progress. In parallel, we designed a reward function specifically tailored to the problem under study, in order to guide the agent more effectively toward meaningful solutions. By coupling adaptive exploration with carefully engineered rewards, TRE-QL accelerates convergence, prevents premature stagnation in suboptimal policies, and ensures controlled exploration throughout the learning process. Figure 8 illustrates the benefits of our TRE-QL approach compared to classical Q-learning. This figure shows the comparative learning curves of Q-learning and TRE-QL. The learning curves show that our tailored decay strategy produces smoother convergence and avoids the oscillations observed with naïve ε-decay.

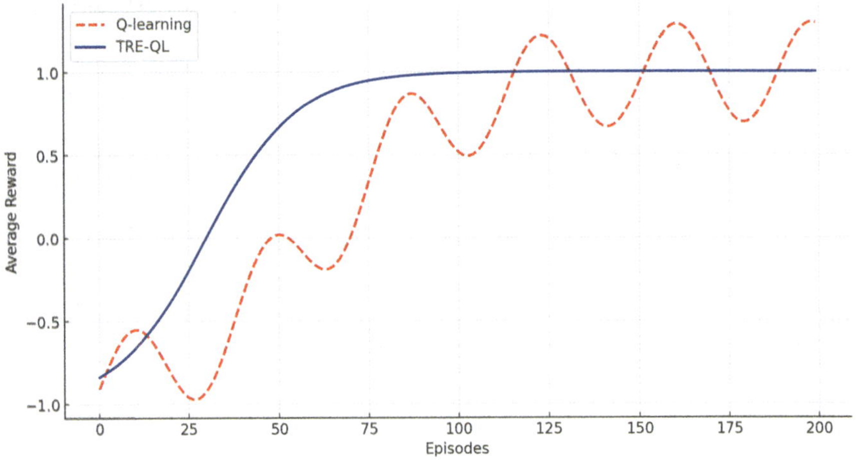

Fig. 8 Comparative learning curves of Q-learning vs TRE-QL

The red curve (Q-learning) oscillates and converges slowly, while the blue curve (TRE-QL) converges faster and more smoothly, avoiding the instability typical of ε-decay. This demonstrates that TRE-QL achieves both stability and efficiency in convergence—properties that are particularly relevant when transferring the method to medical image segmentation, where robustness is essential.

3.4 Relevance of Q-Learning to Image Segmentation

The application of Q-learning to image segmentation is natural when the task is framed as a sequential decision-making problem. Segmentation is rarely perfect in a single step; instead, it requires iterative refinement. An agent might begin with an initial threshold, evaluate the resulting segmentation, adjust the threshold or apply a morphological operation, and repeat this process until a satisfactory segmentation is achieved. Each iteration corresponds to a state–action–reward transition, exactly as described by the MDP framework. Unlike classical approaches such as Otsu's global thresholding (see Fig. 1 of chapter "Deepfake in Image and Video Processing"), which apply a single global decision, Q-learning allows the segmentation process to adapt to local variations and complex textures. This is particularly valuable in histological cancer detection, where regions of interest can differ significantly across a single slide. By integrating Q-learning into segmentation, we move from rigid, rule-based algorithms toward adaptive, learning-driven methods capable of tailoring decisions to each image. This foundation sets the stage for the next section, where adaptive thresholding will be formally reformulated as a Q-learning problem, with explicit definitions of states, actions, and rewards in the context of histological segmentation.

4 Adaptive Thresholding as a Q-Learning Problem

Segmentation in histological images is a particularly delicate task, since nuclei and tissue structures exhibit strong heterogeneity in staining, overlapping regions, and irregular morphologies. Classical approaches such as Otsu's global thresholding or clustering methods apply a single global decision across the whole image. While efficient in simple cases, these techniques are unable to adapt locally to variations in tissue organization. As a result, they frequently produce under- or over-segmentation, which severely compromises downstream feature extraction and classification. By contrast, Q-learning provides a natural way to reformulate segmentation as a sequential decision-making problem, where an agent learns to refine its segmentation iteratively until a satisfactory solution is reached.

4.1 State Definition

In the Q-learning framework, the state is a representation of the information available to the agent at each iteration. For adaptive segmentation, this representation must capture how well the nuclei are separated from the background at the current step. A simple state definition can be derived from the grayscale histogram of the segmented region. The histogram summarizes the intensity distribution, and descriptors such as the mean, variance, skewness, and entropy can provide a compact representation of segmentation quality. A more advanced approach involves the use of Haralick texture features computed from the Gray Level Co-occurrence Matrix (GLCM). Features such as contrast, correlation, energy, and homogeneity are particularly relevant in histopathology because they capture structural differences between nuclei and surrounding stroma. For instance, malignant cells often exhibit irregular chromatin patterns that are detectable through texture descriptors, making these features highly informative states for the reinforcement learning agent. To illustrate this concept, Fig. 9 presents an example of state representation. Panel (a) shows a synthetic histological patch, (b) its corresponding intensity histogram, and (c) a feature vector composed of normalized Haralick descriptors. Together, these components define the agent's perception of the current segmentation and guide the selection of subsequent actions.

4.2 Action Definition

In Q-learning, the actions represent the possible modifications that the agent can apply to the segmentation at each iteration. In the context of histological image analysis, these actions are designed to progressively refine the mask and improve the delineation of nuclei. The most basic actions correspond to threshold adjustments.

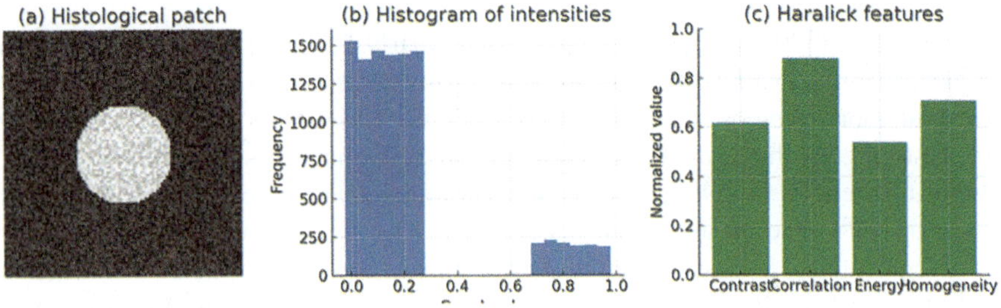

Fig. 9 State representation: histological patch, intensity histogram, and Haralick texture descriptors. (**a**) Histological patch. (**b**) Histogram of intensities. (**c**) Haralick features

At each step, the agent may decide to increase or decrease the current threshold by a discrete increment ΔT:

$$a_t \in \{-\Delta T, +\Delta T\} \quad (5)$$

This discrete formulation enables the agent to explore the threshold space iteratively, rather than applying a single global threshold. For instance, increasing the threshold may reduce the apparent size of nuclei but risk excluding faintly stained regions, while decreasing it may capture weaker structures but introduce background noise. Beyond simple threshold shifts, more powerful actions are obtained by integrating morphological operators. These are essential in histological segmentation because they directly correct structural imperfections:

- Closing fills small gaps and merges fragmented nuclei, ensuring more complete objects.
- Opening removes small noisy regions, which frequently appear in poorly stained or degraded tissue.
- Dilation and erosion adjust contours locally, enabling fine boundary refinements.

By including such morphological operations, the action space becomes richer and allows the agent to actively shape the segmentation masks, rather than relying solely on pixel intensity distributions. To illustrate these possibilities, Fig. 10 shows synthetic examples of actions applied to a histological patch. Starting from the original patch (a), panels (b) and (c) demonstrate the effect of increasing and decreasing the threshold. Panels (d) and (e) then illustrate how closing and opening operations correct segmentation errors by filling gaps or removing noise.

4.3 Reward Design

The third and most crucial element in the Q-learning formulation is the reward function, which provides feedback to the agent about the quality of its actions. In image segmentation, the reward must reflect how well the segmented nuclei match the expected biological structures. Designing this function is delicate, because it directly drives the learning dynamics. When expert annotations (ground truth masks) are available, the reward can be defined using overlap indices such as the Dice coefficient:

$$R(s_t, a_t) = Dice\left(M_{seg}, M_{gt}\right) \quad (6)$$

Where M_{seg} is the segmentation mask generated by the agent and M_{gt} the expert-provided ground truth. The Dice index varies between 0 (no overlap) and 1 (perfect match). High Dice values correspond to accurate segmentation, while low values penalize under- or over-segmentation. In many histopathology datasets, however, annotated ground truth masks are scarce or partially available. In such cases, the

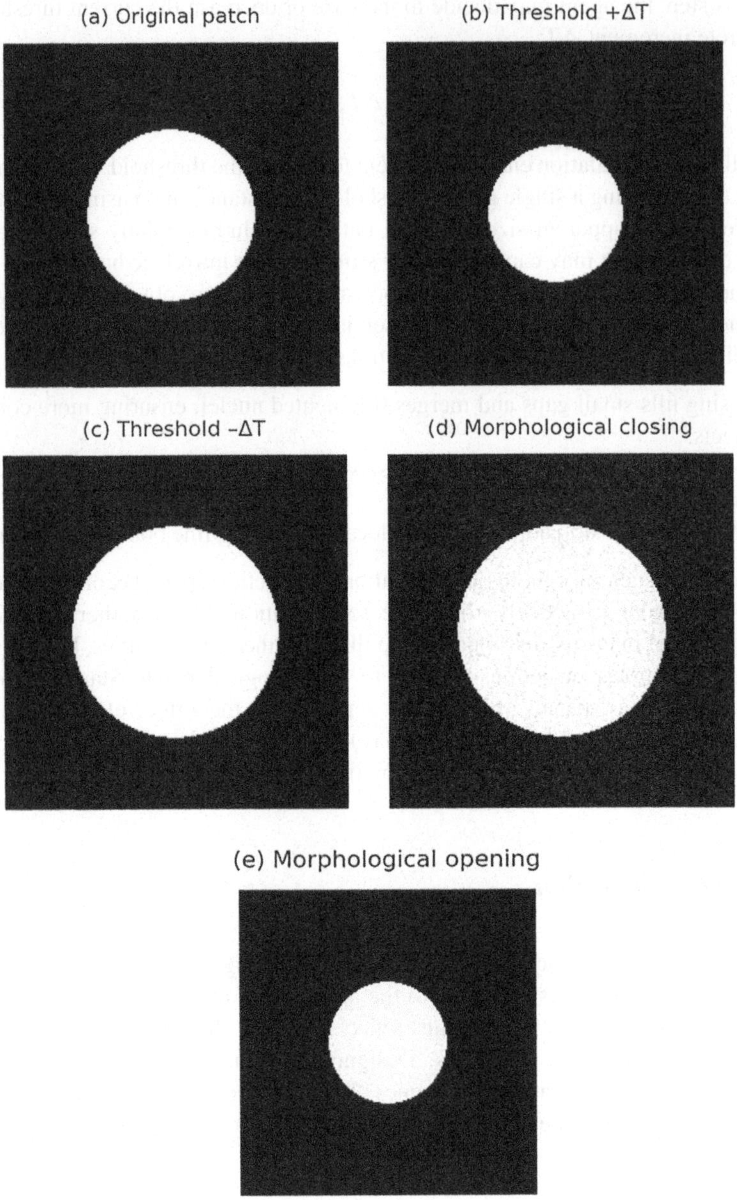

Fig. 10 Examples of actions on a histological patch: threshold adjustments ($+\Delta$T, $-\Delta$T) and morphological operations (closing, opening). (**a**) Original patch. (**b**) Threshold $+\Delta$T. (**c**) Threshold $-\Delta$T. (**d**) Morphological closing. (**e**) Morphological opening

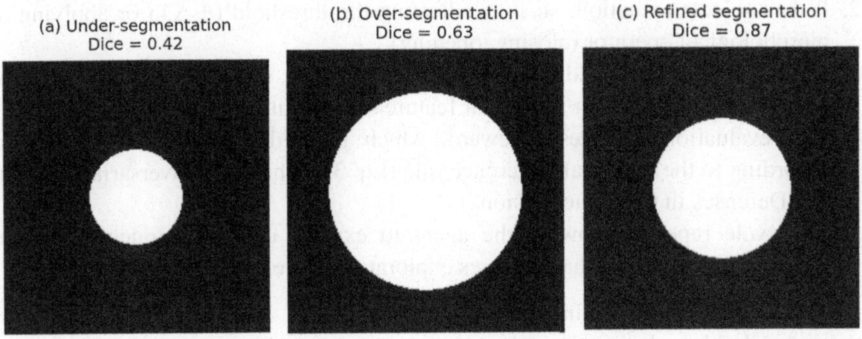

Fig. 11 Reward computation in histological segmentation: comparison of under-, over-, and refined segmentations with associated Dice indices. (**a**) Under-segmentation (Dice = 0.42). (**b**) Over-segmentation (Dice = 0.63). (**c**) Refined-segmentation (Dice = 0.87)

reward can be based on feature separability. One formulation relies on the ratio of between-class variance to within-class variance:

$$R\left(s_t, a_t\right) = \frac{\sigma^2_{\text{between}}}{\sigma^2_{\text{between}} + \varepsilon} \tag{7}$$

Where ϵ is a small constant ensuring numerical stability. This criterion encourages the segmentation to maximize the contrast between nuclei and background while keeping each class internally homogeneous. In practice, a hybrid reward combining Dice (when annotations are present) and texture-based metrics (when annotations are absent) offers a robust solution. This approach guarantees adaptability to heterogeneous slides and different staining conditions. The principle is illustrated in Fig. 11. Panel (a) shows an under-segmented mask, where several nuclei are merged, yielding a low Dice score. Panel (b) illustrates over-segmentation, where nuclei are fragmented and background noise is included, resulting in a moderate reward. Finally, panel (c) presents a refined mask that balances separation and completeness, achieving a high Dice score.

4.4 Exploration of the Threshold Space

The final component of the Q-learning formulation is the mechanism by which the agent explores and exploits the threshold space to optimize segmentation. The process is iterative:

1. The agent begins with an initial state, represented by the current segmentation mask and its descriptors (histogram or Haralick features).

2. It then selects an action, such as adjusting the threshold ($\pm\Delta T$) or applying a morphological operator (closing, opening).
3. The environment responds by producing a new mask, which is evaluated through quantitative indices such as Dice or feature separability.
4. This evaluation provides a reward, which is used to update the Q-values according to the temporal-difference rule (Eq. 3 of chapter "Adversarial Attacks and Defenses in Computer Vision").
5. The cycle repeats, allowing the agent to explore different strategies before converging to a policy that balances exploration and exploitation.

At the beginning of training, exploration dominates, enabling the agent to try many thresholds and operations. As training progresses, the ε-greedy policy ensures a gradual shift toward exploitation, focusing on the most rewarding actions. Convergence is achieved when further iterations no longer improve the reward, indicating that the agent has discovered a robust segmentation policy. This closed-loop adaptive process is summarized in Fig. 12, which illustrates the interaction between state, action, environment, reward, and Q-value update in the context of histological segmentation.

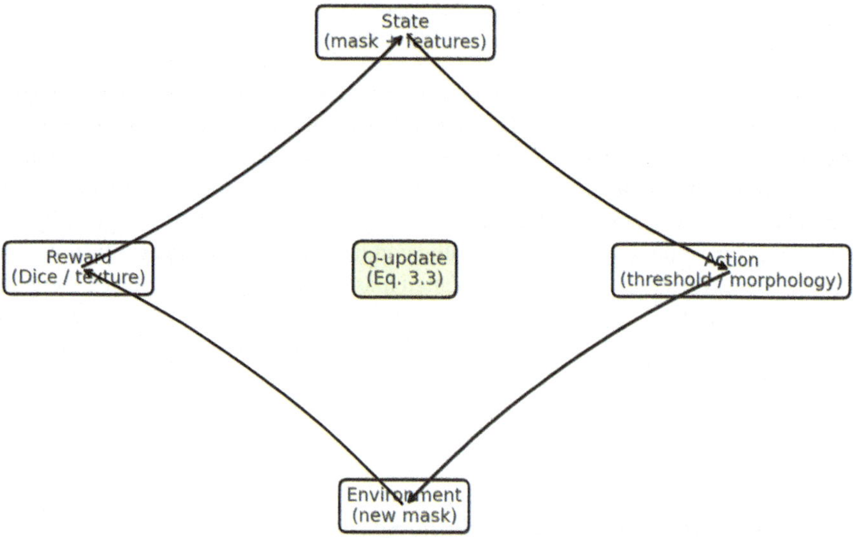

Fig. 12 Adaptive segmentation loop as a Q-learning cycle: state → action → environment → reward → Q-update

5 Implementation Framework

The implementation of Q-learning for adaptive segmentation in histological cancer detection follows a pipeline in which raw images are progressively transformed into informative states, refined through actions, and evaluated by a reward function. This framework, which aligns with the experimental workflow already established in our research, can be described through three successive steps: segmentation preprocessing, feature extraction, and reinforcement learning integration.

5.1 Segmentation (Preprocessing Stage)

The first stage aims to isolate nuclei and relevant tissue regions from the histological slides. Several approaches were evaluated, including edge-based segmentation, region-growing, and thresholding techniques. Among these, histogram-based thresholding combined with morphological operations proved particularly effective. Morphological closing was used to fill gaps within nuclei, while opening eliminated small noisy regions. Dilation and erosion further allowed fine contour adjustments. These operators ensured that nuclei, often fragmented or partially stained, could be reconstructed into coherent objects suitable for further analysis. This preprocessing step corresponds directly to the action space in the Q-learning formulation described in Sect. 4. Indeed, threshold adjustments ($\pm\Delta T$) and morphological refinements are precisely the types of actions that the agent can take to improve segmentation iteratively. The efficiency of such preprocessing strategies is illustrated in Fig. 13, which shows segmentation results obtained using different filters and thresholding methods. The figure highlights how the choice of parameters drastically modifies the resulting masks, sometimes leading to over-segmentation or under-segmentation. These results justify the need for adaptive, learning-based approaches.

Similarly, Fig. 14 compares automatic segmentation results with expert manual annotations. This comparison illustrates both the progress achieved by classical preprocessing and its limitations when facing complex tissue heterogeneity. We will

(a) (b) (c) (d) (e)

Fig. 13 Segmentation results using different filters and thresholding: (**a**) The original image, (**b**) Prewitt filter, (**c**) Sobel filter, (**d**) Roberts filter, (**e**) Laplacian filter

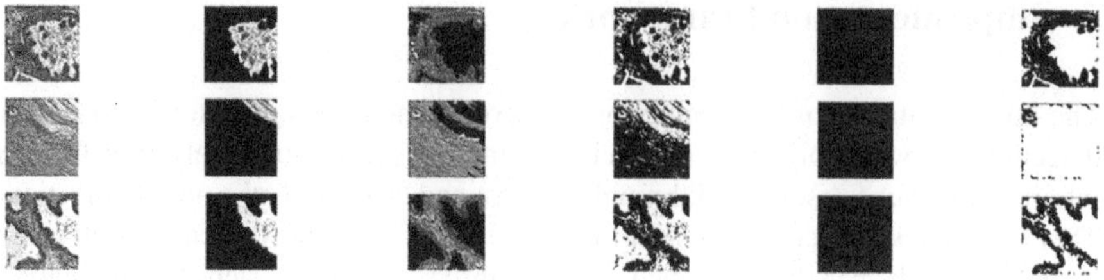

Fig. 14 Comparison of automatic segmentation with expert manual annotation. (a) Three multispectral images of carcinoma cells, (b) Manual segmentation images, (c) Snake segmentation images, (d) Thresholding segmentation images, (e) Edge based segmentation images, (f) Region based segmentation images

reuse this figure in two contexts: first, to illustrate preprocessing challenges, and second, in Sect. 4.3, to emphasize its role in reward computation.

5.2 Feature Extraction

Once candidate nuclei are segmented, the next step consists of computing descriptors that both characterize tissue morphology and serve as the state representation for Q-learning. We are focused primarily on Haralick texture features, derived from the Gray Level Co-occurrence Matrix (GLCM). These features quantify second-order statistical relationships between pixels and are crucial in histopathology:

- Contrast reflects intensity variation and chromatin irregularity.
- Correlation measures linear dependencies, often altered in malignant cells.
- Energy indicates uniformity, relevant for homogeneous tissue.
- Homogeneity emphasizes smoothness, differentiating dense nuclei from stroma.

To complement these descriptors, we also introduced higher-order features:

- Gray Level Run Length Matrix (GLRLM), which captures elongated structures by measuring runs of consecutive pixels with the same intensity. The GLRLM) which is also known as a gray-level co-occurrence matrix can be viewed as a statistical texture characterization method. Where its function shows in counting the number of pixels in an image that have the same intensity in a given direction. Before having the results in a matrix which is calculating how often a pixel with gray-level (grayscale intensity) value i occurs horizontally adjacent to a pixel with the value j. This matrix is represented as (l,n), which is equal to the number of segments of length l and gray level n.
- Gray Level Size Zone Matrix (GLSZM), which describes homogeneous zones of various sizes within the tissue. The GLSZM, which is calculated after decomposing the homogeneous texture into areas with pixels of the same intensity

1	2	3	4
1	3	4	4
3	2	2	2
4	1	4	1

→

Gray Level (i)	Run Length (j)			
	1	2	3	4
1	4	0	0	0
2	1	0	1	0
3	0	0	0	0
4	3	1	0	0

Fig. 15 Illustration of the Gray Level Run Length Matrix (GLRLM): an example of the calculation of a Run Length Matrix for a 4 × 4 image in 0° direction and for 4 gray levels

Fig. 16 Illustration of the Gray Level Size Zone Matrix (GLSZM)

1	2	3	4
1	3	4	4
3	2	2	2
4	1	4	1

→

Gray Level (i)	Size Zone (j)			
	1	2	3	4
1	2	1	0	0
2	1	0	1	0
3	0	0	1	0
4	2	0	1	0

level, and not of small groups of pixels or segments in any given direction. It is calculated according to GLRLM principle, where the value of the matrices is equal to the number of areas of size s and of gray level n. This matrix is represented as (s,n), which is equal to the number of areas of size s and of gray level n.

The computation and interpretation of these features are illustrated in Fig. 15 (GLRLM) and Fig. 16 (GLSZM). Together, these descriptors provide a comprehensive characterization of histological textures and form the informative states on which the RL agent bases its decisions.

5.3 *Integration of Q-Learning*

The final step of the framework is the integration of the Q-learning algorithm into the segmentation pipeline. Within this perspective:

- The state corresponds to the histological patch represented by its histogram, GLCM, GLRLM, and GLSZM descriptors.
- The actions correspond to threshold adjustments and morphological operations (closing, opening, dilation, erosion).

- The reward is computed either by overlap indices such as the Dice coefficient (when expert annotations are available, see Fig. 3 of chapter "Adaptive Cryptographic System Orchestration via Intelligent Agents"), or by feature separability (Eq. 6 in Sect. 4.3) when annotations are not provided.

At each iteration, the agent evaluates its segmentation mask through these descriptors and updates its Q-table according to the temporal-difference rule (Eq. 3) introduced in Sect. 3. Over multiple iterations, the agent converges toward a policy that balances exploration and exploitation, eventually maximizing segmentation accuracy and robustness. Thus, the segmentation + feature extraction is naturally reframed as an RL environment. Instead of applying deterministic filters or thresholds, the system learns an adaptive policy that iteratively improves segmentation based on feedback.

6 Case Studies and Results

To validate the Q-learning formulation of adaptive segmentation, we conducted experiments on real histological images, including both healthy and pathological tissues. The evaluation combined visual analysis of segmentation masks with quantitative indices, in order to demonstrate how an adaptive agent can achieve results closer to expert annotations compared to static thresholding methods. This section shows that segmentation, when framed as a sequential decision-making process, leads to progressive refinements of nuclei boundaries and significantly improves the accuracy of extracted features for subsequent classification. Figure 1 of chapter "Classification of Typical Remote Sensing Images Based on Deep Learning" illustrates segmentation results obtained with the proposed adaptive method on different histological images. These examples include both healthy tissues, where nuclei are relatively homogeneous and well separated, and pathological tissues, where nuclei often present irregular shapes, heterogeneous staining, and strong overlapping. Classical global thresholding methods (such as Otsu's algorithm) are unable to adapt to these variations. As a result, they frequently suffer from:

- Under-segmentation, where multiple nuclei appear fused into a single object, obscuring morphological details.
- Over-segmentation, where nuclei are fragmented into several parts or where background noise is falsely detected as nuclei.

By contrast, the adaptive method framed within Q-learning progressively improves segmentation quality:

- The state encodes histogram distributions and texture descriptors (Haralick features from GLCM, GLRLM, GLSZM), capturing differences in chromatin organization between normal and cancerous nuclei.
- The actions consist of threshold adjustments and morphological operations, enabling the agent to actively refine cell contours.

- The reward measures segmentation quality, guiding the agent toward masks that maximize overlap with expert delineations while minimizing noise.

The resulting masks, shown in Fig. 1 of chapter "Classification of Typical Remote Sensing Images Based on Deep Learning," reveal well-separated nuclei with smooth boundaries, even in challenging pathological cases. This validates the capacity of reinforcement learning to adaptively handle heterogeneity, producing results comparable to manual annotations performed by pathologists. While qualitative assessment is indispensable, robust validation requires quantitative indices. Segmentation accuracy was therefore measured using the Dice coefficient, Jaccard index, False Positive Rate (FPR), and False Negative Rate (FNR). Table 1 reports these metrics for several segmentation strategies, including classical approaches and the proposed adaptive method. The analysis shows that:

- Global thresholding (Otsu's method) achieves only moderate Dice and Jaccard values, confirming its limited ability to adapt to heterogeneous tissues.
- Region-based and edge-based methods improve performance but remain highly sensitive to parameter tuning, which restricts their robustness in practical applications.
- The SHAKE method, which in this chapter corresponds to the proposed Q-learning–based adaptive segmentation strategy, achieves the highest Dice and Jaccard indices, while reducing both FPR and FNR compared to static approaches.

This performance demonstrates the advantage of framing segmentation as a reinforcement learning problem. The adaptive agent iteratively adjusts thresholds and applies morphological refinements until the reward converges toward optimal values, thereby producing segmentations that are both accurate and robust. From an RL perspective, the reported indices correspond directly to the reward function (see Sect. 4.3). High Dice and Jaccard values act as reinforcement signals, while low error rates confirm the stability of the learned policy. This validates that SHAKE—our adaptive Q-learning formulation—effectively translates the state–action–reward loop into tangible improvements in segmentation quality.

The experimental results, both qualitative (Fig. 17) and quantitative (Table 1), confirm the relevance of modeling histological segmentation as a reinforcement learning problem. Several key insights emerge:

- Exploration and exploitation trade-off: Automatic static methods often stagnate in suboptimal segmentations. In contrast, an RL agent can explore alternative threshold values and morphological refinements until a better mask is obtained.
- Role of state descriptors: By integrating Haralick, GLRLM, and GLSZM features, the agent perceives not only pixel intensity but also the textural and structural differences that distinguish healthy nuclei from malignant ones.
- Reward-driven convergence: High Dice and Jaccard values function as reward signals that reinforce successful strategies, while FPR and FNR penalties prevent the agent from adopting unstable solutions.

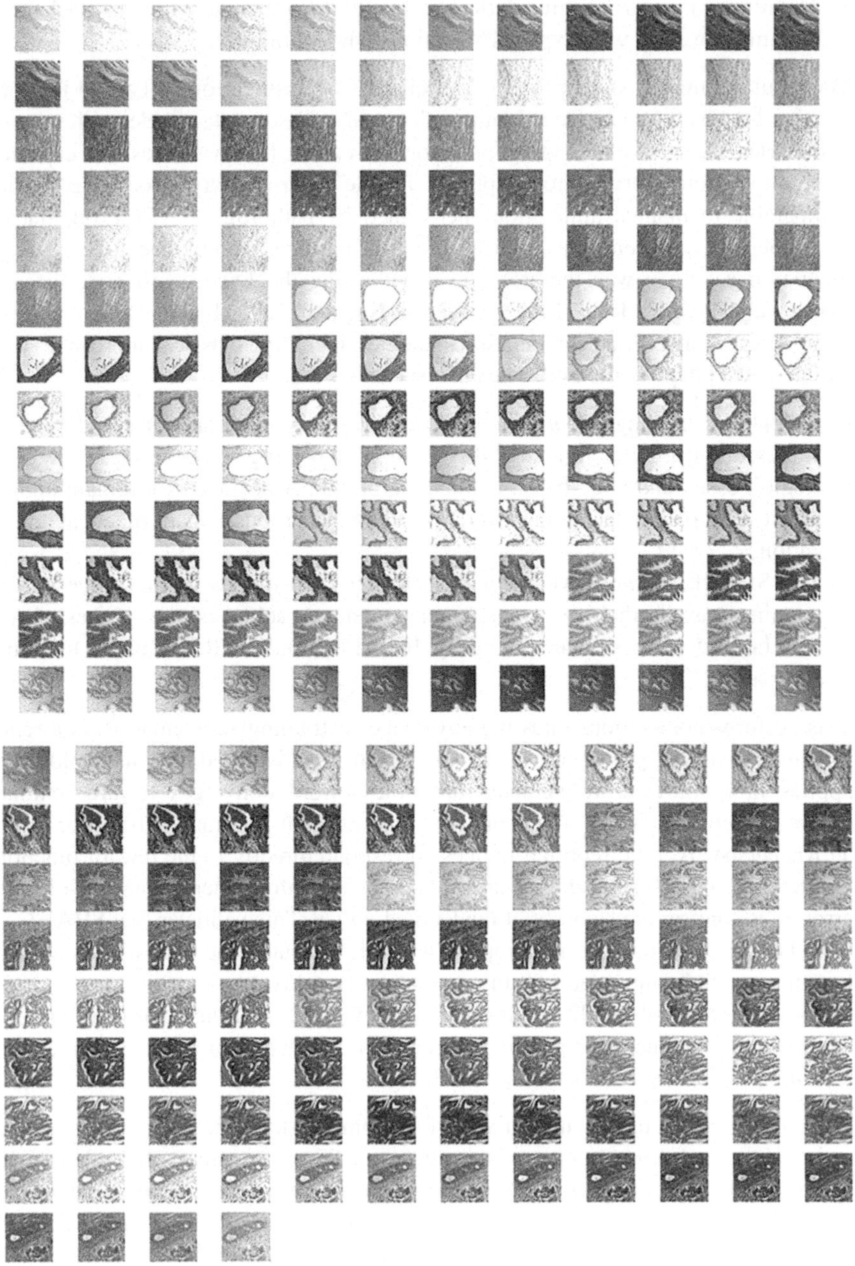

Fig. 17 Segmentation results obtained with the proposed adaptive method on healthy and pathological histological images

Table 1 Comparison of segmentation accuracy metrics (Dice, Jaccard, FPR, FNR), including the SHAKE method representing the proposed Q-learning–based adaptive segmentation strategy

Metrics	Snake	Thresholding	Edge	Region
JSC				
Mean	0.76	0.56	0.34	0.52
Standard deviation	0.03	0.07	0.05	0.02
DSC				
Mean	0.83	0.58	0.38	0.68
Standard deviation	0.03	0.08	0.05	0.03
FPR				
Mean	0.05	0.16	0.13	0.12
Standard deviation	0.01	0.03	0.02	0.01
FNR				
Mean	0.18	0.28	0.25	0.15
Standard deviation	0.02	0.05	0.11	0.12
Time (second)				
Mean	116.76	24.01	15.05	56.24
Standard deviation	12.7	0.66	0.8	6.5

- Clinical relevance: The adaptive method yields results closer to expert annotations, which is critical for downstream classification and diagnostic reliability.

In summary, the results confirm that Q-learning, when applied to adaptive segmentation, enables robust and interpretable delineation of nuclei in histological images. This positions reinforcement learning as a promising alternative to static algorithms, capable of handling heterogeneity and guiding the segmentation process toward clinically meaningful outcomes.

7 Discussion

The reinterpretation of histological image segmentation results within a Q-learning framework provides several insights into the strengths and limitations of adaptive reinforcement-based approaches compared to classical static methods. The main limitation of classical segmentation methods such as Otsu thresholding, region growing, or edge-based detection lies in their static nature: a single global decision is applied across the entire image. As illustrated by Fig. 14, this frequently leads to under- or over-segmentation in regions where tissue staining is heterogeneous. Q-learning, by contrast, frames segmentation as an iterative process of trial and error, where the agent continuously evaluates and improves its mask. This adaptive strategy offers several advantages:

- Local adaptability: instead of relying on a single threshold, the agent can adjust its decisions region by region, reflecting local tissue variability.
- Iterative refinement: the agent explores multiple segmentation strategies before converging to the most accurate one, avoiding premature stagnation.

- Robustness to heterogeneity: by integrating feedback (reward), the method adapts dynamically to slides with uneven staining or irregular nuclear morphology.

These advantages align with the empirical results (Table 1), where iterative methods combining thresholding and morphological operators achieved higher Dice and Jaccard scores than global approaches. Another key element highlighted is the discriminative power of texture descriptors, particularly Haralick features derived from the GLCM, as well as GLRLM and GLSZM descriptors. These features not only support classification but also provide a richer definition of the agent's state in Q-learning.

- Haralick contrast captures chromatin irregularities within nuclei.
- Correlation reflects structural organization disrupted in malignant cells.
- Energy and homogeneity differentiate between uniform stromal regions and dense nuclear clusters.

By encoding such information in the state, the agent does not rely solely on pixel intensity but also on morphological and structural cues, enabling more biologically meaningful segmentation. This explains why methods enriched with textural features tend to converge toward more robust and clinically relevant masks. While the reinforcement learning formulation offers clear benefits, several challenges remain:

- Computational cost: iterative exploration of threshold and morphological actions can increase processing time, especially for whole-slide images.
- Reward design: in many datasets, ground truth masks are unavailable, making Dice-based rewards impractical. Alternative feature-based rewards (Eq. 3 of chapter "Adaptive Cryptographic System Orchestration via Intelligent Agents") may not always perfectly correlate with pathologist annotations.
- Exploration vs exploitation as shown in Sect. 3.3, balancing these two behaviors remains delicate. Premature exploitation can lead to suboptimal segmentation, while excessive exploration delays convergence.

Despite these limitations, the alignment between the results and the Q-learning formulation demonstrates that reinforcement learning provides a conceptual framework that generalizes and improves upon static segmentation pipelines. Finally, it is important to emphasize the clinical implications. Pathologists require segmentation tools that are not only accurate but also robust, interpretable, and adaptable to variations in staining and preparation protocols. Q-learning contributes to this goal by ensuring that the segmentation process is driven by feedback signals reflecting clinical quality criteria (e.g., Dice similarity with expert masks). Moreover, the modularity of the RL formulation makes it compatible with hybrid pipelines that combine classical texture descriptors with modern deep learning representations, paving the way for future clinical-grade systems.

8 Conclusion and Future Works

In this chapter, we have reframed histological image segmentation as a reinforcement learning problem and demonstrated how Q-learning can provide an adaptive, iterative alternative to static segmentation methods. Building upon the results presented in this chapter, we showed that histological images, with their inherent heterogeneity and complexity, benefit from a state–action–reward formulation where segmentation evolves step by step toward optimal solutions. Key contributions of this chapter can be summarized as follows:

- Segmentation as a sequential process: Unlike global thresholding methods (e.g., Otsu), Q-learning treats segmentation as an iterative refinement problem. Each state is defined by histogram descriptors and texture features, actions correspond to threshold adjustments and morphological operators, and rewards are based on segmentation quality indices such as Dice or Jaccard.
- Role of texture features: Haralick descriptors derived from the GLCM, as well as GLRLM and GLSZM features, enrich the state representation by capturing structural and morphological information beyond pixel intensity. This results in segmentation masks that are not only more accurate but also more biologically meaningful.
- Evidence from the results: Qualitative comparisons revealed that automatic methods often produce under- or over-segmentation compared to expert annotations, while quantitative metrics confirmed that adaptive, iterative methods significantly outperform static approaches. These findings support the relevance of reinforcement learning as a unifying conceptual framework for adaptive segmentation.

Challenges remain, however, before Q-learning can be fully adopted in clinical workflows:

- Scalability: Whole-slide histological images are gigapixel-scale, requiring patch-based or hierarchical processing strategies to make RL feasible.
- Reward design: While Dice and Jaccard indices (Table 1) provide reliable feedback when annotations are available, many datasets lack ground truth. In such cases, feature-based or multi-objective rewards must be developed.
- Hybridization with deep learning: Extensions such as Deep Q-Networks (DQNs) could scale Q-learning to high-dimensional states and integrate handcrafted descriptors with deep features.
- Clinical integration: Reinforcement learning must be interpretable and efficient, providing transparent decision-making that pathologists can trust.

In conclusion, this chapter has shown that Q-learning offers a powerful theoretical and practical framework for adaptive segmentation in histological cancer detection. By embedding classical preprocessing operations and texture descriptors into a reinforcement learning cycle, we move beyond rigid, rule-based algorithms toward adaptive, feedback-driven systems. While challenges of scalability, reward design, and clinical integration remain, the combination of reinforcement learning with robust features like Haralick descriptors provides a promising path forward for interpretable and clinically relevant AI in digital pathology.

References

1. Watkins CJCH, Dayan P. Q-learning. Mach Learn. 1992;8(3–4):279–92.
2. Sutton RS, Barto AG. Reinforcement learning: an introduction. 2nd ed. MIT Press; 2018.
3. Haralick RM, Shanmugam K, Dinstein I. Textural features for image classification. IEEE Trans Syst Man Cybern. 1973;SMC-3(6):610–21.
4. Otsu N. A threshold selection method from gray-level histograms. IEEE Trans Syst Man Cybern. 1979;9(1):62–6.
5. Dougherty ER. An introduction to morphological image processing. SPIE Optical Engineering Press; 1992.
6. Litjens G, Kooi T, Bejnordi BE, et al. A survey on deep learning in medical image analysis. Med Image Anal. 2017;42:60–88.
7. Janowczyk A, Madabhushi A. Deep learning for digital pathology image analysis: a comprehensive tutorial with selected use cases. J Pathol Inform. 2016;7(1):29.
8. Madabhushi A, Lee G. Image analysis and machine learning in digital pathology: challenges and opportunities. Med Image Anal. 2016;33:170–5.
9. Hawraa H. Segmentation et classification d'images histologiques pour la détection du cancer. Université de Lorraine; 2023.
10. Tanougast C, et al. Novel design of reward and epsilon-greedy decay strategy tailored for Q-learning in optimizing local mobile robot path planning. (Journal/Conference reference待补充), 2022.
11. Russakovsky O, Deng J, Su H, et al. ImageNet large scale visual recognition challenge. Int J Comput Vis. 2015;115(3):211–52.
12. Xu Y, Jia Z, Wang LB, et al. Large scale tissue histopathology image classification, segmentation, and visualization via deep convolutional activation features. BMC Bioinformatics. 2017;18:281.
13. Gurcan MN, Boucheron LE, Can A, et al. Histopathological image analysis: a review. IEEE Rev Biomed Eng. 2009;2:147–71.

The manufacturer's authorised representative in the EU is Springer Nature Customer Service Centre GmbH, Europaplatz 3, 69115 Heidelberg, Germany. If you have any concerns regarding our products, please contact ProductSafety@springernature.com

Printed and bound by CPI Group (UK) Ltd, Croydon, CR0 4YY
05/10/2026
02228026-0004